BEAUTY AND THE BRAIN

BEAUTY AND THE BRAIN

The Science of Human Nature in Early America

RACHEL E. WALKER

The University of Chicago Press
CHICAGO AND LONDON

The University of Chicago Press, Chicago 60637
The University of Chicago Press, Ltd., London
© 2023 by The University of Chicago
All rights reserved. No part of this book may be used or reproduced in any manner whatsoever without written permission, except in the case of brief quotations in critical articles and reviews. For more information, contact the University of Chicago Press, 1427 E. 60th St., Chicago, IL 60637.
Published 2023
Paperback edition 2024
Printed and bound by CPI Group
(UK) Ltd, Croydon, CR0 4YY

33 32 31 30 29 28 27 26 25 24   1 2 3 4 5

ISBN-13: 978-0-226-82256-3 (cloth)
ISBN-13: 978-0-226-83678-2 (paper)
ISBN-13: 978-0-226-82257-0 (e-book)
DOI: https://doi.org/10.7208/chicago/9780226822570.001.0001

Library of Congress Cataloging-in-Publication Data

Names: Walker, Rachel E. (Historian), author.
Title: Beauty and the brain : the science of human nature in early America / Rachel E. Walker.
Other titles: Science of human nature in early America
Description: Chicago ; London : The University of Chicago Press, 2023 | Includes bibliographical references and index.
Identifiers: LCCN 2022015065 | ISBN 9780226822563 (cloth) | ISBN 9780226822570 (ebook)
Subjects: LCSH: Physiognomy—United States—History—19th century. | Physiognomy—United States—History—18th century. | Phrenology—United States—History—19th century. | Phrenology—United States—History—18th century. | Physiognomy—Social aspects—United States. | Phrenology—Social aspects—United States.
Classification: LCC BF851 .W255 2023 | DDC 138—dc23/eng/20220513
LC record available at https://lccn.loc.gov/2022015065

♾ This paper meets the requirements of ANSI/NISO Z39.48-1992 (Permanence of Paper).

CONTENTS

Introduction 1

ONE
Founding Faces 17

TWO
A New Science of Man 44

THREE
Character Detectives 80

FOUR
The Manly Brow Movement 115

FIVE
Criminal Minds 144

SIX
Facing Race 166

Conclusion 193

Acknowledgments 203   *Notes* 207
Selected Bibliography 257   *Index* 265

INTRODUCTION

When modern Americans use terms like "highbrow" and "lowbrow," they are typically trying to distinguish between the leisurely amusements of elite society and the practices of the broader masses. Going to the Metropolitan Opera, sauntering through an exclusive art gallery, or thumbing through *The New Yorker* are highbrow pursuits. Bingeing reality television? Decidedly lowbrow. We say these words but rarely question them, instinctively associating "highbrow" with cultural and intellectual sophistication while using the term "lowbrow" to reference an alleged lack of refinement.

In 1949, *LIFE* magazine even devised a chart to visually codify these distinctions (fig. 1.1). In meticulous detail, this chart linked highbrow culture with images of Bach, ballet, and wine, while depicting Western movies, comic books, and beer as lowbrow enjoyments. Winthrop Sargeant, a self-identified "high-brow" and Senior Writer for the magazine, then followed up with a tongue-in-cheek screed against his "duller-witted contemporaries." Lamenting the fact that "backward children" made up "some 90% of the population," Sargeant decreed that "low-brows may be bank presidents, pillars of the church, nice fellows, good providers, or otherwise decent citizens, but, culturally speaking, they are oafs."[1]

Nowadays, magazine publishers would probably think twice before dismissing the vast majority of Americans as childish "oafs." Yet terms like "highbrow" and "lowbrow" have not gone away. They have become entrenched in our shared cultural lexicon—words so ordinary that we barely wonder where they came from. These terms, though, have a vibrant history. For more than two hundred years, Americans have spoken of brows high and low. When they did so, they meant something far more literal than most of us realize today. In the eighteenth and nineteenth centuries, people were—to put it frankly—*obsessed* with each other's brows. They measured their height,

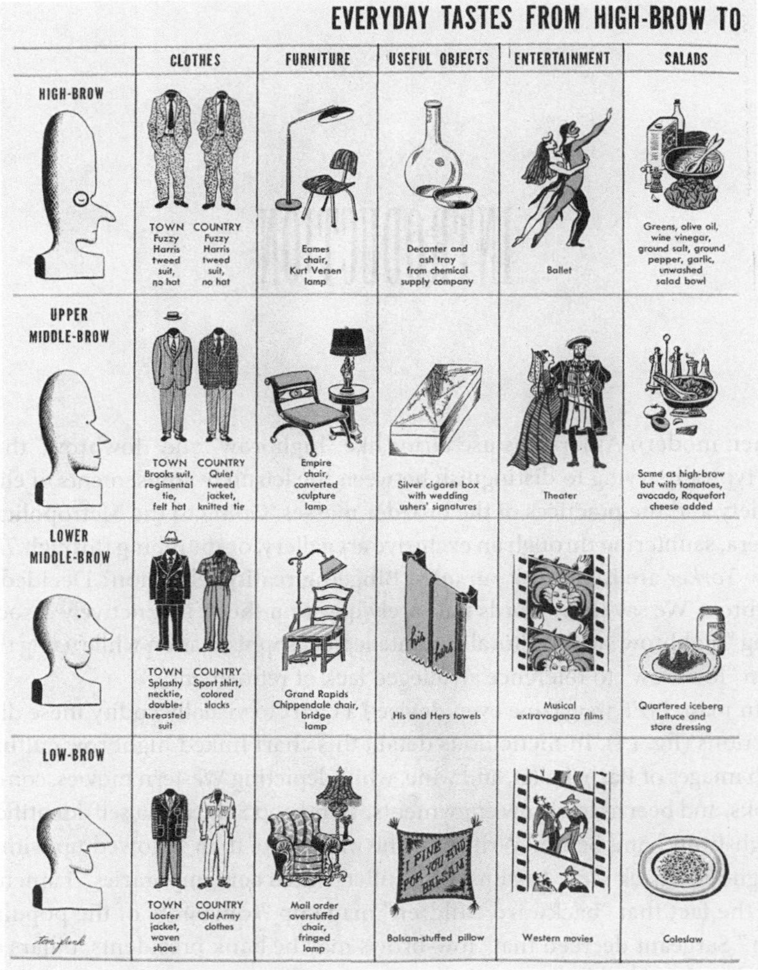

FIGURE I.1. "Everyday Tastes from High-Brow to Low-Brow Are Classified on Chart," *LIFE* (New York), April 11, 1949. Neither this chart nor the accompanying article makes any references to physiognomy or phrenology, but the lingering impact of these sciences is clear in the images used to signify the different groups. The representative of "high-brow" culture wears a monocle and sports a soaring forehead, while the "low-brow" figure has barely any forehead at all.

LOW-BROW ARE CLASSIFIED ON CHART

DRINKS	READING	SCULPTURE	RECORDS	GAMES	CAUSES
A glass of "adequate little" red wine	"Little magazines," criticism of criticism, avant garde literature	Calder	Bach and before, Ives and after	Go	Art
A very dry Martini with lemon peel	Solid nonfiction, the better novels, quality magazines	Maillol	Symphonies, concertos, operas	The Game	Planned parenthood
Bourbon and ginger ale	Book club selections, mass circulation magazines	Front yard sculpture	Light opera, popular favorites	Bridge	P. T. A.
Beer	Pulps, comic books	Parlor sculpture	Jukebox	Craps	The Lodge

analyzed their shape, and pored over scientific treatises about the intricacies of the human forehead. They did so because they believed that countenances and craniums could be scientifically plumbed for hidden clues about the human mind.

Between the 1770s and the 1860s, physiognomy (the study of the human face) and phrenology (the study of the human skull) took the transatlantic world by storm. These were popular sciences rooted in a deceptively simple premise: people's heads and faces revealed their intelligence, personality, and character. Physiognomists and phrenologists suggested that eyes, noses, cheeks, and lips could all convey important information about the human mind and soul, but they viewed the forehead as the most significant marker of a person's mental capacity. High, expansive, and well-shaped brows signaled both wisdom and intelligence. Short, misshapen, or low brows were the anatomical indicators of a degraded mind. Within this mental universe, terms like "highbrow" and "lowbrow" were not abstract indicators of hazy cultural divides. They were bodily traits that signified deeper truths about human nature.[2]

While modern Americans have held onto the terminology that physiognomists and phrenologists first devised, they have lost sight of the dynamic cultural and scientific universe that first gave those words meaning. This book excavates that history, revealing how scientific assessments of the human body shaped debates about social and political hierarchy in the early American republic. In doing so, it tackles questions that have long captivated scholars: How have people in the United States historically made sense of inequality while inhabiting an allegedly egalitarian country where it's not supposed to exist? And how have they used science to do so? By exposing how Americans used physiognomy and phrenology to understand human nature, *Beauty and the Brain* reveals how science has historically functioned as a tool of oppression. More importantly, this book uncovers the histories of marginalized people who adopted science for their own political aims, creatively interpreting human minds and bodies in their efforts to fight for racial justice and gender equality.

PHYSIOGNOMY, PHRENOLOGY, AND THE FOUNDING OF A NATION

When colonial leaders declared independence from Britain in 1776, they proclaimed that "all men are created equal." Yet these elite white men faced a troublesome conundrum.[3] Despite their symbolic commitment to liberty, justice, and equality, few of the nation's founders envisioned a world with white women and people of color as full citizens. They rationalized this inequitable reality through a belief in meritocracy. If some Americans were inherently smarter, more virtuous, and more skilled than their less fortunate counterparts, then those illustrious citizens *deserved* more expansive social and political privileges. As Thomas Jefferson once wrote in a letter to John Adams, "I agree with you that there is a natural Aristocracy among men. The grounds of this are Virtue and Talents."[4] Like Jefferson, many of the nation's white male founders believed that only certain individuals should be entrusted with political leadership. This was not egalitarianism, of course, but it was superficially logical. If there were going to be hierarchies, why not base them on merit, rather than on an antiquated aristocratic system that rewarded (and penalized) the accidents of birth? Still, this left a major question unanswered: What was the best way to determine merit?

Physiognomy—and eventually phrenology—provided some solutions. These disciplines were alluring because they promised a way of identifying "Virtue and Talents" through external features. If a rational process of empirical observation found that wealthy white men had better brains and bodies than their compatriots, then it was not discriminatory to treat them as superior beings—it was scientific. This, at least, was the theory. In practice, social prejudices shaped scientific findings. When privileged people looked around, they managed to find the most merit in the heads and faces of those who already held social and political advantages. Physiognomy and phrenology often provided new justifications for old hierarchies, giving people a scientific framework for grappling with the ideological inconsistencies of their unequal republic.

The intriguing thing about these sciences, though, was that they did not appeal only to social and political leaders. Physiognomy and phrenology were initially elite affairs, and many of the rules were crafted by white male thinkers who were convinced of their superiority and ready to defend it scientifically.[5] By the mid-nineteenth century, though, physiognomy and phrenology had become accessible knowledge systems with broad appeal among the public. Unlike other sciences, facial and cranial analysis were practiced not in laboratories, but in private homes and public spaces. Requiring no specialized equipment, massive skull collections, or university training, these sciences were accessible to working people and highly educated intellectuals alike. Established practitioners also made a concerted effort to disseminate their doctrines to as broad an audience as possible. Physiognomists proclaimed that everyone—including babies—had an innate capacity for reading visages. Phrenologists likewise marketed their science to the broader public, churning out cheap pamphlets, almanacs, and "people's editions" of more expensive treatises. Claiming that anyone could "Know Thyself" and others, they explicitly tried to make these sciences attractive to the masses.[6]

Their crusade was successful in large part because they preached a hopeful message that aligned with the nation's meritocratic ethos. Although popular science enthusiasts assumed that everyone was born with a particular set of moral, mental, and physical attributes, they also believed that with enough dedication, people would be able to alter both their minds and their bodies. According to physiognomic and phrenological understandings of the world, a person's body did not determine their destiny. Moral and intellectual cultivation could quite literally stamp itself on the external form, making people more physically attractive as their minds improved. A better brain led to a more beautiful body. Such an understanding of human nature distinguished physiognomy and phrenology from other sciences at the time, which increasingly described minds and bodies as both inherited and immutable.

The optimism and intellectual flexibility of physiognomy and phrenology made these disciplines appealing, and by the late 1830s and 1840s the ranks of popular scientists were no longer populated primarily by lawyers, doctors, and elite naturalists. They also included ordinary men and women, as well as enslavers, formerly enslaved people, and abolitionists. Farmers, factory workers, and early feminist activists enthusiastically analyzed heads and faces. So did Quakers, evangelicals, Mormons, and members of utopian communities. Many of these people were powerful defenders of the status quo, but many others were radical reformers who believed they could use science to build a better, more equitable America. The United States' most privileged citizens might have used physiognomy and phrenology to uphold inequality, but men of color and women of all races adopted and co-opted these sciences to undermine existing hierarchies and to craft alternative methods for measuring human worth. By recovering this largely forgotten history, *Beauty and the Brain* focuses on the imaginative experimentation that characterized the development of modern science. In doing so, it paints a vivid picture of how a diverse group of Americans tried—and often failed—to live up to their founding principles.

This book begins by taking physiognomy and phrenology seriously—as both sciences and cultural practices—not because they are legitimate tools for analyzing human nature, but because early Americans believed they were. Between the 1770s and the 1860s, these disciplines shaped the transatlantic debates of elite scientific thinkers while simultaneously infiltrating popular culture and influencing people's daily lives. These were not silly pseudosciences that people occasionally discussed at dinner parties. They were pervasive practices that suffused almost every aspect of society, culture, and politics. In an era before the advent of modern neuroscience and psychology, physiognomy and phrenology were two of the primary tools that people used to understand the human mind.[7]

Taking that reality as its starting point, this book intentionally uses the word "science," rather than "pseudoscience," when referring to physiognomy, phrenology, and other discredited disciplines of the past. Ideas about what counts as real science—and what does not—have shifted dramatically over time. Dismissing physiognomy and phrenology as pseudosciences might make us feel good because it assures us that we are smarter, better, and more enlightened than people in the past. In the end, though, the word "pseudoscience" merely allows scholars to condemn problematic discourses of bygone eras while maintaining an uncritical veneration of science itself. It also draws artificial distinctions between the science that worked and the science that didn't, anachronistically implying that historical actors were not practicing real science at all. Such a division simply is not useful for the late eighteenth and early nineteenth centuries, a time when both elites and ordinary people considered physiognomy and phrenology to be "legitimate knowledge systems."[8]

Before the consolidation, professionalization, and segmentation of scientific disciplines in the late nineteenth century—and especially before the founding of the American Medical Association in 1847—the meaning of science was both capacious in character and constantly in flux. In fact, while it now seems both timeless and unremarkable, the term "scientist" was not even part of the English lexicon until 1833. For most of the nineteenth century, science was an eclectic enterprise with diffuse aims and porous boundaries. It attracted college-educated intellectuals, lay practitioners, and amateur dabblers with a wide array of interests, specialties, and skill sets. Elite thinkers had not yet clearly demarcated boundaries of belonging and exclusion in their respective disciplines, meaning that the line between "real" science and pseudoscience was always blurry and sometimes nonexistent.[9]

Nor was the distinction between science and popular culture clearly established. As one scholar has argued, "science rubbed shoulders with a wide range of literary modes" during the early decades of the nineteenth century, "and its promoters helped themselves to these modes in their own writing." Artists read scientific treatises, and naturalists engaged in philosophical meditations on the nature of beauty. In the early American republic, art, science, and literature were not clearly delimited disciplines. They were mutually supportive tools that people simultaneously used to make sense of the world.[10]

These tools were wielded by a large and diverse group that ranged from physicians and intellectuals to educated women, artists, novelists, moral reformers, and Black abolitionists. This is not to say that all Americans were enthusiastic body interpreters. Some were true believers, while others were curious but skeptical dabblers. Still others decried physiognomy and phrenology as "humbuggery." But those individuals never constituted a monolithic or coherent social group. It was not as if all the shrewdest physicians, politicians, and lawyers presciently recognized these sciences as quackery while a much bigger and more gullible public brainlessly embraced them. There was considerable overlap between expert and lay knowledge. As Carla Bittel has argued, there was also a "continuum of belief" that ran the gamut "from total advocacy to absolute denunciation, with lots of room for acceptance and rejection in between." Even when people were skeptical of physiognomy and phrenology, they often admitted that their doctrines contained a kernel of truth. Because physiognomy and phrenology were malleable discourses without a clear set of reproducible rules, individuals could experiment with their precepts to develop idiosyncratic philosophies about the human brain and body. If these sciences did not garner universal approval, they did become culturally ubiquitous languages that almost all literate Americans would have known how to speak.[11]

Many of these insights will be familiar to historians of science, whose very discipline presumes that science itself has a history. These scholars have produced a treasure trove of quality work on physiognomy and phrenology in Europe and across the globe, much of it focusing on Britain in the Victorian era.[12] Historians of the early American republic, by contrast, have been more reluctant to view physiognomy and phrenology as real sciences or important intellectual philosophies. Sometimes US historians have dismissed these disciplines as quirky forms of quackery that only briefly captured the American imagination.[13] More often, they've ignored them entirely. The field's leading academic journals and textbooks almost never mention physiognomy and phrenology. If they do, they dispatch with them in a sentence or two before getting back to more well-studied topics: the legacies of the American Revolution, the emergence of the market economy, the bitter conflicts over slavery and freedom, the rise of moral reform crusades, or the struggles and successes of the women's rights movement.[14] While there are specialized books that examine how physiognomy and phrenology influenced American ideas about art, literature, crime, or education, there is still no comprehensive cultural history of these sciences during the country's foundational decades.[15]

Physiognomy and phrenology were not just silly fads, nor were they epiphenomenal to the economic, political, and legal developments that US historians typically emphasize. They were pervasive social practices and intellectual philosophies that people used to better understand their own brains, bodies, and behaviors. In fact, it was often through the language, imagery, and daily practice of popular sciences that people made abstract concepts like "race," "gender," "beauty," and "intelligence" feel real on the ground. To overlook these sciences, then, is to forget the kaleidoscopic system of cultural codes that early Americans used to make their world seem both legible and comprehensible. As the late historian E. P. Thompson famously argued, doing good history requires scholars to abandon "the enormous condescension of posterity" and instead take the people we study seriously. Physiognomy and phrenology might strike modern observers as unfamiliar, unintelligible, or perhaps even absurd, but that's mostly because we've forgotten just how ubiquitous these disciplines were in the past.

In the late eighteenth and nineteenth centuries, physiognomy and phrenology not only shaped how Americans understood their own brains and bodies; they also gave people a practical way of evaluating others and understanding social and political hierarchies. Heads and faces became physical evidence for why certain men rose to national prominence while others committed crimes—or why some women made bad mothers and others made good wives.[16] These sciences allowed people to discriminate according to preconceived notions of race, class, gender, and ethnicity. At the same time, they gave people hope that bodies did not determine destiny, and that individuals might overcome their hereditary endowments through sheer force of will. For this reason, physiognomy and phrenology appealed to scientific racists and social conservatives, but also to reformers who believed the world could be perfected through a mix of good science, grit, and principled idealism. Without understanding this history, scholars risk misinterpreting the very people we seek to study—or at least disregarding the scientific logic that made their worlds make sense to them. Focusing on these disciplines allows us to better understand how Americans thought about human nature and how they used scientific knowledge to navigate political debates about race, class, and gender.

The history of physiognomy and phrenology can also help us reexamine some of our most fundamental assumptions about the relationship between science and social hierarchy in early America. It is enormously tempting to look back at physiognomy and phrenology and dismiss them as racist, sexist, and classist disciplines. This narrative has been pervasive in both scholarly and popular works—and for good reason. It's accurate. Because these sciences were predicated on the idea that countenances and craniums revealed people's inner capacities, they provided scientific explanations for inequality while laying the foundation for later, more quantitative disciplines like craniometry, ethnology, physical anthropology, criminology, and eugenics—all of which would be used to justify the political exclusion of white women, people of color, and other marginalized groups. But that's only part of the story.[17]

Many of the nation's most enthusiastic popular scientists were not political reactionaries seeking to defend the status quo. They were radicals and reformers: abolitionists like Frederick Douglass, Lucretia Mott, and William Lloyd Garrison; women's rights activists like Susan B. Anthony, Elizabeth Cady Stanton, and Margaret Fuller; and a zealous group of moral reformers who sincerely believed they were advocating for the rights of the dispossessed. These women and men did not see physiognomy and phrenology as instruments of inequality. Instead, they used these disciplines to fight back against oppression and undercut emerging theories of biological determinism. As inhabitants of an intellectual world in which many different groups could craft unique forms of anatomical knowledge, both white women and people of color transformed the very discourses that defined them as inferior and instead used science to advocate for racial justice and gender equity.[18]

THE SCIENCE OF MAN

When Americans studied heads and faces, they were participating in a broader Enlightenment project: the application of human reason to the study of human nature itself. As the Scottish philosopher Francis Hutcheson explained in 1747, the most pressing objective for naturalists, intellectuals, and artists of the era was to "search accurately into the constitution of our nature to see what sort of creatures we are." What made people human, Enlightenment thinkers wondered, and what made humans different from other creatures? To address these quandaries, they crafted elaborate classificatory systems.[19]

By the middle decades of the eighteenth century, European intellectuals indulged in what Stephanie Camp has called an "obsessive compulsion for categorization." In the 1750s, Carolus Linnaeus developed his system of binomial nomenclature, while naturalists like Petrus Camper, Johann Friedrich Blumenbach, and the Comte de Buffon studied skeletal structures in an attempt to empirically determine the differences between the world's diverse populations. As Europeans embarked upon brutal crusades to colonize the globe, they fashioned new frameworks to make sense of the people they intended to conquer. Natural history became a way of cataloging unfamiliar populations—and of justifying imperial violence.[20]

Physiognomy and phrenology emerged as modern forms of scientific observation within this larger context. Europeans had long analyzed faces in an attempt to learn about character, but in the 1770s, a Swiss cleric named Johann Caspar Lavater catapulted to international fame by claiming it was possible to do so *scientifically*. Proclaiming that the countenance was both the "index of the mind" and the "mirror of the soul," he devised a physiognomical method that was both observational and reflective. In some ways, there was nothing new about his theory. Lavater was not the first facial analyzer. He was simply the first to insist that physiognomy could be a legitimate science—a system of "fixed principles" that could be almost mathematical in their predictive power—rather than a haphazard art of first impressions. His books enthralled people because they promised a way of rationally interpreting entities which had previously seemed unknowable: the human mind and soul.[21]

People all over the world were soon using physiognomy to analyze themselves and others. But Lavater's system left some people dissatisfied. While the German anatomist Franz Josef Gall believed that the head and face revealed important information about the mind, he critiqued Lavater's method as vague and unscientific. Determined to do better, he developed a more elaborate—and allegedly more empirical—way of reading the human mind in the 1790s. While physiognomy focused on the face, phrenology focused on the skull. Gall divided the brain into different "organs," with each part responsible for a different mental function. Then he analyzed cranial shapes for clues about character. Like physiognomy before it, phrenology soon spread across the globe, gaining both critics and admirers. It became especially popular in the antebellum United States, where it mingled with Lavater's system. Rather than abandoning facial analysis for the science of skulls, Americans read countenances and craniums together. Even when they said they were practicing phrenology, much of their focus remained on the face.[22]

By the late 1830s, Americans were hooked on the "science of man." Factory workers got their heads examined after work, while the elite and middling classes scrutinized each other's features in parlors and at public lectures. Popular sciences also infiltrated the nation's expanding print culture. When antebellum readers picked up a newspaper, got lost in their favorite novel, or delved into the latest magazines, they encountered detailed descriptions of human heads and faces. With differing degrees of seriousness, famous authors like Mason Locke Weems, Edgar Allan Poe, Harriet Beecher Stowe, and Walt Whitman deployed physiognomy and phrenology as descriptive tools in their works.[23] Physicians and lawyers debated the merits of these sciences in hospitals, courtrooms, and medical journals. Children learned about them in school, and prison administrators used them to evaluate the redemptive potential of those they incarcerated.[24] Women's magazines urged readers to take up phrenology, and they gave appearance-conscious readers advice about how to attain physiognomic beauty standards.[25] These sciences even influenced people's most intimate decisions, shaping how they searched for spouses and raised their children.[26] Meanwhile, political periodicals consistently evaluated the mental, moral, and physical traits of prominent legislators. By mid-century, it would have been virtually impossible to open a novel, newspaper, or magazine without seeing references to someone's "beaming eyes," "intellectual brow," or "well-shaped head."[27]

Using the body as their signpost, Americans tried to decipher the mental, moral, and physical characteristics that differentiated human beings from one another. For the most part, physiognomic and phrenological practitioners truly believed that they were engaging in a rational, dispassionate process of scientific observation. The goal was to analyze craniums and countenances objectively and discern new information about others in the process. Yet people invariably brought their existing biases and beliefs to their quest to comprehend human nature. Finding what they expected to see in the bodies they scrutinized, practitioners often came to different conclusions depending on their own racial, gender, and class identities.

A REPUBLIC OF CONTRADICTIONS

The rise of physiognomy and phrenology coincided with broader shifts in the American political landscape. In the immediate post-Revolutionary decades, most states limited suffrage rights to property-owning men. By the 1820s, however, many of the nation's non-landholders began demanding that citizenship be based on moral virtue and mental competence, rather than inherited wealth or connection to a plot of soil. At the Virginia Convention of 1829-1830, a group of non-freeholders protested that land ownership did not automatically result in the "moral or intellectual endowments" that were necessary for citizenship. Property, they argued, did not make people "wiser or better." Wasn't it more important for voters to be smart, sensible, and virtuous? As the United States developed into a vibrant market economy in the early nineteenth century, land ownership seemed an increasingly irrelevant litmus test for full citizenship. By the 1830s, most state legislatures had redefined voting as a natural right, unconnected to property ownership—at least for white men.[28]

In extending suffrage rights to non-freeholders, political leaders seemed to validate the idea that inner characteristics such as virtue and intelligence were the primary qualifications for citizenship. In principle, this seemed like a victory for republican notions of liberty and equality. In practice, moral and mental virtue were not enough to get marginalized people access to the ballot box. At the very same moment that politicians extended suffrage to all white men, they developed new strategies for excluding women and free men of color from the polity. Political leaders may have imagined white women as virtuous wives and mothers, swaying their husbands and sons toward good political decisions through gentle persuasion, but they were reluctant to see women as full citizens. Only in New Jersey could property-holding women vote in the years after the American Revolution. In 1807, they lost even this privilege.[29]

Free men of color also faced a different set of rules. Just as state legislatures began expanding the rights of poor white men, so too did they revoke suffrage for African Americans. In some states, political leaders set landholding requirements prohibitively high for free Black property holders, making it impossible for them to access the franchise. In other states, legislators simply barred Black men from voting entirely. Although many free people of color had previously exercised suffrage rights, they now confronted a new legal and political structure that questioned their capacity for citizenship and denied their right to participate in political decisions.

By mid-century, race and gender had replaced class as the primary dividing line between those who "counted" as full citizens and those who did not. To rationalize this new reality, many political thinkers insisted that both white women and Black men—no matter how rich and no matter how educated—were unequivocally and intrinsically unqualified for the responsibilities of citizenship. When doing so, they argued that suffrage should only be conferred upon those who held the inner capacity to exercise it. Qualifications shifted from external factors—like wealth or property ownership—to internal characteristics. Yet all of this happened during a historic moment when transatlantic intellectuals were becoming increasingly convinced that interior traits could be discerned through external features. Perhaps some people truly *were* more capable and more virtuous than others, they posited, and perhaps these differences were visible in the human body. Anyone who needed justification for these beliefs merely had to tap into a vast and growing scientific literature, which evaluated human capacity through elaborate anatomical observations. By the mid-nineteenth century, physiognomy and phrenology were two of the most influential—and popularly accessible—disciplines within this intellectual tradition.

Physiognomy and phrenology seemed to prove that there were "natural" differences between men and women, and between people of different races and ethnicities.[30] As a result, they appealed to elite white men, who saw their own capacious craniums as physical proof of their intellect, power, and fitness for leadership. Yet popular sciences also appealed to many marginalized Americans. As late as 1868, the Bishop Henry McNeal Turner cited these disciplines in a fiery speech before the Georgia State Legislature, which had prohibited him from being sworn in on account of his race. Turner's goal was to "convince the House, today, that I am entitled to my seat here." To make his case, he tackled scientific racism head-on: "A certain gentleman has argued that the negro was a mere development similar to the orangutan or chimpanzee," he scoffed, "but it so happens that, when a negro is examined, physiologically, phrenologically and anatomically, and, I may say, physiognomically, he is found to be the same as persons of different color." This speech is remarkable for its soaring defense of universal equality, but it is also a fascinating example of how different groups used science for competing aims. Turner did not reject the scientific language of the day, nor did he challenge the notion that it was possible to determine human value empirically. Instead, he wielded physiognomy and phrenology as weapons in the battle against white men who denied his humanity. For Turner, popular sciences validated his right to be a political leader in the state of Georgia.[31]

Even though white physicians, enslavers, and political thinkers used facial and cranial analysis to justify slavery and rationalize racial inequities, Black activists like Turner regularly embraced these disciplines, using them to argue for racial justice and political inclusion. Women's rights activists, too, employed physiognomy and phrenology to defend the brains of women and advocate for their social and political advancement. In such cases, marginalized people challenged their oppressors with the same scientific language that had been used against them. This was possible because they inhabited a vibrant cultural and scientific world in which all individuals—regardless of sex, race, or rank—could cultivate their own understandings of human anatomy and character. In early America's physiognomic universe, scientific knowledge emerged simultaneously from the top down and the bottom up.

By recovering this once vibrant but now forgotten scientific world, this book unearths the stories of ordinary Americans who shaped the trajectory of American cultural and intellectual thought. In the process, it not only rethinks what counted as science in early America; it also reimagines who might have counted as a *scientist*. The first half of the book demonstrates how physiognomy and phrenology emerged in a global context and reveals how elite white intellectuals used these disciplines to rationalize existing hierarchies. Yet these chapters also explain why physiognomy and phrenology attained such massive popularity in the United States, appealing to a wide cast of contradictory characters. Chapter 1, "Founding Faces," shows how Americans used physiognomy to craft an idealized—and exclusive—vision of the disinterested republican citizen. Chapter 2, "A New Science of Mind," then tracks the rise of phrenology in the United States, illustrating how it shaped—and was shaped by—the country's meritocratic ideals, and how it affected people's personal identities, interpersonal relationships, and perceptions of social and political hierarchies.

The rest of the book then shows how different groups used physiognomy and phrenology in innovative and often subversive ways. Chapter 3, "Character Detectives," explains how Black and white women co-opted popular sciences and used them to argue for their own social and educational advancement. Chapter 4, "The Manly Brow Movement," not only illustrates why early feminist activists embraced physiognomy and phrenology, but also exposes the cultural backlash they faced from gender conservatives. The fifth chapter, "Criminal Minds," examines the case files and institutional records of penitentiaries and moral reform institutions to demonstrate how bourgeois moral reformers used popular sciences to establish distinctions between themselves and the "rabble," between penitent prisoners and allegedly recidivistic wrongdoers, and between the "virtuous" and "vicious" poor. The final chapter, "Facing Race," then reveals how Black and white abolitionists deployed physiognomy and phrenology to undermine scientific racism. It argues that when used by people of color, popular sciences could be powerful weapons in the fight for racial justice.

By the mid-nineteenth century, many Americans conceived of race, class, and gender as immutable traits. New sciences like physiognomy and phrenology hinted that old hierarchies were not only legitimate but also based on bodily realities. And yet, countless Americans used physiognomy and phrenology to argue that the body and the mind were malleable entities, capable of moral, mental, and physical improvement. In the minds of many abolitionists and women's rights activists, popular sciences were exciting disciplines that supported their claims for universal human equality. Still, these sciences were deeply problematic.

When activists, abolitionists, and ordinary people embraced physiognomy and phrenology, they conceded that the body could be read for signs of inner worth. In doing so, they helped legitimize an insidious regime of biopolitical scrutiny—one that would have devastating consequences for the poor, for Black Americans, and for women of all races. Undergirding the physiognomic project was a troubling assumption: perhaps some individuals had squandered their futures not because they were biologically destined to be inferior, but because they simply had not worked hard enough to overcome their supposed neurological and anatomical inadequacies. Physiognomy and phrenology ultimately enforced existing inequities, but they also convinced many progressive thinkers that hierarchy itself was both unstable and impermanent. The advocates of racial and gender justice who embraced these disciplines were not heroic radicals, nor were they hopelessly deluded dreamers. They were active and equal participants in a vibrant—and yet problematic—

scientific culture. Sometimes, they were subversive proponents of an ethical worldview that undermined existing power structures. At other times, they were inadvertent co-fabricators of a discriminatory system that devalued their own brains and bodies. If white women and people of color were shapers of their ideological climate, then they were the victims of it, too. This book tells their story.

1

FOUNDING FACES

What does a United States citizen look like? As Americans braced themselves for war with the British in 1775, Abigail Adams made a few conjectures. After meeting George Washington for the first time, she told her husband she had been "struck" by his appearance. "Modesty marks every line and feature of his face," she gushed, writing that "the gentleman and soldier, look agreeably blended in him." Months later, she used similar language to describe the famed diplomat, Benjamin Franklin. "You know I make some pretensions to physiognomy," she declared, "and I thought I could read in his countenance the virtues of his heart, among which, patriotism shone in its full lustre." For Abigail Adams, it seemed obvious that moral and mental characteristics would imprint themselves on people's faces. Sketched in the visages of men like Washington and Franklin, she detected traits like modesty, martial strength, patriotism, and virtue. These were necessary characteristics for the model citizens of a new republic.[1]

Abigail Adams styled herself a physiognomist, but she wasn't exactly using this science to discover new information. After all, she had made up her mind about George Washington and Benjamin Franklin long before she scrutinized their visages. In his response, John nevertheless praised her "skill in Phisiognomy," as well as her "Talent at drawing Characters." In these friendly letters, Abigail tried her hand at a new transatlantic science while John signaled his own awareness of the same discourse. But these letters were far more than an informal exchange of cultural and scientific knowledge. When Abigail analyzed the appearances of Washington and Franklin, she was assuring herself that they were honorable men who would bravely lead their compatriots in a grand political experiment. Assuming it was possible to identify the signs of republican virtue through an examination of people's faces, she hinted that some individuals were better qualified to run the country than others. Washington and Franklin, she speculated, had been blessed with an impressive array of inner qualities—a set of moral and mental traits that would manifest themselves in the body. By writing about their countenances, she was outlining the anatomical indicators of disinterested republican citizenship.[2]

Physiognomy began as a European science, but it attained a unique resonance in the United States. After the War for Independence, Americans embarked upon a challenging task: building a new nation from the scattered remnants of a colonial past. Their fledgling republic was fragile, and they feared it might easily succumb to corruption, vice, and aristocratic excess. Only by cultivating personal virtue in the country's citizenry could they forge a viable political system. Precisely because Americans believed the integrity of the country's citizens was critical to the nation's success, they placed immense value on the ability to discern people's inner character. This meant that physiognomy played an important role within the early American republic. Particularly in the 1790s, writers and political thinkers began claiming it was possible to see a man's civic virtue by examining his countenance. Through facial analysis, Americans hoped they might be able to distinguish between those who were truly respectable—genteel individuals with the capacity for republican citizenship—and those who were supposedly undeserving of political rights.[3] Physiognomy proved especially appealing for white Americans in the elite and middling classes because it promised a method for delimiting the boundaries of political belonging. To help people discern who "counted" as a virtuous citizen and who did not, authors, artists, and engravers published portraits and "character sketches" of the United States' most distinguished white men. In doing so, they crafted a textual and visual archive where the countenances of the nation's founders represented republican virtue personified.

American writers and political thinkers likewise used physiognomy to establish class divisions, formulate gender ideals, and construct rudimentary ideas about racial difference. By assuring the public that true republican women had a particular appearance, they argued that proper wives and mothers might be identified by their facial features—and that traits such as modesty, benevolence, intelligence, and virtue would be visible on white women's visages. They also used science to associate mental eminence with the visages of white genteel men, crafting a vision of citizenship that optically excluded people of color from the body politic.

Physiognomy ultimately served three major functions in the early national decades. First, it assured the citizenry that their leaders were illustrious men, perfectly suited to guide their constituents through the novel and tumultuous ordeal of nation building. Second, it helped Americans convince themselves they were just as good as their European counterparts, both in body and in mind. Finally, physiognomy provided mental comfort to society's most privileged members. By scientifically confirming that some people's brains and bodies were better than others', it solidified the notion that certain individuals were uniquely suited for the rights and responsibilities of republican citizenship.

THE BIRTH OF A NEW SCIENCE

At least since the time of Plato and Aristotle, the ancient Greeks had studied human countenances for signs of internal character. Yet for hundreds of years, Europeans imagined physiognomy as more of a mystical art, classifying it alongside disciplines like astrology, necromancy, palm reading, and alchemy. In 1775, a Swiss cleric named Johann Caspar Lavater upended this reality by publishing the first volume of the *Physiognomiche Fragmente* (1775–1778), a series of essays which would spawn over a dozen translations and at least twenty English-language editions by 1810. With rhetorical flamboyance, Lavater insisted physiognomy could be a credible science, and he maintained that the skeletal structure of the human countenance revealed people's innate character. His volumes soon captivated the attention of millions, transforming the practice of reading faces from a marginal form of mysticism into a widespread cultural phenomenon.[4] By the early decades of the nineteenth century, Lavater was a transatlantic superstar. Authors and advertisers referred to him as a "celebrated" or "ingenious" thinker, publishers reprinted his maxims, and people from all over the world traveled to his home in Switzerland for personalized facial readings. After his death in 1801, the *Scots Magazine* dubbed him "one of the most famous men in Europe."[5]

In those early years, Americans gained access to physiognomical theory by reading the French translations of Lavater's treatises or by seeing fragments in British magazines. In the 1780s, American periodicals began printing short translations of Lavater's French and German texts, lamenting that an English version of the *Essays on Physiognomy* still did not exist.[6] British publishers released the first English-language volumes of the *Essays on Physiognomy* in 1789, and the first American edition appeared in 1794. From that point forward, Lavater's treatises appeared in parlors and libraries on both sides of the Atlantic.[7]

The multi-volume *Essays on Physiognomy* comprised enormous tomes with fancy decorative plates, but printers and authors regularly shortened, adapted, and reprinted portions of Lavater's text to reach a wider popular audience. For those who could not afford the elaborate illustrated volumes, there were smaller editions like the "Pocket Lavater," the "Juvenile Lavater," or the "Aphorisms on Man."[8] American editors and publishers also reprinted sections from the *Essays on Physiognomy* in periodicals and newspapers, and they regularly invoked Lavater's methodologies to provide "character sketches" of real and fictional characters.

By the early nineteenth century, most Americans would have been familiar with physiognomic theories. As Dror Wahrman has contended, the *Essays on Physiognomy* were "reprinted, abridged, summarized, pirated, parodied, imitated, and reviewed so often" that it would be "difficult to imagine any literate, semi-literate, or otherwise culturally conscious person remaining unaware of its basic, and deceptively simple, claims."[9] Even people who never gained access to Lavater's tomes would have encountered his ideas in the published texts they encountered daily. As New York's *Lady's Monitor* proclaimed in 1801, the *Essays on Physiognomy* took the world by storm: "In Switzerland, in Germany, in France, in Britain, and in America, all the world became passionate admirers of the physiognomical science of Lavater." It seemed as if "everyone was eager to learn to read his neighbor's heart in his face," and the *Essays on Physiognomy* "were thought as necessary in every family as even the bible itself."[10]

Despite the widespread cultural enthusiasm for physiognomy, Americans did not universally accept Lavater's theories. Scientists, philosophical thinkers, and ordinary people regularly criticized his efforts to frame his science as an empirical form of observation. Yet even as people questioned physiognomy's legitimacy, few rejected the central premise undergirding the discipline: the idea that facial features and expressions conveyed useful information about a person's internal character.[11] This was, in part, because Lavater's method capitalized on a cultural practice with long historical roots. Eight years before the first volume of *Essays on Physiognomy* was published, John Adams had used physiognomical language in a flirtatious letter with his then love interest, Abigail Smith. John described one of Abigail's friends as "a Buxom lass," cheekily confessing that he "longed for a Game of Romps with her." Yet John also analyzed the woman's countenance, saying she had "An Eye, that indicates not only Vivacity, but Fire—not only Resolution, but Intrepidity." His interpretation quickly veered into critique. Adding the parenthetical disclaimer, "(Scandal protect me, Candor forgive me,)" he wrote, "I cannot say that the Kindness, the softness, the Tenderness, that constitutes the Characteristick Excellence of your sex . . . are very conspicuous Either in her Face, Air or Behaviour." Despite the lack of warmth in the woman's countenance, she was apparently appealing enough for a "Game of Romps." Luckily, she escaped Adams's unsavory advances (if not his armchair physiognomical analysis).[12]

The fact that Adams wrote this letter in 1764—several years before physiognomy enraptured the transatlantic world—hints at why the science became so popular in the first place. By the 1770s, Europeans and Americans were already used to analyzing people's faces. Lavater was not the first practitioner of physiognomy, nor was he the first to write a book on the subject. His primary innovation lay not in his originality, but rather in his aptitude for transforming a long-standing cultural pastime into an accessible and pragmatic methodology, one that promised to help people discern character with scientific precision.[13]

In medical discourses, popular literature, and daily practice, a broad-based faith in the readability of the human countenance uncomfortably coexisted with widespread skepticism of physiognomy's scientific legitimacy. This meant that Americans analyzed faces with enthusiasm and regularity, despite a reluctance to believe in the discipline's infallibility. When the Quaker Elizabeth Drinker checked Lavater's *Essays on Physiognomy* out of the library, for instance, she assessed it in her diary: "I believe there is a great deal in what he advances and am not of the oppinion [sic] of those who say, he is a madman, or out of his sences [sic]." Drinker was nonetheless skeptical of Lavater's contention that physiognomy was an unimpeachable discipline. Although she did not distrust the science entirely, Drinker believed Lavater had overstated his own talents. "I think he carries some things much too far, and has rather too much conceit of his abilities," she declared. While Drinker harbored mistrust for physiognomy's most bombastic claims, the person who wrote her obituary was apparently less skeptical. Memorializing Drinker's life, the author said she "possessed uncommon personal beauty, which the gentleness of her temper preserved, in a great degree, to the last; for her countenance was a perfect index of a mind, whose feelings were all attuned to harmony."[14]

By 1799, physiognomy had become so popular that Joseph Bartlett focused his Harvard University commencement speech on the subject. One aspiring Philadelphia physician had so much faith in the science that he decided to write his dissertation on it. His doctoral thesis, "An Essay on the Truth of Physiognomy and Its Application to Medicine," earned him a degree from the University of Pennsylvania's medical school in 1807.[15] Benjamin Rush, the nation's most famous physician, also dabbled in the subject, suggesting that "genius" typically announced itself through "a certain size of the brain, and peculiar cast of features, such as the prominent eye, and the aquiline nose."[16]

Between the 1810s and the 1830s, Nathaniel Chapman, physician, professor of medicine, and editor of the *Philadelphia Journal of the Medical and Physical Sciences*, gave his students lectures on the "Physiognomy of Diseases."[17] John Kearsley Mitchell, another Philadelphia physician, collected physiognomical sketches of the mental patients in residence at the Pennsylvania Hospital.[18] Isaac Ray, an early pioneer in the field of psychiatry, published *Conversations on the Animal Economy* (1827), an educational text with two chapters dedicated to simplifying physiognomical principles for schoolchildren.[19]

Far more than a trivial and evanescent fad, physiognomy was a phenomenon that enraptured the transatlantic world for several decades. Almost sixty years after Elizabeth Drinker first critiqued Lavater's *Essays on Physiognomy*, Catharine Maria Sedgwick was still debating the merits of the science in a letter to her niece. In 1852, a man named "Dr. Redfield, a professor of the art of reading physiognomy," came through Sedgwick's town to do personalized facial readings. She expressed suspicion of physiognomy's legitimacy, saying Redfield "pretends that it is an exact science." Even so, Sedgwick was impressed by his conclusions and was pleasantly surprised when he accurately captured people's characters: "truly his readings here were wonderful," she gushed.[20] In an analysis of the New York Society Library records, one scholar discovered that Lavater's *Essays on Physiognomy* was still one of the most popular books for female borrowers in the late 1840s and 1850s. Puzzling over this fact, he speculated that these patterns signified "an unexpected interest in a rather unorthodox science." For early Americans, though, there would have been nothing "unorthodox" or "unexpected" about this finding. The fact that educated people were still discussing and debating Lavater's theories at mid-century demonstrates just how profoundly his work had suffused the broader culture within the United States.[21]

Americans knew physiognomy was not perfect. Yet even some of Lavater's fiercest critics admitted there was a kernel of truth in the science. As one periodical put it in 1822, "The basis of physiognomy, that the face is the silent echo of the heart, is substantially true."[22] That same year, Louisa Adams articulated a similar sentiment in a letter to her husband, John Quincy Adams. Assessing a woman that she met in Philadelphia, she wrote, "Mrs. Ingersol is a very pretty Woman, and as far I can judge she is intelligent and well educated—There is however something in her eye that repels." Louisa knew she had a "propensity of passing judgment at first sight," and she recognized this habit as a flaw in her own character. She also doubted physiognomy's reliability, joking that "Poor Lavater little knew the mischief he was to do" when people like her "adopted his system without his skill or his combinations to guide them." Adams knew she was not the only person to judge someone at face value and then high-mindedly criticize herself for doing so. She also acknowledged that physiognomy had an impact on interpersonal interactions, "however disinclined we may be to trust it." Through this comment, she showed that it was entirely possible to harbor doubts about physiognomy while continuing to use it. At a time when neither Americans nor their European counterparts knew much about the complexities of human cognition, physiognomy became one of the most efficient ways to penetrate the deepest recesses of the mind and soul. It is hardly surprising, then, that people began using this science to make sense of social and political hierarchies.[23]

THE FACE OF A CITIZEN

When Americans declared independence from the British Empire, they rejected the notion that society should be governed by an aristocracy of birth: an unequal arrangement in which people inherited their social and economic position from their ancestors. Instead, Americans insisted they were forging an aristocracy of merit, in which individuals could earn their social position through their intelligence, virtue, and hard work. As long as someone assiduously cultivated their unique combination of personal talents, they were sure to succeed. Soaring rhetoric aside, this imagined aristocracy of merit was not as egalitarian as the nation's founders might have liked to believe. As Americans forged a new political structure, they did not eliminate the institution of slavery, nor did they grant legal, political, or economic rights to women. In the early national decades, many states did not extend voting rights to poor white men. Even though the nation's founders replaced the British monarchy with a new republican form of government, they maintained the inequities that had plagued the old system.

Scientific facial analysis helped them reconcile these contradictions. When Americans interpreted people's heads and faces, they convinced themselves they were rationally and objectively discerning character. They weren't. In most cases, people were already acquainted with the individuals they tried to analyze. At the very least, they *assumed* they knew them well enough to determine their moral and mental capacities. Still, popular sciences were comforting. They allowed Americans to make broad pronouncements about their contemporaries and then retroactively validate those beliefs by invoking the explanatory power of science. Perhaps some people had been blessed with superior brains and bodies, they posited. If that was true, then the nation's leading citizens deserved the privileges they had accumulated. After all, were they not superior specimens of humanity—people with illustrious minds and virtuous characters who just happened to be thriving in a meritocratic republic? With the help of popular sciences, Americans assured themselves that "true" republican citizens looked a certain way, and that certain people's bodies disqualified them from full membership in the body politic.

As physiognomy grew in popularity during the early national decades, elite and middling white men increasingly used its tenets to rationalize their own privileged position within society and marginalize the figures they deemed less intelligent, virtuous, and capable than themselves. Arguing that civic virtue was both a moral and a physical quality, the country's artists, politicians, and popular authors contended that the capacity for republic citizenship was an internal reality that manifested itself in the visages of the nation's most prominent white men. These efforts constituted a dual attempt to elucidate the eminence of the US population and accentuate the moral, mental, and physical distinction of the country's most exemplary citizens.

During the 1780s and 1790s, artists such as Charles Willson Peale, Gilbert Stuart, Pierre du Simitière, and Charles B. F. de Saint-Mémin began compiling physiognomic galleries of distinguished Americans. Their goal was simple: to give the public access to the likenesses of the nation's most eminent men so that the masses might emulate the mental and moral characteristics of their supposed social betters.[24] In these portrait galleries, figures such as George Washington, Thomas Jefferson, Benjamin Rush, and Benjamin Franklin became embodied versions of abstract political ideals. As the model citizens of a new nation founded on republican principles, their faces became anatomical representations of virtue, gentility, and patriotic disinterestedness. Almost invariably, the people whose portraits graced the pages of printed portrait galleries and adorned the walls of museum exhibits were wealthy, white, and male. These illustrations—and the supposedly scientific descriptions that accompanied them—suggested that the bodies of white women and people of color were simply not capable of displaying the qualities of true republican citizenship.[25]

Charles Willson Peale's "Repository for Natural Curiosities" is perhaps the most famous of these pantheons. In 1786, he opened a museum featuring the busts and visages of influential Americans within a "Gallery of Distinguished Personages." Peale insisted that his portraits were doing important political work. As a believer in physiognomy, he claimed that the examination of profile portraits would "be a very certain means of studying Characters, to determine the measure of Intulects [sic] as well as disposition." In his mind, physiognomical renderings would "instruct the mind and sow the seeds of Virtue" by encouraging the general public to examine the countenances of exemplary figures, discern traits of intelligence and morality in their visages, and then imbibe these characteristics through rational observation and emulation.[26] Peale sought realistic depictions of his sitters' visages. The goal was to use portraiture to convey deeper information about people's inner nature. As he wrote in his diary in December 1817, "It is the mind I would wish to represent through the features of the man, and he that does not possess a good mind, I do not desire to portray his features."[27] Because Peale thought that countenances conveyed character, he tried to highlight the individual particularities of people's faces—a task he viewed as especially important when painting the nation's most prominent leaders.[28]

Peale intended his museum to convey both scientific and political messages. In his worldview, certain people's minds—and bodies—were more distinguished than others'. For this reason, he organized his collections in ways that invoked both the "Great Chain of Being" and the new Linnaean classificatory system. Using portraits, sculptures, and physical specimens, he visually exhibited a divinely sanctioned order. Images of rich, white, and politically powerful men hung near the ceiling, towering over specimens of insects, birds, and other mammals. Every species had a providentially ordained place. At the top were the nation's founders, Revolutionary War heroes, and eminent scientists and intellectuals. Such displays portrayed republican citizenship as a privilege restricted to a certain subset of white men.[29]

These galleries also served an important patriotic function: they confirmed that American citizens were equal to—and perhaps better than—Europeans. In the late eighteenth and early nineteenth centuries, natural historians like the Comte de Buffon had described the American continent as inferior, in its landscape and population as well as in its flora and fauna. Other Europeans likewise argued for the inferiority of American minds and bodies. In his *Sketch of the United States of North America* (1814), the French author Louis-Auguste Félix Baron de Beaujour contended that Anglo-Americans had "little delicacy in their features, and little expression in their physiognomy." Americans were not necessarily "ugly," he conceded, but their "tall forms, ruddy and soft," often concealed an "obtuse mind and soul devoid of energy." This deficiency was readily observable in their countenances. It was apparently due to "this vice in their physical constitution, more than to their geographical position, that the eternal irresolution of their government is owing." Through this statement, the Baron de Beaujour tied the political instability of the early republic to the inferior minds and visages of its citizens. He nonetheless assured readers that Americans were not doomed to perpetual inadequacy. Over time, "their temperament will improve with their climate," and as a result, "the Americans will some day or other acquire more vivacity of mind and more vigour in their character." Relying on an environmentalist understanding of human nature, he described American bodies as simultaneously inadequate and improvable.[30]

Well into the early decades of the nineteenth century, foreign authors continued to suggest that Americans were inferior versions of their European counterparts. In 1824, several US periodicals reprinted an article from *Blackwood's Magazine* that had compared the American physiognomy unfavorably with the British. If British faces were "full of amplitude, gravity, and breadth," then American countenances had "more vivacity, and a more lively character." At first glance, this description does not seem particularly negative. But the author had more to say. Although he admitted the "English face" conveyed a "pompous expression," he claimed this smugness was justified. After all, the English were an "extraordinary people," members of "an empire that never had—has not, and never will, have a parallel on earth." The English might be arrogant, but they had every right to be. They were the greatest, most powerful people on the planet. Americans, by contrast, were "exceedingly vain, rash, and sensitive." They, like the British, were proud of their country. They, too, exuded confidence. But American self-assurance was artificial at best and delusional at worst. Residents of the United States had a bad habit of peevishly insisting upon their own greatness, but they did not—and simply could not—compare to the British, either in body or in mind. The writer implied that, deep down, Americans already knew this. Since they were "less assured of the superiority" of their country, they had developed a "waspish and quarrelsome" appearance, "like diminutive men, who, if they pretend to be magnanimous, only make themselves ridiculous." "And so," the author continued, "in the keen, spirited, sharp, intelligent, variable countenance of an American, you will find a correspondent indication of what he is." With a frankness bordering on brusqueness, he proclaimed that Americans were not fooling anyone. Despite their bravado, their inferior countenances betrayed their inadequacies as a people and a nation.[31]

As early national artists and political thinkers saw it, the only way to combat such discourses was by creating a visual archive of American distinction. In response, they patriotically highlighted the superior brains and bodies of the nation's most distinguished citizens. When the Philadelphia publisher Joseph Delaplaine published his *Repository of the Lives and Portraits of Distinguished American Characters* in 1815, he complained that the "writers of Europe" had unfairly tried "to degrade the character of the natives of America." Delaplaine insisted that the people of the United States were "the immediate descendants of European ancestors," excluding both African Americans and Native Americans from his vision of national identity. He then lamented how his countrymen had "been declared to be inferior, both in body and intellect, to those who are born in the eastern hemisphere." He nevertheless hoped his "galaxy of genius" would bely the claims of European authors by exhibiting the "transcendent greatness" of white Americans.[32] Was there any better moment "to pass by these [European] dotards in refinement, and to become their tutor in the true art of political greatness?" According to Delaplaine, this feat could be accomplished by publishing the "correct and striking likenesses" of America's most "distinguished characters."[33]

Between 1820 and 1827, John Sanderson embarked upon a similar crusade, printing a multi-volume *Biography of the Signers to the Declaration of Independence*, which included physiognomical sketches of the nation's founders. His work went through numerous editions, with magazines and newspapers repeatedly reviewing and reprinting excerpts from the volumes.[34] These books included a detailed physical description of Benjamin Rush, the nation's most influential physician. His "uncommonly large" skull and "prominent forehead" were signs of his mental acumen, as were his "aquiline nose, highly animated blue eyes, with a chin and mouth expressive and comely." Sanderson argued that Rush's "thoughtful" visage "and the general traits of his physiognomy bespoke strength and activity of intellect." His books made similar claims about other politicians. Thomas Jefferson's countenance was "remarkably expressive, his forehead broad . . . and the whole face square and expressive of deep thinking." Gouverneur Morris "possessed a fine, open, and benevolent countenance" with "strong" features, which appeared "austere" when he was in the midst of "deep meditation." Edward Rutledge, for his part, had a heart "so well expressed in his fine countenance, that the dullest physiognomist could scarcely mistake the delineation of its feelings." All these descriptions crystalized into an unmistakable message: the nation's founding fathers were a superior set of beings, with mental and physical attributes that separated them from the general population.[35]

Sanderson's *Biography* was followed by a series of printed pantheons. James Thacher released the *American Medical Biography* in 1828, a tribute to the nation's greatest physicians, followed by James Herring and James Barton Longacre's *National Portrait Gallery of Distinguished Americans* in 1834. In 1845, Stephen W. Williams released the *American Medical Biography*, and in 1856, Rufus Wilmot Griswold released a memorial of *The Republican Court; or American Society in the Days of Washington*. Each collection provided visual and textual "evidence" of America's greatness by highlighting the biographies and physiognomical portraits of its most accomplished white men. Despite its European beginnings, physiognomy attained a singular importance in the early republic as people used it to argue for the exceptionalism of the new nation and the distinction of its leaders. In doing so, they invoked a European science to craft a uniquely American identity.

What, then, did a respectable US citizen look like? The answer was not always clear. After all, physiognomy was not a reliable system with invariable or reproducible rules. Despite his insistence on physiognomy's exactitude, Lavater had suggested a tautological method for learning the mysteries of the human visage. He instructed newly minted face decoders to begin their studies with the easiest examples. If one was trying to identify the physiognomical symbols of a degraded mind, his advice was simple: "visit hospitals for idiots," and sketch "the most remarkable traits of the most stupid." Eager to identify the physical symbols of intellectual power? Seek out "men of wisdom and profound thought, and proceed as before." In other words, Lavater argued that budding physiognomists did not have to begin with the facial features at all. They should start by making broad assumptions about entire groups of people. Then, after observers had *already* come to a decision about a person's mind, they could retroactively discern those internal traits in the external features. It was a circular form of logic, which undercut Lavater's claim to have created a rational and dispassionate form of anatomical observation.[36]

Within such a system, early Americans could draft physiognomical rules that conveniently confirmed the mental and moral distinction of their leading scientific and political thinkers. The public debates about George Washington's visage provide an example of how this process unfolded. Of all the faces they highlighted, the nation's authors, artists, and writers were most captivated by the countenance of their first President. One periodical contributor, for instance, waxed poetic about the genius and physical appearance of "the greatest man America had ever produced."[37] Similarly, before launching into a detailed description of Washington's character, the author John Bell insisted it was incumbent on him to sketch "the person of this distinguished man," because it "bears great analogy to the qualifications of his mind." In Washington's "sensible, composed, and thoughtful" physiognomy, Bell argued, viewers could see "the striking features of his character," which included affability, simplicity, and sincerity. He claimed that "no man ever united in his own person a more perfect alliance of the virtues of a philosopher with the talents of a general."[38]

Bell was not alone in his obsession with Washington's body. Samuel Powel, one of Washington's closest friends, decided to sketch his own silhouette of the general during a tea party. Powel was undoubtedly influenced by physiognomy, as evidenced by the fact that he copied several paragraphs from the French edition of Lavater's *Essays on Physiognomy* into his account book around the same time.[39] Even if he had not been reading his Lavater, though, it would have been hard to avoid physiognomical discussions of Washington's visage in the 1790s. Periodicals like the *Philadelphia Monthly Magazine* were claiming that his "countenance carries the impression of thoughtfulness," while the *American Museum* suggested that the "general's goodness beams in his eyes."[40] Emphasizing both his martial prowess and his cheerful disposition, the *Rural Magazine* proclaimed that Washington's "whole aspect pronounces him the hero," even as "the mildness and benignity of his countenance" revealed that he was more peaceful than pugnacious.[41] Joseph Delaplaine likewise suggested Washington's "corporeal majesty" was so bewitching that whenever viewers gazed upon "the lineaments of his face," they lost themselves in "the impress of his spirit and the expression of his intellect."[42]

Writing in 1824, Rembrandt Peale explicitly articulated what was at stake when memorializing Washington's visage. Even if "his features had been more ordinary, and his expression less distinguished, the rising generation would still wish to know his own peculiar look." Yet Washington was far from ordinary. He was the corporeal manifestation of republican virtue: "his aspect was as noble as his character," and "his countenance corresponded with his conduct." Peale intended to give posterity "the true and impressive image" of Washington. His concern was both artistic and scientific. If the President's face conveyed his inner nature, then it was critically important to commemorate his appearance properly. Peale wanted future generations to gaze upon Washington's image, optically absorb his sublime genius, and be inspired to "emulation and patriotism."[43]

Such portraits of political luminaries did not just provide object lessons in republican virtue for American citizens; they also allowed Americans to show off for European readers. People in the United States must have been particularly proud to see that both Washington and Benjamin Franklin made appearances in the *Essays on Physiognomy*. Before these works were translated into English, Philadelphia's *Columbian Magazine* reprinted Lavater's engravings of the two men (fig. 1.1). Henry Hunter's English translation of the *Essays on Physiognomy* (1789) then provided a portrait and detailed description of Washington's visage. Noting the President's military distinction, Lavater asked, "would not the Physiognomist be eager to know the features of him, whom Providence selected as the instrument of effecting a revolution so memorable?" Answering the question in the affirmative, the book emphasized Washington's "expression of probity, wisdom, and goodness" (Fig. 1.2).[44]

Lavater's analysis, while flattering, exposed the slippery logic undergirding his science. Before scrutinizing Washington's face, Lavater had already concluded that his subject was a Revolutionary hero. He faced a conundrum,

FIGURE 1.1. "Portraits from Lavater's celebrated essay on Phisyognomy [sic]," *The Columbian Magazine* (Philadelphia), March 1788. National Portrait Gallery, Smithsonian Institution.

however. Washington's forehead simply did not live up to expectations. Although the engraving showed "uncommon luminousness of intellect," it portrayed a brow "deficient in point of depth," which seemed to "exclude penetration." His eyes were "replete with gentleness and goodness, but they possess neither that benevolence, nor prudence, nor heroic force, which are inseparable from true greatness." How were viewers supposed to reconcile this contradiction? Fortunately, the ever-inventive physiognomist devised a solution. Whenever Washington's portraits failed to exhibit the stunning intellect and "noble boldness" that Lavater expected, he simply blamed the engraver. Since Washington had already illustrated his worthiness by spearheading the Revolution, his distinction was not in dispute. Surely, then, it was "the Designer" who had failed. This allowed Lavater to maintain his faith in physiognomic principles (which he never systematically defined, despite writing hundreds of pages on the subject), even as he made exceptions for individuals whose portraits did not measure up to the discordant standards he delineated.[45]

What, then, were the corporeal indicators of moral and mental distinction? The answer was rarely clear. The rules of physiognomy were murky, leading both Americans and Europeans to contest its legitimacy as an "exact science" from the moment the *Essays on Physiognomy* achieved international acclaim in the 1770s. Still, even the most diverse practitioners managed to settle on a few general rules. Physiognomists largely agreed the human body could be

FIGURE 1.2. Portrait of General Washington, in Johann Caspar Lavater, *Essays on Physiognomy: Designed to Promote the Knowledge and Love of Mankind*, trans. Henry Hunter, vol. 3 (London: Murray and Highley, 1798). Courtesy of the Winterthur Library: Printed Book and Periodical Collection.

divided into three major components: mental, moral, and animal. According to Lavater, the face and head represented the mental capacities, the chest signified the moral capacities, and the digestive and reproductive organs embodied the animal propensities. But he also believed the face was a miniature visual distillation of all these properties, displaying a mix of moral, mental, and animal traits. Following Lavater's lead, European and American physiognomists typically divided the head into three parts. The brow and forehead supposedly connoted intellectual capacity, with the eyes being the gateway to the soul. The cheeks, nose, and middle portions of the face indicated morality. Finally, the mouth and jaw indicated one's animal propensities. This tripartite division of the human countenance provided a general guideline for both amateur dabblers and serious practitioners of physiognomy.[46]

Beyond these general precepts, the consistency of physiognomic doctrine broke down quickly. The "rules" were vaguely defined and poorly outlined from the discipline's very inception. Christopher Lukasik has shown that Lavaterian physiognomy purported to "read moral character from unalterable and involuntary facial features."[47] And yet, as the literary scholar Christopher Rivers has demonstrated, Lavater's method was both "contradictory and incoherent." His "system" was hardly a system at all. Even if we "suspend our disbelief for a moment and try to imagine ourselves as earnest seekers of physiognomical method," argues Rivers, "we come up against the undeniable problem that there simply is no method in Lavater's work."[48] Lavater claimed the bones of the countenance invariably indicated one's permanent and unalterable nature, but he often broke his own rules and adapted the guidelines he laid out for his readers. As the *New-England Galaxy* complained in 1818, "no certain rules can be found for the practice of this science." In the end, Lavater's tomes were full of ambiguous pronouncements, idiosyncratic facial readings, and a veritable onslaught of conflicting advice.[49]

Americans nonetheless embraced Lavaterian ideas, using them to craft physiognomical standards for republican citizenship. Beyond the basic requirements of whiteness, masculinity, and gentility, the most important trait was a large and expansive forehead: the symbol of intellectual refinement. Over and over again, newspapers, magazines, and books described the nation's leaders as men with high brows. Americans also tended to associate dark eyes with intellectual profundity. This sometimes presented a problem. Many of the nation's leaders—including such luminaries as George Washington and Thomas Jefferson—had light complexions with fair hair and light eyes. Luckily, the rules were easily adaptable. "Blue eyes are generally more significant of weakness, effeminacy, and yielding, than brown and black," Lavater wrote, even as he contended there were plenty of "powerful men with blue eyes." Adding a bigoted coda to these broad pronouncements, Lavater noted what was (to him at least) a curious fact: people of Asian descent almost never had blue eyes, even though there were "no people more effeminate, luxurious, peaceable, or indolent than the Chinese." He quickly moved on from this inconsistency without addressing the racist and contradictory assumptions of his own logic. As such examples make clear, it was easy for physiognomists to believe that they were decoding people's minds by analyzing their appearances, even if they were subconsciously using facial analysis to rationalize and confirm their preconceptions.[50]

Physiognomists insisted that their science was capable of determining people's character with meticulous exactitude, but it never was. Readings were personalized and idiosyncratic, and a person's public identity invariably mattered more than their features. Everyone could agree that George Washington had a striking physiognomy because they had already collectively decided that he was an impressive man. When eminent figures did not live up to established archetypes, people simply reformulated the rules or carved out exceptions. George Washington's eyes might be mild, observers admitted, but they were also "pensive," "benevolent," and "noble," with an "air of gravity." Practitioners could personalize their interpretations, finding a wide array of character traits, depending on what they were looking for.[51]

Free to draw their own conclusions, Americans sometimes wielded physiognomy as a weapon against their political enemies. Resentful that Alexander Hamilton had tried to "secretly" and "cunningly" thwart her husband's election to the presidency, for instance, Abigail Adams turned to physiognomy to discredit her rival. "O I have read his Heart in his wicked Eyes many a time. The very devil is in them," she wrote. "They are lasciviousness itself, or I have no skill in Physiognomy." Knowing that she was engaging in indecorous behavior, she followed her insults with a request: "Pray burn this Letter. Dead Men tell no tales. It is really too bad to survive the Flames." In a victory for historians, John never burned the letter—Abigail's physiognomic barbs are now safely preserved in the archives.[52]

The Adamses were not just *practitioners* of facial analysis, though. They were also the *targets* of physiognomic warfare. At some point between 1820 and 1880, a young Quaker woman named Emma Howard Edwards copied physical descriptions of eminent men into her scrapbook. One excerpt was an analysis of John and Abigail's son, John Quincy Adams, taken from a Virginia newspaper. No fan of the Massachusetts politician, the paper described him as a "suspicious," "malignant," "unforgiving," and "revengeful" man who had few friends and betrayed everyone who trusted him. Turning to a description of Adams's appearance, the author wrote that the "furtive weariness of his small gray eye, his pinched nose, receding forehead, and thin, compressed lips, indicated the malignant nature of his soul." Clearly, the paper had other reasons to resent Adams. He was anti-slavery in sympathy, and his father had been an arch political rival of Virginia's favored son, Thomas Jefferson. Still, the paper mostly relied on physiognomical tropes to make its point. If Adams's "pinched nose" and thin lips indicated his secretiveness and viciousness, then his "receding forehead" signaled a brain in decline. Emma Howard Edwards then copied the description into her scrapbook, pairing it with character sketches of other powerful men. Her scrapbook reveals how Americans used physiognomy to navigate political conflict, but it also shows how easily scientific knowledge could move between elite treatises, popular prints, and ordinary people's manuscripts. Both in public and in private, Americans used physiognomic language to evaluate their political leaders.[53]

THE FACE OF HIERARCHY

In the wake of the Revolutionary War, the nation experienced a series of transformative political and economic developments. As the market economy expanded, young men moved into urban centers, undermining older methods of maintaining the patriarchal order. Moreover, as cheap clothing became available and social mobility increased, it became harder to tell genteel individuals from the clever tricksters and confidence men who were waiting to corrupt an otherwise virtuous American citizenry. Physiognomy gave more privileged Americans hope that they would be able to read faces properly, and in the process, discern the dissimulators in their midst. By rooting social status in the countenance, rather than clothing or behavior, facial analysis provided a seemingly stable method for separating distinguished individuals from "the rabble."[54]

Readers of physiognomic books and articles would have learned how to tell a "man of business" or "lawyer" from a "scoundrel" or "rogue," regardless of how they disguised their appearance. Since the skeletal features were supposedly inherited and immovable, a careful physiognomist would never be fooled by a well-dressed crook. Authors likewise instructed elite and middling Americans to consult the science of facial analysis when hiring servants, forging friendships, and "choosing matrimonial or mercantile partners."[55] Noah Webster, for instance, proclaimed that he would never elect a politician with "Lavater's strait nose," nor would he hire "a man or a woman with a monstrous under lip, to labour for me." Warning his readers to study visages "before you make a bargain," he drew an analogy between the emerging capitalist marketplace and the science of physiognomy: "Nature has been kind enough to hang, upon every man's face, the sign of the commodities for market within." There was no excuse to make bad decisions—in business, in politics, or in everyday encounters—because the human countenance was "the best letter of recommendation."[56]

Early national authors also used physiognomy to distinguish between the "virtuous" and the "vicious" poor. A careful observer, physiognomists argued, would be able to discern refinement in even the most hapless characters. In 1796, Mathew Carey published a story about a fictional, poverty-stricken heroine whose husband had just died. Consumed by sorrow and preoccupied with the care of her young children, the woman did not notice the man who was furtively staring at her. "I had therefore full leisure to exercise my skill in physiognomy," Carey's narrator mused. The woman's dress was both "homely and coarse," which "excited the idea of an uncultivated owner." But her face was different. According to the narrator, "Less skill than that of Lavater would suffice to discover marks of refinement and tenderness there, that might have done honor to elevated situations." Carey encouraged his readers to revel in their own distress: "Check not your tears, tender reader—Let them flow freely." After all, this moralistic tale was intended to evoke sympathy in its readers, assuring them of their own refined sensibilities. The story also had a deeper function: to conjure up a tangible image of virtuous poverty and provide readers with a template for identifying it in the human body. Carey's character was clad in ragged clothing, but her homely attire could not obscure her refined visage. She was so virtuous, in fact, that her neighbors raised over sixty dollars for her support. A woman like this, the tale suggested, deserved the benevolence of others.[57]

Just as physiognomy served as a scientific justification for class hierarchies, it also became one of the most important precursors to—and foundations for—the rise of scientific racism. Between the 1770s and the 1820s, Americans primarily used physiognomy to draw distinctions between white men. As Christopher Lukasik has shown, the vast majority of physiognomic depictions before 1825 focused not on people of color but on people of European descent.[58] Yet as soon as physiognomy made its earliest incursions into American culture, at least some authors began using it to rationalize white supremacy. In February 1792, a writer calling himself "Africanus" employed Lavaterian ideas to suggest that people of color were incapable of scientific learning, success in the arts, or "civilization" more broadly. If there was "any truth in the science" of physiognomy, he argued, then Black degradation was more than his own treasured opinion; it was, instead, a provable scientific fact. Africanus insisted "the physiognomy of the white man excels that of the black, as much as the appearance of the liveliest sagacious horse exceeds the stupid aspect of the dullest ass in the creation." Because of the implicit assumption of white male superiority that suffused physiognomic doctrines, this author quite easily adapted Lavaterian language to justify unequal race relations.[59]

Lavater himself did not spend much time discussing racial difference in the *Essays on Physiognomy*, but he built on the work of someone who did: a Dutch naturalist, artist, and anatomist named Petrus Camper. Camper's most important—and insidious—contribution to physiognomic racial science was his theory of the facial angle. According to Camper, the different "nations" of the world all had different skull structures and unique facial features. Camper preferred to analyze skulls in profile (a technique that Lavater enthusiastically adopted). To calculate the facial angle, Camper drew a horizontal line between a person's nose and ear, as well as a vertical line that stretched from the forehead to the chin and passed through the lips. Then he measured the angle where the two lines intersected (figs. 1.3 and 1.4).

Although Camper claimed that he was not trying to hierarchically rank human beings, he arranged his skulls on a continuum, with Europeans and the ancient Greeks conspicuously on one side and dogs, apes, and Africans together on the other. Camper argued that the faces of the ancient Greeks were the most beautiful, with facial angles approaching 100 degrees. This meant that their foreheads extended far beyond their chins: the mark of a large brain and small animal propensities. Caucasians, he claimed, were closest to the Grecian standard, with facial angles between 80 and 90 degrees. The other "nations" of the world—groups Camper referred to as "Moors," "Calmucks," and "Negroes"—had facial angles between 70 and 80 degrees. Camper believed in the universal humanity of all mankind, but he nonetheless devised a framework that white people would later use to suggest that people of African descent were a different species entirely. Scientists, politicians, and popular authors continued using Camper's facial line well into the late nineteenth century. In this way, physiognomic ideas contributed to one of the earliest mathematical justifications for white supremacy in the transatlantic world.[60]

"GENERAL REMARKS ON WOMEN"

In addition to using physiognomy to establish racial categories, Americans turned to this discipline when making sense of gender differences and prescribing social roles for men and women. From reading physiognomical works, Americans would have learned a few things. All white women were born with a unique capacity for motherhood; all were "weak," "beautiful," and "tender" beings; and all should be "subject to man." Lavater, for instance, argued a "comparison of the external and internal make of the body, in male and female, teaches us that the one is destined for labour and strength, and the other for beauty and propagation." He stated that the "bones in the female are more tender, smooth, and round; have fewer sharp edges, cutting and prominent corners," and he claimed that the "light texture of [women's] fibres and organs" made them "ready of submission to the enterprise and power of the man."[61] American periodicals reprinted Lavater's words throughout the early national decades, though they mostly failed to identify the famous physiognomist by name. Instead, they repackaged his words under broader titles such as "Characteristic Differences of the Male and Female of the Human Species" or "General Remarks on Women." At a time when plagiarism laws were virtually nonexistent, Lavater's comments on the nature of women were reproduced in cities as distant as London, Paris, Edinburgh, New York, Philadelphia, and Boston.[62]

Because Lavater's work traveled easily between elite intellectual circles and popular culture, his ideas about sexual difference shaped popular conceptions of gender, beauty, and human nature in the early American republic. His statements on gender difference soon became so influential within Anglo-Atlantic intellectual circles that early editions of the *Encyclopaedia Britannica* simply extracted his comments, parroted them directly, and substituted them for the encyclopedia's entry on "Sex." The first two editions provided only a brief definition: "SEX, something in the body which distinguishes male from female."[63] Later editions kept some version of this first sentence, but then followed it with Lavater's chapter on women, lifting the text straight out of the *Essays on Physiognomy*. Readers hoping to find a positive evaluation of the female mind in the encyclopedia were out of luck. Instead, they would have discovered Lavater's provocative contention that women were incapable of "deep inquiry," easily manipulated, and flightier than their male counterparts.[64]

Insisting on the mediocrity of the female brain, Lavater argued that women suffered from "extreme sensibility" due to "the irritability of their nerves." This resulted in an "incapacity for deep enquiry and firm decision" and a troubling tendency to "become the most irreclaimable, the most rapturous enthusiasts." As Lavater put it, "The female thinks not profoundly; profound thought is the power of the male." Although he framed these assumptions about women as the product of physiognomic observation, he only made

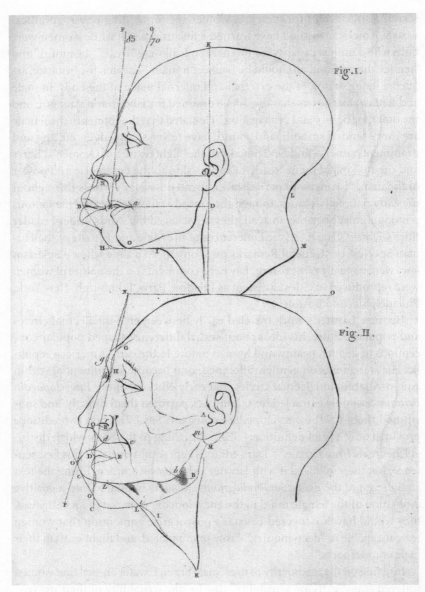

FIGURE 1.3. An illustration of Camper's facial angle, from *Dissertation sur les variétés naturelles qui caractérisent la physionomie des hommes des divers climats et des différens ages* (Paris: Chez Francart, 1792). The Library Company of Philadelphia. Scientific racists adopted the facial angle as a mathematical technique for defending white supremacy.

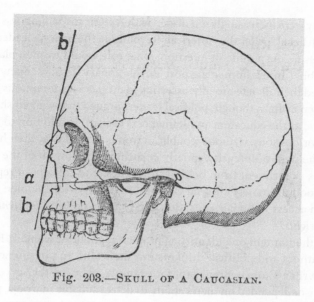

Fig. 203.—SKULL OF A CAUCASIAN.

FIGURE 1.4. "Skull of a Caucasian," from Samuel Wells, *New Physiognomy; or Signs of Character, as Manifested through Temperament and External Forms* (New York: S. R. Wells, 1870). Courtesy of the American Antiquarian Society. The facial angle was calculated by drawing a horizontal line between the nose and ear, and then intersecting it with a perpendicular line that passed from the forehead to the lips. Well into the late nineteenth century, American physiognomists, phrenologists, and ethnologists used Camper's facial angle to claim that white people were intellectually superior to people of color.

superficial attempts to tie these internal traits to physical characteristics. Engaging in a form of reverse physiognomy, he began with an assumption of female inferiority and then described this inferiority in physiological terms. His words eventually spread throughout the transatlantic world. In 1798, when Thomas Dobson issued the first encyclopedia to be published in the fledgling United States, he copied from the *Encyclopaedia Britannica*'s third edition, including the exact text of Lavater's "General Remarks on Women." This ensured that when early national readers looked for a standard definition of "sex," they instead found a series of excerpts from the *Essays on Physiognomy*. Between the 1790s and the 1860s, then, Lavater's theories effectively stood in as the definitive statement on sex and gender in the Anglo-Atlantic print universe. Only in the *Encyclopaedia Britannica*'s eighth edition, published in 1860, did publishers finally remove his maxims from the text.[65]

Americans continued to reprint Lavater's comments throughout the early national decades, even though his bald avowals in favor of men's mental superiority were increasingly out of step with Americans' willingness to value female intellect.⁶⁶ After the American Revolution, the nation's leaders found themselves facing a difficult question: what role would women play in the new republic? In the immediate post-Revolutionary decades, many women interpreted their dictate broadly, asserting identities as autonomous political actors. Increasingly, though, political leaders argued that women should instead serve as the educators and influencers of men. In addition to mothering meritorious sons, virtuous republican women would sway their husbands toward the path of integrity, quietly ensuring the triumph of the nation's ambitious experiment from behind the scenes. The concept of republican womanhood ultimately allowed elite and middling white women to claim increased access to higher education and to exert political *influence*, if not political *rights*.⁶⁷

How, then, might one identify a "proper" republican woman? Early national authors and political thinkers put their faith in physiognomy. The prolific writer Thomas Branagan, for example, believed the success of republicanism depended on men's ability to detect suitable marriage partners, and he imagined scientific facial analysis as a weapon that men could wield against "female policy." By scrutinizing women's faces for signs of virtue, men could avoid being matched with "tyrannical female beauties." Branagan was a strong advocate for female education who disagreed with Lavater's contention that women were the intellectual inferiors of men. He was nonetheless convinced by Lavater's instructions on how to find an honorable and appropriately submissive wife. "I do solemnly declare," he penned with rhetorical flourish, that "a competent knowledge of the science of physiognomy" would be the best way to avoid "matrimonial misery." If only men could learn to properly read women's visages, he asserted, they would be able to see past the wiles and affectations of coquettes.⁶⁸

Branagan believed that no republican citizen should have to marry a "she fiend." Luckily, the physical markers of proper femininity were visible on every worthy woman's countenance. Lavater had said a "noble spotless maiden" would have a "large arched forehead," in which "all the capacity of immeasurable intelligence which wisdom can communicate be visible." Branagan parroted this description. Paraphrasing Lavater, he described his archetypal republican bride: "In her black, her brilliant, her smiling eyes may be read the candour and generosity of her heart, her large arched forehead, plainly denotes a capacious mind." Painting his sketch of the perfect wife, Branagan suggested that "her compressed eyebrows bespeak her understanding, her gentle outlined nose shews refined taste, her placid and pleasing lips point out the complacency and docility of her nature." Through these statements,

Branagan proclaimed that men could achieve matrimonial bliss by looking for the anatomical identifiers of virtuous republican womanhood.[69]

Mason Locke Weems similarly physiognomized the perfect republican wife. Scholars usually remember Weems as the brilliant book peddler who concocted a story about a youthful George Washington who cut down a cherry tree and could not tell a lie. Less well known is a story that Weems published in Virginia's *Lover's Almanac* in 1798, where he pondered the importance of republican womanhood and painted a picture of a fictional young heroine named "Miss Delia D," a paragon of feminine virtue and "the most attractive form that ever charmed the world to love." In addition to being "tall and majestic," Delia had an elegant form and an "exquisite shape." Her "fine, soft blue eyes" resembled "the bright azure of an unclouded sky," and her arching eyebrows rivaled the "bow of heaven." Weems assured his readers that Delia was stunningly beautiful—the ultimate catch for any United States bachelor. But then his story took a curious turn. After meticulously delineating the "beauties of Delia's person," Weems suddenly dismissed the importance of her bewitching face and form. After all, he argued, "The body is but the casket, the mind is the *juice*, the diamond that gives it all its loveliness and worth." What truly made Delia exceptional was the "far superior beauties of her mind." This was a fine sentiment, at least at first glance. It rested on the idea that women should be valued for their intellect and character, rather than their bodies alone. But it was also dishonest. Even as Weems insisted on the insignificance of Delia's physical beauty, he obsessed over her corporeal features. Delia's "eye sparkles with benignity," and her countenance exhibits "features animated by a show of devotion." Her "lips are the temples of truth, her cheeks are the emblems of modesty," and her "soft melting eye bespeaks a soul that's united to all around her." According to Weems, Delia was an exemplary republican woman—an identity that was visible in her visage. Weems chided young bachelors for their infatuation with female bodies, only to furnish them with careful instructions for reading women's physiognomical features.[70]

If facial analysis helped men identify wives of quality, it was also useful for unmasking malicious or coquettish women. In a different story, *The Bad Wife's Looking Glass*, Weems urged men to shun the "*pretty statues*" who only pretended to be virtuous.[71] Despite what "certain Physiognomists have stated," a pleasing face did not always correlate with a virtuous heart. To illustrate his point, Weems analyzed the appearance of "the pretty Miss Becky Kannady," a ruthless woman who hatched a plot to pulverize her husband's skull while he slept. Intriguingly, Kannady's "form and features were unrivalled," a trait that might have fooled a reckless observer. A thoughtful physiognomist, by contrast, would see that her face was "like a fine casket without a jewel." Although she had "the vain attractions of a little skin deep beauty, aided with the frivolities of dress," Kannady's countenance hinted at her vicious nature: "her eyes, though bright, lacked the all-animating intelligence which kindles the beholder into admiration," and "her face, though faultlessly featured," lacked "those divine charms of innocence and goodness." Kannady was beautiful on the surface, but she lacked a beautiful mind. This reality could only be revealed with a thoughtful physiognomic examination.

Like Weems, Thomas Branagan warned that vain or frivolous women might be alluring at first, but he assured men that physiognomic scrutiny could disclose vicious interiors. Quoting directly from Lavater's *Essays on Physiognomy*, he beseeched America's men to avoid disreputable women, who would be identifiable by "the haughtiness of their eyes" and "their languid unmeaning lips." To demonstrate his point, Branagan told his readers a cautionary tale: "Walking one day in the street, I inadvertently was struck with a most beautiful female countenance." On closer inspection, he "recognised the most prominent traits of boisterous irritability." The woman's "dictorial [sic] eyebrows, authoritative nose, arrogant and piercing eyes, the affectation of her measured steps, all conspired at once, to convince me she was a beautiful tyrant." Several months later, he encountered the same woman, who had since married an amiable young man. Sure enough, she had turned her new husband into a miserable wretch. He now took refuge in a tavern, descending into drunkenness and abdicating his responsibilities as a sober and respectable citizen.[72]

Branagan's book purported in its title to be about "*The Rights and Privileges of Republicanism*," but he dedicated one-fourth of his work to the study of physiognomy and female beauty. This might seem surprising to modern readers. And yet, for Branagan, beauty, politics, and facial analysis were intimately connected. As he saw it, men needed physiognomy if they wanted to forge stable marriages. Reading faces was important, not simply because it disclosed women's characters, but also because it helped men pick suitable life partners. By shoring up the institution of marriage, physiognomy ensured the success of republicanism itself.

Through the science of physiognomy, Americans assured themselves that respectable republican citizens—and proper republican women—*looked* a certain way. Physiognomy was enticing because it gave people a method for rationally and objectively identifying the physical signals of republican virtue. The reality, of course, was more complex. Physiognomists insisted they were empirically analyzing the character of their contemporaries, but that wasn't quite true. In many cases, they had already decided who was an honorable republican citizen and who was not. Facial analysis was merely a way to rationalize the inequities Americans had previously decided were legitimate.[73]

Physiognomy ultimately depended far more on the opinions of the analyzer than on the face of the person being observed. The rules were murky, the system was not a system, and the methodology was nothing more than a disjointed collection of individualized interpretations. Powerful white thinkers tried to convince people that their way of seeing was the right one—but they did not always succeed. Even though physiognomy suffused elite discourses and penetrated some of the nation's most prominent educational institutions, it was mostly practiced in public spaces. Because physiognomy was a popular science, white male physicians, politicians, and intellectuals never exerted absolute control over the practice of science in early America.

This left open an intriguing possibility. If scientific facial analysis was fundamentally dependent upon the individualized interpretations of its practitioners—and if it was theoretically accessible to all Americans—then marginalized people, too, might use it to their own advantage. Despite their best efforts, society's most privileged members could not unilaterally control the practice of science, nor could they dictate the lessons that Americans would draw from it. This became especially true during the antebellum decades, as a new "science of mind" entered the scene: phrenology. In the 1830s, this discipline would revolutionize how ordinary people made sense of their own bodies and brains, allowing an increasingly diverse group of Americans to craft alternative models of social and political belonging.

2

A NEW SCIENCE OF MAN

It was the winter of 1854 and John Brown Jr. was worried. He wasn't good at his job. Or so he feared. Luckily, he believed he was finally starting to improve, a sentiment confirmed by his boss. "You have always questioned your own ability," Samuel Wells reminded him, despite "Substantial Evidence of Success." In addition to being a fervent anti-slavery activist like his more famous father, Brown Jr. was a traveling lecturer for the firm of Fowler and Wells. His task? Tour the American West spreading knowledge of phrenology, an intriguing practical science which captivated the American public in the middle decades of the nineteenth century. Phrenology, like physiognomy, was rooted in a simple assumption: people's bodies revealed secrets about their character. While physiognomists focused on the face, phrenologists turned their attention to the skull, proclaiming that it was possible to read people's minds by analyzing the bumps and crevices of their craniums. Brown's job was to lecture about the science, analyze heads, and solicit subscribers for the *American Phrenological Journal*. Pleased by the money Brown was bringing in, Wells sent his employee an encouraging message: "You are, and have long been, capable of teaching Phrenology. And I am glad to know that you have finally found it out." The senior phrenologist encouraged his protégé to "buckle on the armor and with truth of your tongue, & righteousness in your heart & head, go on, *on, and on!*" In his employer's imagination, Brown was an "apostle" on a heroic mission to dispel "ignorance" and help ordinary Americans learn the truth about their minds and bodies.[1]

Brown was not a phrenological lecturer for long, but he maintained a friendly relationship with his employers in the years that followed. When his father, John Brown Sr., was executed for trying to violently overthrow enslavers in 1859, Nelson Sizer—another phrenologist at Fowler and Wells—praised the radicalism of his former employee's "martyred & immortal father." Boasting that John Brown had taken the time to write him a letter from jail as he awaited his execution, Sizer lauded the abolitionist's courage and excitedly spoke about the Underground Railroad. Then, he gave Brown Jr. a nervous warning: "You must not let the pro slavery sharks get hold of you." Although the two men no longer worked together, they still had certain things in common: they both despised slavery and they were both trained skull interpreters.[2]

At first glance, this might seem surprising. Historians have long recognized how white supremacists used disciplines like physiognomy and phrenology to justify racial inequality and defend slavery—and for good reason. Scientific thinkers regularly deployed physiognomic and phrenological language to suggest the brains and bodies of white Americans were superior to those of their Black counterparts. They also used these disciplines to argue for the inferiority of white women, working people, and immigrants. Despite this fact, social conservatives, white supremacists, and advocates of the status quo were not the only ones to embrace popular sciences. In fact, it was often radical activists—as well as more moderate reformers—who became especially enthusiastic about these disciplines.

Henry Ward Beecher, the nation's most celebrated minister, publicly sang the praises of cranial analysis, even asking his students to read phrenological texts as part of their religious training.[3] His sister, Harriet Beecher Stowe, similarly embraced popular sciences in her literary works, skillfully reading the heads and faces of fictional characters in anti-slavery novels like *Uncle Tom's Cabin*.[4] Horace Mann, known for spearheading the common school movement, believed phrenology could be used to revolutionize the nation's educational system.[5] William Lloyd Garrison not only commissioned a reading of his own skull but also published it in *The Liberator*, the anti-slavery newspaper where he regularly advertised and defended phrenology.[6] The abolitionists and women's rights activists Lucretia Mott and Abby Kelley Foster became so interested in the science that they used it to understand their own bodies and make sense of their commitment to abolitionism.[7] Elizabeth Cady Stanton, Susan B. Anthony, Sojourner Truth, and John Brown were all intrigued enough to submit their craniums for examination. John Brown even spent some of his final moments drafting a letter to a famous phrenologist. When viewed within this broader context, John Brown Jr. was not unique. He was just one of many activists who drew connections between popular science and progressive reform.

Physiognomy and phrenology were invidious disciplines that reified hierarchies of race, class, gender, and ethnicity. Yet they were also culturally pervasive sciences that diverse groups of people deployed for a wide variety of practical purposes. Why, then, did such a wide variety of people so fervently embrace facial and cranial analysis? What was it about these sciences that appealed to both the privileged and the powerless—to social conservatives, scientific racists, feminist thinkers, Black abolitionists, bourgeois reformers, and laboring Americans? And why did marginalized people sometimes end up endorsing disciplines that marked their bodies and brains as defective?

In many ways, it was the ideological flexibility of physiognomy and phrenology that made these sciences so broadly appealing. On one level, they enshrined the notion that people's bodies revealed their innate capacities. Suggesting that each person inherited a set of unique mental and physical features, physiognomists and phrenologists claimed it would be difficult—if not impossible—for someone to overcome their predetermined neurological destiny. At the same time, popular scientists declared that every brain could be cultivated and that every body could be beautified. Brains were inherited, and they molded the body according to their image. Yet brains could improve, and bodies could change. By claiming mental cultivation led to physical enhancement, facial and cranial analyzers crafted a capacious vision of corporeal mutability that different groups of people could—and did—deploy for opposing purposes.

Physiognomy and phrenology were ultimately appealing because they helped Americans justify existing inequities while giving them hope that their own bodies and brains were fungible and fundamentally improvable entities. And yet, even as they dangled the promise of personal betterment before the United States population, popular scientists never really provided a coherent roadmap for eliminating the country's most entrenched structural inequalities. Instead, they reproduced the inconsistencies at the heart of American democracy, nurturing a hopeful vision of meritocratic possibility while scientifically rationalizing the very hierarchies that existed within nineteenth-century society.

SCIENCE FOR THE PEOPLE

Phrenology, like physiognomy, began as a European science. It was the brainchild of a German anatomist named Franz Josef Gall, a man who came of age in a Lavaterian world where it was common to draw connections between external features and internal characteristics. When narrating the origin story of his science, Gall claimed it all began when he made a strange observation as a student: all his classmates with good memories also had "protruding eyes." Was there a connection? Might the physiognomists be right? Could "mental qualities be externally recognizable?" To hear Gall tell it, his juvenile curiosity blossomed into a program of sustained empirical research during the 1790s. Eager to resolve the craniological mysteries of his youth, he started collecting skulls from across the globe. He enlisted the help of an earnest young medical student named Johann Gaspar Spurzheim in 1804. Together they would promulgate an exciting new system for reading the human mind.[8]

Gall and Spurzheim built on the conclusions of the physiognomists, but they also complained that facial analysis was simply not scientific enough. The "language of Lavater is obviously always vague," they grumbled. As a result, they tried to develop a more empirical methodology for gauging people's intelligence, personality, and character. Rather than focusing on the face alone, they studied the size and shape of people's craniums. Their science was rooted in a few fundamental assumptions: 1) all cognition happened in the brain, 2) different parts of the brain controlled different cerebral functions, and 3) the brain imprinted itself upon the skull from the inside out, giving every individual a unique cranial shape.[9]

After years of collaboration—and plenty of personal drama—the two men parted ways, with Gall pursuing scientific legitimacy among the European intelligentsia and Spurzheim choosing to disseminate phrenology on the public lecture circuit. Spurzheim eventually made his way to the United States in 1832, where he dazzled audiences with live brain dissections and anatomical demonstrations. Americans described him as an "excellent," "eminent," and "indefatigable" scholar who revolutionized global scientific practices. Harvard invited him to speak, and at Yale "the professors were in love with him." Then disaster struck. Just three months after his arrival, Spurzheim unexpectedly died in the city of Boston. By that point, he was the world's most famous phrenologist. Now, at just fifty-five years old, he was gone.[10]

Bostonians were shaken by Spurzheim's death, calling it both "startling" and "appalling." Eager to carry on his legacy, they resolved to keep the phrenological flame burning.[11] Although phrenological discourses had been percolating in the United States since the early nineteenth century, the 1830s marked a clear turning point. The Scottish phrenologist George Combe had bolstered phrenology's popularity with his global bestseller, *The Constitution of Man*, in 1828, and Spurzheim's tour further elevated American interest in the science. Soon after his visit, phrenological societies "sprang up throughout the country." Combe then embarked upon his own US lecture tour in 1838, additionally enhancing phrenology's profile. Spurzheim might have died in 1832, but his science experienced an American rebirth.[12]

The earliest adopters of phrenological theory in the United States were physicians, lawyers, and urban professionals—society's intellectual elites. By the late 1830s, though, some elite thinkers were beginning to turn their noses up at the science. This coincided with a rapid explosion of cheap print culture, which increasingly made scientific knowledge accessible to broader swaths of society. Around the same time, a new group of "practical phrenologists" also entered the scene, popularizing the science and forever altering how ordinary people thought about their craniums. The most famous was Orson Fowler, a homegrown skull interpreter and the first US resident to match the popularity of European practitioners. He was cantankerous and flamboyant—an evangelist who dedicated his life to the science of skulls after abandoning his initial dream of becoming a minister. Teaming up with his siblings, Charlotte and Lorenzo, Orson opened workshops in New York, Philadelphia, and Boston. In these establishments, inquisitive visitors learned how to read minds and bodies.

To enter the Fowler phrenological cabinet was to be visually assaulted by a panoply of plaster heads (fig. 2.1). Entry was free, giving people of all backgrounds an opportunity to gawk at the "sculls of murderers, assassins, rogues, warriors, chiefs, idiots and various destructive animals."[13] For a small fee, visitors could even get their own heads examined. By the 1840s, the Fowlers were dispatching teams of "practical phrenologists" on national tours, and their "Repository of Curiosities" in New York rivaled P. T. Barnum's Museum as one of the city's most popular attractions. Such innovations made scientific knowledge accessible to both elite and working people.[14]

European phrenologists might have sparked people's interest in cranial analysis, but it was the Fowlers who helped make phrenology a profitable business and ubiquitous social practice in the United States. The Fowlers were not just amateur scientists and savvy entrepreneurs; they were also successful publishers. In 1841, they took over the *American Phrenological Journal and Miscellany*, which would become the world's longest-running phrenological magazine. More resilient than most nineteenth-century periodicals, it had more than 20,000 subscribers by 1847 and stayed in print until 1911. The Fowlers published a mix of material, including books by luminaries such Walt Whitman, Edgar Allan Poe, and Margaret Fuller. Their press trafficked in self-help manuals and cheap advice leaflets, targeting a popular audience and embracing a wide variety of topics, ranging from beauty, sex, love, and marriage to fashion, gambling, temperance, and personal hygiene.[15]

Phrenology did not supplant physiognomy so much as it amplified—and capitalized on—its popularity. In fact, phrenology became so influential because it built upon the preexisting cultural practice of analyzing countenances. Phrenology was a tactile discipline, encouraging practitioners to touch the

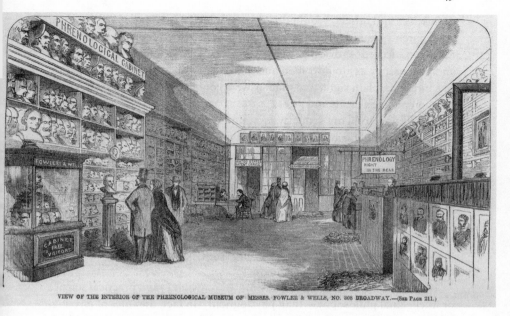

FIGURE 2.1. "View of the Interior of the Phrenological Museum of Messrs. Fowler & Wells, No. 308 Broadway," wood engraving, *New-York Illustrated News*, February 18, 1860. New-York Historical Society, 98766d.

protuberances of people's skulls. But the face remained critically important to cranial interpreters, who continued to rhapsodize about mouths, cheeks, foreheads, and eyes.[16] In one 1845 almanac, the Fowlers proudly declared that "the system of Physiognomy is *founded in* Phrenology." Had they been more concerned with chronological accuracy, they would have reversed the order of this sentence. In any case, the Fowlers believed people could—and should—pair the two disciplines. "Neither, without the other," they wrote. "*Both together*, complete our knowledge of human nature." Facial analysis continued to be a widespread practice, in large part because it was easier to interpret people's visages than to poke and prod their craniums. In fact, Samuel Wells—the phrenologist, physiognomist, and husband of Charlotte Fowler—argued that facial analysis had "greater practical availability" since the visage was "open to observation at all times, whether the head be covered or not." In any case, Wells expressed confidence that physiology, physiognomy, and phrenology were "in perfect harmony from beginning to end." Rather than being in competition, facial and cranial analysis coexisted symbiotically.[17]

Together, these disciplines gave people a code for unlocking secrets about themselves and others. Would this man be a loving husband? Will I make a good parent? Should I trust this stranger? Have I chosen the right career? Am I likely to succeed? Providing "easy answers to hard questions," physiognomy and phrenology assured women and men that knowledge of human nature was hiding in plain sight, and that answers could be deciphered from visual clues on people's heads and faces.[18]

In addition to being seductively simple, physiognomy and phrenology were uniquely accessible disciplines that did not require expensive training or equipment. This meant that laypersons could be active creators—not simply passive consumers—of knowledge. As the minister Henry Ward Beecher argued, phrenology in particular "brought the human mind within the reach and comprehension of persons of ordinary intelligence." Previously, it had been mostly "mental philosophers and metaphysicians" who studied the human brain. But the "simple and sensible" nature of the phrenological method made it "the philosophy of the common people." As Beecher saw it, this was an extraordinary victory for democracy.[19]

It wasn't that phrenology was any more accessible—as a practical science or intellectual philosophy—than physiognomy. In fact, the rules of physiognomy were even more flexible and even more dependent on the individualized interpretations of the observer. Simply by staring someone in the face, people could develop their own methods for studying the human mind. Yet the success of both physiognomy and phrenology by mid-century was rooted not only in the "simple and sensible" character of their doctrines, but also in innovations in print technology and distribution that made scientific knowledge accessible to a broader swath of society. During the 1830s and 1840s, a "cheap literature revolution" ensured that scientific books, magazines, pamphlets, and newspapers could be rapidly produced and disseminated. Because literacy rates were particularly high in the United States, magazines became especially important tools for the transmission of science. The rapid proliferation of museums, lending libraries, and lyceum circuits in the United States further ensured that people had consistent access to information. This put science at people's fingertips.[20]

Traveling lecturers further enhanced the popularity of physiognomy and phrenology, bringing them to populations in large cities and rural outposts alike. On the public lecture circuit, scientific knowledge became available to individuals—particularly women—who were otherwise excluded from the nation's colleges and universities. Lecture attendees rarely made distinctions between subjects that we now see as "real" sciences—like astronomy, chemistry, geology, and comparative anatomy—and disciplines that we now classify as pseudosciences: mesmerism, magnetism, and phrenology. Instead, they described all these lectures as mutually supportive educational experiences that collectively enhanced their knowledge of the natural world. After seeing a phrenologist analyze her friends and neighbors, the young Georgiana Barrows proclaimed that he "gave their characters exactly." Another woman, Lucy Chase, was so intrigued by a series of phrenological lectures that she got her own head examined. She also used physiognomic language to evaluate the intelligence, virtue, and sincerity of subsequent lecturers. Her favorite speaker was a "Polish Exile" named Major Tochman, whose "frank, open countenance" bore a "benign expression." According to Chase, the "expression of sincerity upon his countenance" not only allowed the audience to "clearly read the soul within," but also stimulated her own phrenological organ of "Ideality." Another lecturer, she claimed, had a "pleasant intelligent countenance," while yet another surprised her with his eloquence, giving the audience "much more than the *head* promised." For amateur cranial analyzers like Chase and Barrows, the process of going to lectures—and studying the faces and skulls of the speakers—was part of a broader quest for rational enlightenment.[21]

Popular sciences also captivated Americans for a more basic reason: they flattered people. When the upstate New Yorker Mary Ferguson brought her toddler to get his head examined, she was thrilled to learn that her young Josie was "precocious" and "by far too smart." The phrenologist declared Ferguson's son to be "one of the brightest and most knowing children I have examined," someone who was "born to command." Ferguson was "so pleased with Phrenology" that she and her sister then went to have their heads examined. Though she had gleefully reported on Josie's cranial reading, she was more hesitant when it came to her own skull. "I think he flattered us a little," she admitted, even though she continued to find phrenology "very interesting."[22]

Sometimes, people balked at their cranial readings. Harriet Robinson, a laborer at the Lowell Mills, had particularly unpleasant memories of a phrenological visit. When "Professor Fowler" examined her cranium, he informed her she was lacking in the organ of "Veneration." She didn't know precisely what this meant (Fowler was likely commenting on her alleged lack of religiosity), but "supposed was an awful thing, because my teacher looked so reproachfully at me when the professor said it."[23] Many other young women, though, embraced phrenology. Harriot Curtis, another employee in the Lowell Mills, became so enamored with the science that she solicited Orson Fowler's mentorship and eventually became a traveling phrenologist in her own right.[24] Facial and cranial readings were ultimately beguiling—not just because they provided insights about seemingly unknowable entities like the mind and soul, but also because they often confirmed people's suspicions about their own exceptionality. This seemed painfully obvious to Lucy Larcom, yet another operative at the Lowell Mills. In her memoirs, she wrote of the various phrenologists who "came around" analyzing heads. In Larcom's telling, these "professors" gave positive readings to "almost everybody," meaning that "many very commonplace people were flattered into the belief that they were geniuses, or might be if they chose." As Larcom saw it, phrenology was not a legitimate science but a clever monetization of disingenuous praise.[25]

Even though Larcom was critical of phrenology, her comments help to explain why this science achieved such remarkable popularity. It persuaded ordinary people they "were geniuses, or might be if they chose." This fundamentally optimistic outlook was embedded within the very methodology of phrenological analysis, which validated the notion that everyone was capable of change. This had not always been the case. When Gall first developed his science, he envisioned it as a rather deterministic discipline. His method left little room for the possibility of mental and physical betterment. Everyone would inherit a unique set of mental traits, he believed, and those heritable characteristics would be reflected in the corporeal features.[26] Future practitioners would tell a more hopeful tale. Both Spurzheim and Combe suggested that individuals might strategically augment certain characteristics while subduing others. The Fowlers capitalized on this claim, transforming phrenology into a "quintessentially American" discipline that closely aligned with the United States' meritocratic ethos of self-improvement.[27]

When Americans read the Fowlers' phrenological books, articles, and almanacs, they learned that they could take charge of their own neurological destinies. If they were plagued with a weak constitution and an overactive brain, all they needed to do was "stop those sedentary or mental pursuits," exercise, and "take things easily." Be sure to get enough sleep, Orson Fowler proclaimed. Adopt a vegetarian diet. And stop drinking coffee, tea, and alcohol. Then, "the equilibrium" would "be restored." But what of those who had the opposite problem? "Are your muscles strong, but mind dull, and feelings obtuse . . . ?" Never fear. Simply "work less, but read, and think, and feel more." For individuals who labored in physically demanding jobs, this was easier said than done. The Fowlers nonetheless stressed that personal grit could overcome structural inequalities. Anyone might rehabilitate a sluggish mind by exercising their "brain more and muscles less," they argued, but people had to do it for themselves. Self-betterment took time—and there were no quick fixes for a faulty temperament. With hard work and a bit of patience, though, anything was possible. This message was both empowering and condemnatory. Preaching the gospel of meritocratic individualism, the Fowlers hinted that those who failed to succeed had only themselves to blame. At the same time, they appealed to working people with a clear-eyed critique of economic inequality.[28]

The Fowlers disdained rapacious capitalism almost as much as they loved the science of skulls. At a time when the United States was gripped by the market revolution's jarring upheavals, *Fowler's Practical Phrenology* (1840) proclaimed that "The labouring classes, instead of consuming their whole existence in working, should be better paid for their labour, and thus allowed time to cultivate their intellects, and exercise their finer feelings." This was more than a book about how to read craniums. It was a searing critique of capitalism. "The present arrangements of society tend to make the rich man richer, and the poor man poorer," Orson Fowler grumbled. "This is certainly not the order of nature. The possession of great wealth is not right." After all, what gave the "capitalist" the right to "sacrifice upon the altar of his selfishness, all the lives of all these human beings?" Certainly not science. Rather than hoarding money for themselves, employers should pay their employees better. "Let the wages of the labouring classes be doubled, and trebled, and quadrupled," Fowler proclaimed. Then, he demanded that employers grant workers more time for "reading and mental culture." This was particularly important in "a republic like ours," he said, "where every thing depends on the intelligence of the people." It was a strikingly egalitarian rallying cry.[29]

Above all else, the Fowlers self-identified as reformers, and they encouraged ordinary people to be "self-reformers," too.[30] Throwing their weight behind crusades such as the temperance movement, penitentiary reform, and women's rights, they rejected the notion that people's personalities were fixed and unchangeable. They did not think of the skull as a permanent, hard-and-fast indicator of a person's essential character. Instead, they claimed that craniums could adjust and develop over time, due to education, employment, and other external factors. If a person could understand their own flaws—whether inherited or acquired—then they could change them. A good-faith engagement with the phrenological method necessitated a cold, hard look at one's own strengths and weaknesses. With enough effort and enthusiasm, anyone could become a respectable and well-rounded citizen. Self-improvement was not only possible, but probable. At least that was the promise. It was the same ideological contradiction that had characterized physiognomy from its inception.

Physiognomy, like phrenology, had been based on a series of antithetical arguments about the heritability of human difference. Lavater claimed appearance and character were determined by parental inheritance, and thus were unaffected by environmental forces. Yet he also hinted that faces could transform over time, responding to stimuli as people cultivated certain characteristics. This left physiognomists with some vexing questions to iron out: Was character innate and unalterable? Or could it develop over time? Would people's faces maintain a consistent appearance throughout their lives? Or did internal developments leave indelible marks upon the human countenance? Lavater insisted that "true" physiognomists would be able to distinguish the difference "between original form and deviations." But how, exactly, was one supposed to differentiate between "original" character and cranial variations? What should observers be looking for, anyway? Was it more important to discern someone's *inherited* traits or to study their *current* appearance and character? Lavater himself admitted that physiognomy could only be a real science if observers were consistently able to evaluate people's intrinsic and unchangeable features. But if character, intellect, and appearance were all flexible, then what was the point of the whole enterprise? The answers were rarely clear.[31]

Phrenologists did not solve any of these quandaries. Sometimes, they made grim proclamations about the heritability of mental and physical traits. Individuals should rely on science to select well-suited spouses, they warned. Otherwise, they might saddle their progeny with undesirable propensities. This was a real concern for some people. In 1823, a young woman wrote a note to her friend in which she poked fun at her cousin Francis for seeking a spouse with a "grecian face" and "all the proper bumps on her head." Clearly, Francis thought his future wife's skull would have important ramifications for his own life.[32] The radical abolitionists Abby Kelley and Stephen Foster likewise used popular science to think through the consequences of marriage and child-rearing. Before their wedding, the couple exchanged thoughts about the various phrenological texts they were reading. Abby had recently finished Fowler's "Phrenology Applied to Marriage" when she penned a letter to Stephen asking, "Will you read it as soon as you can procure it?" In this book, Fowler gave advice on how to spot a spouse with faculties and propensities that complemented one's own.[33]

Emphasizing the connection between craniums and connubial bliss, the Fowlers published hundreds of articles with titles like "Hereditary Descent," "Idiocy and Superior Talents, Hereditary," and "Intellect Hereditary—As to Both Kind and Amount."[34] Many of these pieces read like early forms of eugenics, disguised as helpful advice on marriage and parenthood. As such texts often decreed, the only way to ensure your children's success was to procreate with the proper spouse. Yet phrenologists and physiognomists also printed titles like "Culture of the Mental Faculties," "Self-Improvement," and "Education, Phrenologically Considered."[35] Asserting that each and every person could improve their mind and beautify their body, popular scientists provided a puzzling mix of deterministic messages and helpful hints about how to attain a well-balanced brain. Sometimes they suggested that every individual would inherit a distinctive set of propensities which could not be fundamentally changed. On other occasions, they described science as an empowering aid in everyone's self-help journey.

Even though most leading physiognomists and phrenologists maintained that everyone was reformable, they made some caveats. Children, they suggested, could more easily alter their destinies (and their craniums) than adults. In one children's story, first published in 1788 and republished in schoolbooks well into the 1830s, a fictional father gave his daughter advice on how to be beautiful. If she and her friends wanted better faces, he lectured, they should focus not on their bodies, but on their minds. Why? Internal makeovers would result in physical changes. He insisted that his daughter should embark upon her moral metamorphosis immediately. As long as she was young, there was still a chance for mental developments to permanently influence her features. If she waited too long, though, her visage might be forever marked with the visible deformities of vice. Why? Because faces were harder to change in adulthood.[36]

Phrenologists articulated a similar theme. While they claimed everyone was improvable, they also hinted that it was easier to course-correct in youth. For this reason, they placed great importance on early childhood education (and the role mothers played in it). Still, they suggested there was hope for everyone. One phrenologist, for instance, wrote that "intellectual cultivation" resulted in "an evident change in the expression of the eye, a softening of the lines of the eyebrows, and a lateral expansion of the nose from the bridge downward." In addition to these shifts, "the lips become more gracefully arched and firmer; the chin more delicate and clearly defined, and the lines of the face, as a whole, more diversified and beautiful." Binding mental enrichment to physical attractiveness, he claimed that mental "culture" led to physical beauty. Still, he waffled. Intelligence was innate, but education was important. Brains were inherited, but also changeable. Body shape was determined by nature, but also by nurture. And the skulls of children were, apparently, more mutable than those of adults.[37]

At least some readers were irritated by these mixed messages. Sylvester Graham—the man who made a name for himself as an anti-masturbation crusader, vegetarian activist, and inventor of graham crackers—once wrote a letter to the editor of the Fowler press in which he lauded the *American Phrenological Journal* for synthesizing a wide variety of scientific information for the general public. Still, Graham complained that "a large majority of its readers need a competent editorial authority to which they can safely look for a thoughtful verdict between contradictory testimonies." Noting that the magazine provided readers with ambiguous and incompatible messages, he urged the family to exert a stronger editorial hand. His advice went unheeded. The truth was, the intellectual diversity of the *American Phrenological Journal* was good for business.

It wasn't that the Fowlers were mere money-grubbers who amassed a fortune by commodifying quackery. They profited from phrenology, yes, but they were true believers in the science. They simply practiced a version of bodily analysis that simultaneously allowed for both flexible and deterministic understandings of human nature. This ensured that just about every reader could find something that appealed to them. Gender conservatives who yearned for scientific evidence of women's intellectual inferiority found plenty of it. White supremacists likewise uncovered encouraging conclusions about their own racial superiority. At the same time, anti-slavery activists, health reformers, and early feminists easily found material that validated their political agendas. Because these sciences encouraged people to conceptualize their subjective interpretations as objective discoveries made by empirical observation, individual practitioners were free to draw different conclusions from the same bodies. To practice facial and cranial analysis was to search for certainty while embracing ambiguity. Still, physiognomy and phrenology became popular not in spite of these inconsistencies, but because of them.[38]

HIGH BROW, LOW BROW

Despite the variabilities in physiognomic and phrenological doctrines, facial and cranial analyzers tended to agree on a few general rules. For one, they largely concurred that the ancient Greeks had been the most intelligent and attractive beings ever to walk the earth. They did not come up with this idea on their own; rather, they built on the work of anatomists, art critics, and natural philosophers such as Johann Joachim Winkelmann. Since the era of the Renaissance—but particularly in the eighteenth century—European artists and thinkers had invoked Grecian statuary as the standard against which all people should be measured. Physiognomists and phrenologists largely adopted this standard, a reality showing how aesthetic traditions could shape scientific conclusions. Yet science also shaped artistic expression, influencing people's ideas about what was—and was not—beautiful. Physiognomists and phrenologists did not merely reproduce European beauty ideals; they provided ostensibly empirical evidence for them. The result was a new science of beauty that validated old assumptions through its encomiums to Grecian heads with aquiline noses, thin lips, and towering brows.[39]

Physiognomists and phrenologists also agreed that human beings were governed by a tripartite system of mental, moral, and animal traits. Physiognomists argued that the forehead and eyes were the "mirror, or image, of the understanding," while the "nose and cheeks" were "the image of the moral and sensitive life," and the "mouth and chin" were the signifiers of "animal life." Phrenologists extended these assumptions to the skull, claiming that each cranial region revealed something unique about human nature (fig. 2.2). Intellectual power supposedly manifested itself in the forehead; the "moral sentiments" and spiritual faculties resided at the top of the skull (which was, after all, closest to God); and the back and side portions of the cranium disclosed the "animal propensities," including the love of sex, spouses, and children.[40]

In addition to shaping scientific discourses, this tripartite understanding of the human cranium shaped nineteenth-century artwork. On one occasion, the American businessman William Aspinwall commissioned a "bust of our Savior" from the famous sculptor—and avid phrenologist—Hiram Powers. The client asked Powers to decrease the size of the statue's mouth, complaining that the large lips had given Jesus Christ an "expression of haughtiness or voluptuousness." Powers bristled. Scribbling a principled (if transparently petty) retort, Powers informed his patron that his criticism had actually been a "high compliment," writing, "if you have found in the expression of the mouth a certain fullness indicative of the passions of men, though subdued, and under the restraint of the intellectual power shown in the upper part of the face, you have found precisely what I wished to reveal." According to Powers, the eyes were the "windows of the soul" and "vehicles of intelligence." The mouth, by contrast, was the seat of human fallibility. Because Christ "combined in himself all the evils of our fallen nature" with "the Divine," it followed that both spiritual and animal traits should be visible in his countenance. Why gloss over the truth? Aspinwall later responded that he was willing to trust the artist's "discretion," but he repeated his request. He wanted smaller lips.[41]

If nothing else, this testy exchange over the marbleized mouth of Jesus Christ reveals that heads and faces had consequences in the nineteenth century. Like Powers and his disgruntled patron, Americans largely took for granted that the "upper part of the face" revealed the moral and mental

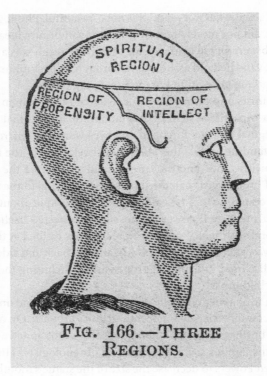

FIGURE 2.2. "Three Regions," in Samuel Wells, *How to Read Character: An Illustrated Handbook of Phrenology and Physiognomy* (New York: Fowler and Wells, 1882). Courtesy of the American Antiquarian Society.

capacities, while the mouth and jaw exhibited the baser passions. Physiognomic practitioners would also have agreed on something else: the smartest and most sophisticated individuals had large and well-developed foreheads. As early as the 1770s, Lavater contended that a short or "lowering forehead" signified "manifest dullness, and debility." He went so far as to say that a decent brow was a necessary attribute for all aspiring physiognomists, claiming that no one should "enter the sanctuary of physiognomy who has a debased mind, an ill formed forehead, a blinking eye, or a distorted mouth."[42] Phrenologists refined and popularized this idea, arguing that the height of the brow indicated one's perceptive capacities, while breadth signaled the strength of one's reasoning faculties.

Phrenologists assured people that high foreheads usually appeared in creative types with imaginative and vivacious minds. Broad foreheads, by contrast, indicated intellectual power, and therefore were most common in analytical thinkers known for their profundity, good judgment, and solidity of mind. According to these guidelines, artistic thinkers would most likely exhibit the "poet's forehead" (high and narrow), while forceful individuals with dogged personalities would have the broadest brows. The latter category included military generals and combative politicians, who sometimes lacked the high brows signifying genuine mental delicacy. Skilled mathematicians and serious philosophers, of course, would require both acuteness of perception (height) and the capacity for rational deliberation (breadth). These traits would be visible in the "mathematical forehead, broad and profound."[43]

By mid-century, high foreheads were not merely metrics in the pages of scientific treatises—they were beauty ideals that ordinary men and women alike sought to emulate. In both published works and personal manuscripts, Americans demonstrated a fascination with the forehead, regularly complimenting each other for "high brows" and "intelligent countenances." Clara Crowninshield's journal, for instance, described numerous men as having "an intellectual forehead."[44] When describing the author Nathaniel Hawthorne, Charlotte Forten, a young African American woman from Philadelphia, emphasized his "splendid head," writing that his "noble, expansive brow bears the unmistakeable impress of genius and superior intellect."[45] Another woman wrote a letter depicting the young Louisa May Alcott as a child with "a high forehead, and altogether a countenance of more than usual intelligence."[46] Elizabeth Blackwell, a British physician and the first woman to earn a medical degree in the United States, analyzed her contemporaries in similar ways. In one letter, she christened someone a "stupid fox" with a "retreating forehead." By contrast, she pronounced an Italian violinist a musical "genius" with a large cranium and "fine forehead." Blackwell also analyzed the appearances of "work-women" in a British textile factory, noting how they seemed "brutified by their toil; their physiognomies were assuming the projecting mouth of the lower animals."[47]

Such class-based assumptions could infiltrate people's self-perceptions in ways they did not always consciously admit. On a tour through the streets of Rome, one American woman was startled when she passed by a window and saw a portrait of a peasant who resembled her. When her friends saw the painting, "they both cried out, 'Why it is exactly like you.'" The woman agreed, writing, "The eyes, nose, mouth, cheek and shape of the head are as like me as two peas." She nonetheless thought it necessary to point out that the peasant's forehead was "a trifle lower" than her own. Otherwise, "the rest of the face might pass perfectly for me." While this woman did not launch into a scientific discussion of the forehead's physiognomical significance, she clearly thought it was important to note that her forehead was higher than that of an unidentified peasant. Why? She fancied herself a sophisticated and intelligent member of the refined classes, and within the cultural and intellectual universe of the nineteenth century, a high brow signified mental cultivation.[48]

Even people who did not self-identify as physiognomic practitioners often imbibed these ideas from the cultural stew in which they marinated. The Lowell worker Harriet Robinson, remember, had a rather adversarial encounter with "Professor Fowler," but this didn't stop her from using physiognomic language to explain her first encounter with her friend Margaret Foley (who later became a famous sculptor). She remembered Foley as "very attractive," with "a high, broad forehead, which, in connection with her refined features, gave her the stamp of intellectual power." Robinson then emphasized Foley's whiteness, saying she had "merry blue eyes, and a head as classic and a skin as white as her own beautiful marbles." Such racialized—and ostensibly scientific—descriptions appeared thousands of times in the published and private writings of nineteenth-century Americans. Within this context, big heads and high brows were not merely incidental characteristics. They could also be empirical justifications for existing social and political hierarchies.[49]

"THE HEADS OF THE LAWMAKERS"

Just as early national artists and authors had used physiognomy to elevate the nation's leaders as corporeal specimens of republican virtue, antebellum Americans used science to paint "political portraits" of the country's most distinguished politicians. These descriptions were so ubiquitous that the English author Charles Dickens complained about people repeatedly asking him if he was impressed by the "*heads* of the lawmakers" in the United States. By this, they apparently did not mean the "chiefs and leaders, but literally their individual and personal heads, whereon their hair grew, and whereby the phrenological character of each legislator was expressed." Whenever he told his questioners that he was not impressed, they were allegedly dumbstruck. Yet Dickens's dismissiveness could not quash American enthusiasm for corporeal analysis. Whether in published works, private manuscripts, or ordinary conversations, it was common for people to comment on the craniums of eminent men.[50]

In addition to being objects of physiognomic scrutiny, some of the nation's politicians directly engaged with facial and cranial analysis. When the phrenologist George Combe visited the United States, he met with President Martin Van Buren in the White House and attended a dinner party with several senators and representatives. He was evidently unimpressed by John Quincy Adams's cranium (which perhaps had something to do with the fact that Adams had previously described "Craniology" as a discipline with "no foundation in truth," despite his own family's tendency to engage in physiognomic analysis).[51] Other politicians were similarly skeptical, but they were willing to be convinced. In 1826, James Madison corresponded with the phrenologist Charles Caldwell, telling him he had "obviated some of the most popular prejudices against the Science, and enabled the uninformed, like myself to take an instructive view of it."[52] Henry Clay even brought phrenology to the floor of the United States Senate in 1832, wielding it against his rival, Andrew Jackson. Sarcastically lamenting the deaths of Gall and Spurzheim, Clay regretted that phrenology's "most eminent propagators" were no longer around to "examine the head of our Illustrious Chief Magistrate." He was convinced that the phrenologists would have located an overly developed "organ of destructiveness" in the region above Jackson's ears (see fig. 2.3). After all, the president was trying to exterminate Indigenous Americans, demolish the National Bank, and obliterate the country's most treasured institutions. As Clay saw it, Jackson had "built up nothing, constructed nothing." His platform was "destruction, universal destruction." Why wouldn't this trait be reflected in his cranium?[53]

Unlike Clay, US newspapers and magazines provided more measured critiques of Jackson, simultaneously emphasizing his "sternness," "self-esteem," and "inflexible resolution."[54] They did the same with other politicians, publishing hundreds of articles about the physiognomical characteristics of figures such as John Quincy Adams, Henry Clay, and Martin Van Buren. Yet there was one man—above all others—who garnered unique attention. If George Washington was the physiognomical archetype of the early national decades, then the New England senator Daniel Webster became the model for mid-nineteenth-century Americans. The nation's periodicals waxed poetic about his "high, bold, and majestic *forehead*," as well as the "lines of intellectual labor" imprinted upon his "extraordinary face." Even Webster's political rivals grudgingly admired his capacious cranium, acknowledging it as "perhaps the largest head of any man in the country." As Richmond's *Southern Literary Journal and Magazine of Arts* stated, Webster's "noble forehead reminds one of some massive castle, armed with veteran troops, and impregnable to the assaults of the enemy." A reluctant admirer of the New England politician, the Virginia magazine described him as a taciturn figure who sometimes resembled a "demon." The author nevertheless mustered the magnanimity to praise Webster's mental gifts, proclaiming, "Nature has cut

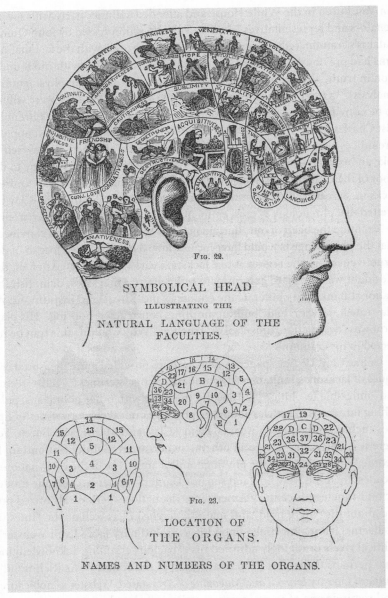

FIGURE 2.3. "Symbolical Head Illustrating the Natural Language of the Faculties," in Samuel Wells, *How to Read Character: A New Illustrated Handbook of Phrenology and Physiognomy* (New York: Fowler and Wells, 1882). Courtesy of the American Antiquarian Society. The organ of "Destructiveness" was positioned directly above the ears (Organ #7).

FIGURE 2.4. Eastman Johnson, *Daniel Webster* (1848), lithograph by William Sharp. National Portrait Gallery, Smithsonian Institution. This portrait emphasized the height and breadth of Webster's forehead and highlighted the prominence of his brow.

him out, body and soul," for leadership. It wasn't just that Webster's features explained his political success. His skeleton quite literally confirmed the legitimacy of physiognomy and phrenology. Who could deny the "truths" of these sciences when looking at a man whose "Shakespearean forehead," "strongly marked mouth," "large dark eyes," and "aquiline nose" (fig. 2.4) provided tangible proof of his superiority?[55]

If national papers could be believed, Webster was a special sort of genius who attained national prominence through a combination of heritable traits and hard work. Commentators made a point of saying he was the "poor son of a poor New Hampshire farmer," a man of humble beginnings, "brought up, as all New England country-boys are, to earn his bread by the sweat of his brow." Webster, they insisted, had not been born into wealth. Inheriting his body and brain from hardy New England parents, he was the product of a different and seemingly more legitimate aristocracy: an aristocracy of merit, where brain power determined destiny. Over and over again, the press rhapsodized about Webster's countenance and cranium, describing them as corporeal markers of his "giant intellect." Even as a young child, he had supposedly been smarter and more talented than his contemporaries. Sparks of genius enlivened "his infant eye," and "his infant forehead" conveyed "the mind of a man." Such articles hinted that Webster was merely claiming his biological birthright by ascending to the Senate.[56] Despite overwhelming evidence to the contrary, political thinkers fostered the fiction that anyone could attain success through intellect and virtue alone. Without an archaic aristocracy of birth, uniquely gifted people might naturally rise to distinction. Daniel Webster's body seemed to buttress this faulty narrative, making him the prototypical American success story—an example of what could happen when a republican government worked precisely as it was supposed to.

By holding up distinguished intellectuals and famous politicians as model citizens, physiognomists and phrenologists helped people rationalize hierarchy in a country that was increasingly unequal. During the early decades of the nineteenth century, Americans experienced unprecedented social and economic changes. The growth of industrial capitalism altered individuals' relationship to labor and capital, transforming how they exchanged and purchased goods. More and more people moved to cities, and the country's population skyrocketed due to a combination of natural increase and an influx of European immigrants. Roads, canals, steamboats, and railroads revolutionized the nation's transportation network, while new print and communication technologies connected Americans across vast distances. This emerging market economy nonetheless left people behind, devaluing the work of skilled artisans, confining people of color in a set of exploitative positions, and creating an underclass of poorly paid factory workers who labored long hours to build wealth for the privileged and powerful.[57]

Americans also formulated increasingly restrictive gender ideologies as the capitalist economy solidified. Even though many poor white women and women of color labored for wages during this period, popular writers and scientific thinkers increasingly argued that "true" women would confine themselves to the domestic "sphere," where they could serve God and family with cheerful resignation. Naturally, these happy homemakers would not require legal, political, or economic rights of their own.[58]

If anyone gained increased access to political power during the 1820s and 1830s, it was white men. Yet their advances almost invariably came at the expense of white women, African Americans, and Native Americans. As state legislatures expanded suffrage to non-landowning white men, they stripped voting rights from free people of color (and in New Jersey, from property-holding women, too). Meanwhile, enslavers ripped Black families apart, forcing approximately one million enslaved people from mid-Atlantic plantations to the cotton fields of the Deep South. The federal government facilitated this process by violently expelling Indigenous nations from their sovereign lands in the Southeast, clearing land for the rise of the Cotton Kingdom. By the 1840s, jingoistic expansionists were insisting it was their "Manifest Destiny" to control the North American continent "from sea to shining sea." To some people, it might have seemed as though the country was experiencing unparalleled territorial expansion and technological advancement. Yet this alleged progress depended upon white supremacy, gender discrimination, and Native dispossession. Rhetorically, white Americans embraced the revolutionary ideals of liberty, justice, and equality. In practice, they crafted a strikingly unequal society.

How were people supposed to reconcile these two visions of the United States—the productive, virtuous, and egalitarian country that lived in the minds of idealistic Americans and the exploitative and inequitable nation that people saw on the ground? Physiognomy and phrenology provided some answers. In theory, intelligence and virtue were the only legitimate prerequisites for citizenship in the United States. But these were invisible traits. How was one supposed to measure mental power? What was the metric for morality? Who counted as a "true" American? And how would one be able to tell? Physiognomy and phrenology allowed Americans to visualize abstract characteristics and, in the process, delineate the bodily signifiers of national belonging. Perhaps unsurprisingly, they crafted artificial standards that marked women and people of color as the mental and physical inferiors of white men.

SCIENTIFIC RACISM

Until the 1820s, physiognomists and phrenologists had primarily focused on the heads and faces of white people. That began to change in the 1830s, as facial and cranial analyzers increasingly turned their attention to Indigenous and Black Americans. This shift coincided with broader political developments in the United States. As the federal government enacted the genocidal policy of Indian removal, popular science practitioners presented "evidence" that Indigenous Americans were "indomitable" and "ferocious" figures who could never be part of the country's political community. In an especially egregious form of phrenological imperialism, United States military officers "dug up and decapitated the Indian dead" during the Second Seminole War, collecting Native skulls and displaying them as specimens in phrenological cabinets. As Cameron Strang has argued, "White officials and phrenologists looked to the remains of Florida Indians for proof that the Seminoles were a unified ethnic group, one that lacked meaningful attachments to Florida and whose brains were so geared toward destruction that their extermination seemed inevitable." Through such practices, the US army enacted the scientific principles that were circulating in both elite treatises and popular texts.[59]

It was also during the 1830s that white Americans began more aggressively invoking science to justify the exploitation of enslaved people and free people of color. In the South, enslavers crafted inventive ideological justifications for human bondage. Meanwhile, in the North, politicians disenfranchised Black voters and white mobs perpetrated violence against Black communities. Within this context, physiognomy and phrenology provided superficially logical justifications for racist policies and beliefs. In one tract from 1833, the pro-slavery author Richard H. Colfax provided what he saw as scientific "proof" of the "natural inferiority" of African Americans. He maintained that "we are not believers in physiognomy, (as a science,) yet we cannot avoid making a remark upon the negro's face." He then launched into a diatribe about Black visages, belying his own professed skepticism about the legitimacy of physiognomic analysis. Black people, he pontificated, had thick lips, slanting foreheads, projecting jaws, and retreating chins. Insisting his bigotry was "consistent with science," he suggested that it was "improper and impolitic" for Black men to be "allowed the privileges of citizenship in an enlightened country!"[60] As Colfax saw it, physiognomy demonstrated the reasonableness of racial hierarchies. If external beauty signified mental fitness, then Black people's alleged physical deficiencies disqualified them from citizenship. Such arguments were effective because, as Douglass Baynton has argued, one of the most common arguments for slavery "was simply that African Americans lacked sufficient intelligence to participate or compete on an equal basis in society with white Americans." For some white thinkers, racism was not just logical; it was the natural consequence of aesthetic and anatomical realities.[61]

Physiognomical and phrenological discourses developed into new—and particularly pernicious—forms of scientific racism during the 1830s. In the first few decades of the nineteenth century, anatomists in the United States began collecting human skulls from across the globe, measuring the cranial capacity of various "races," and insisting that the "inferiority" of people of color was a quantifiable anatomical reality, rather than a social or political decision.[62] The most influential among them was Samuel Morton. In 1839, he published *Crania Americana*, a voluminous work that became the foundation for a new "American School" of ethnology. Pioneering the discipline of craniometry, Morton tried to determine the mental capacity of different "races" by measuring the relative size of their brains. To accomplish this goal, he meticulously filled skulls with various substances and then recorded how much they could hold. Bigger craniums meant bigger brains. Bigger brains signified more powerful minds. Morton then used his findings to suggest that people of European ancestry were more intelligent than people of color. Unlike most physiognomists and phrenologists, he ultimately came to a disturbing conclusion: people from different races were different species entirely.[63]

Morton circulated around the edges of an emerging—but still insecure—scientific establishment, which encompassed a group of "serious" thinkers who tried to establish their legitimacy by distinguishing themselves from the peripatetic lecturers who brought facial and cranial analysis to the masses. These new craniometrists emphasized more quantitative methods than the physiognomists and phrenologists who came before them, but they had much in common with the popular scientists they increasingly tried to dismiss as quacks. Morton, for instance, not only incorporated physiognomic ideas into his work but also collaborated with the phrenologist George Combe, who penned the appendix for *Crania Americana*. While ethnologists insisted that they practiced a more responsible and empirical form of science than the physiognomists and phrenologists, they rarely acknowledged the extent to which their own work built upon—and reinforced—the conclusions of both amateur and professional cranial analyzers.[64]

If Morton was publishing expensive tomes that impressed elite thinkers throughout the transatlantic world, then other writers were articulating many of the same ideas, but for a popular audience. Josiah Nott and George Gliddon, for instance, became two of the United States' most notorious white supremacists by synthesizing Morton's ideas for general readers in their racist magnum opus, *Types of Mankind* (1854). Printing sketches of Black countenances alongside detailed physiognomical descriptions, they claimed it was not just skull shape that distinguished the races, but physical beauty itself.[65] The Fowlers made similarly offensive claims. Despite viewing himself as a progressive reformer and an advocate of universal human equality, Orson Fowler claimed people could easily distinguish between the skulls of "Africans," "Indians," and "Caucasians" by examining the brow: "the Caucasian Race is superior in reasoning power and moral elevation to all other races, and accordingly, has a higher and bolder forehead."[66] *The Illustrated Annuals of Phrenology and Physiognomy* likewise described the "Ethiopian" race as having a "small but long and narrow skull, low and retreating forehead, high cheekbones, projecting teeth, thick lips, and large mouth." These features supposedly signaled that "the scale of intellectuality is low among" people of African descent.[67]

By mid-century, physiognomy, phrenology, and craniometry had melded messily together in both popular culture and scientific treatises. Merging with white people's aesthetic prejudices, these three scientific traditions combined into a racist ethnological system, predicated on the faulty assumption that external beauty conveyed internal worth. As white Americans expelled Native peoples from their ancestral homelands, dramatically expanded slavery in the South, and stripped political rights from free people of color in the North, corporeal analysis provided an ostensibly scientific explanation for social and political inequality.

"DIFFERENCE BETWEEN THE SEXES"

In addition to validating white supremacy, physiognomists and phrenologists turned to the skull in their efforts to naturalize and defend gender hierarchies. Since the early modern era, European naturalists had been trying to distinguish between the bodies and brains of men and women. Using a mix of Aristotelian ideas, Galenic humoral theories, and social prejudices, they differentiated the feminine mind from the "manly" mind. European intellectuals argued that female bodies were "cold and humid," which rendered them "weak, timid, indiscreet, impatient, babbling." Unlike men, women supposedly lacked "solidity" of mind. Instead, their brains were delicate and vivacious. European thinkers argued that women might have brilliant bursts of mental energy, "which dazzle for a moment," but they insisted that female minds were not strong and stable. Male minds, by contrast, were apparently capable of long and laborious deliberation. Although some thinkers challenged this hierarchical conception of gendered intellect, these ideas persisted throughout the eighteenth and nineteenth centuries.[68]

One of the prevailing theories in transatlantic medical literature was that women's nervous fibers were softer and more impressionable than those of their male counterparts. This meant women were superior in "perception" and "sensibility," while men excelled in "reason." Physiognomists found evidence for these theories in the human forehead. They not only argued that women's bodily fibers were delicate (and thus unsuited to serious study); they also suggested that female skulls exacerbated the problem. In both Europe and the United States, physiognomists contended that the most intelligent women would have high and narrow foreheads. Since high foreheads signaled superior perceptive abilities, it was not surprising that certain elite and middle-class white women were quick thinkers and witty conversationalists, even if they were unable to dedicate themselves to rigorous study over sustained intervals. For such intense mental deliberations, they apparently needed broader brows.[69]

The consensus seemed to be that women had greater "intensity" of thought, but less "permanence." When an idea or sensation hit a woman's brain, her flexible fibers would vibrate quickly and intensely through her narrow forehead, producing a brilliant exertion of mental energy. Then, just as quickly as the sensation appeared, it would vanish, causing women to be capricious and easily distracted. If women's impulsive unpredictability resulted from the shape of their skulls—as well as the delicate texture of the filaments in their nervous system—then it ultimately did not matter if they were blocked from the highest echelons of the educational system. After all, women's narrow foreheads and malleable brain fibers had already cursed them with a diminished capacity for sustained intellectual activity.[70]

Phrenologists took these ideas and expanded upon them during the early decades of the nineteenth century. By the late 1830s, American readers could access simplified versions of transatlantic theories, reformulated for popular audiences in the pages of phrenological almanacs, magazines, and pamphlets. Lorenzo Fowler's discussion of the "Difference between the Sexes," for instance, built on the ideas of eighteenth-century anatomists and physiologists, as well as on the forehead doctrines of the physiognomists. Yet Fowler did not plunge readers into the scientific debates taking place in transatlantic intellectual circles. Instead, he authoritatively stated that the female head was "narrower" while the male head was "broader from ear to ear." Because the "male has a higher and deeper forehead," Fowler asserted, "the male sex possess stronger intellect," while "the female sex have stronger feelings and moral sentiments." Supplementing physiognomic forehead doctrines with insights of their own, phrenologists argued that the back of the skull indicated a person's parenting abilities. Women whose skulls protruded in the back were sure to be good mothers, but if this region of their skull was underdeveloped, they would likely flounder as parents (fig. 2.5). Phrenologists dubbed this trait "philoprogenitiveness" (love of progeny), and argued that it was visible in people's craniums.

Physiognomists and phrenologists were typically willing to admit that *certain* women had high brows and refined intellects, even as they insisted male brains were *generally* larger and stronger than female brains.[71] When speaking of working women, immigrants, and women of color, however, they used physiognomical descriptions of female beauty to paint a racially and ethnically exclusive vision of human capacity. Popular scientists, for instance, often pointed to the supposedly "retreating" foreheads of African Americans to prove their minds were no match for those of their white counterparts. If a high and broad forehead was the ultimate symbol of intellectual distinction, then a "low and retreating forehead" was a marker of mental inferiority. The pro-slavery phrenologist Charles Caldwell, for instance, declared it was infeasible "for an individual with a low, narrow, and retreating forehead, to be intellectually great." Predictably, he claimed people of color were more likely to exhibit this trait.[72]

In the eyes of many white thinkers, true beauty simply did not—and could not—extend to Black women's bodies. For this reason, they spent very little time talking about Black female beauty, even as they expostulated at length about white women's beauty. When they did turn their attention to Black women's physical features, white scientists usually insisted women of color had inferior foreheads, prognathous jaws, and course features.[73] In 1841, for instance, the *American Phrenological Journal* described what the magazine dismissively referred to as "the skull of a stupid negro woman," saying it displayed a "low forehead" with projecting eyes. "This form is incompatible with intellect," the author said, providing no evidence for the accusation.[74]

What many of these theories had in common was a focus on the forehead. Seeing this trait as the most reliable metric of mental capacity, physiognomists and phrenologists insisted that the forehead indicated one's mental faculties while the mouth and jaw indicated one's animal propensities. When proving white women's intellectual inferiority to white men, scientists focused on the narrowness of the female brow. When arguing in favor of white supremacy, they used a different strategy. Insisting people of African descent had retreating foreheads and large protruding jaws, they claimed that Black women's animal propensities predominated over their moral and mental faculties. Whenever they could not come up with a method for definitively quantifying Black and female inferiority, scientific thinkers reverted to an ideologically facile claim: native-born white women were simultaneously the intellectual inferiors of white men and the aesthetic superiors of everyone else. Facial beauty and cranial conformations thus provided tangible "evidence" of racial distinctions.

Embedded within these scientific discussions of female bodies was an undercurrent of erotic desire. The physiognomist Alexander Walker, for instance, peppered his treatises with images of nude female bodies. He assured readers that these images were scientific and not salacious in nature. Responding to potential critics, Walker wrote: "some will tell us that the analysis of female beauty . . . is indelicate.—I shall, on the contrary, show that decency demands this analysis." After all, it was men who held the "power of selection" when forging sexual partnerships. How else would they learn to identify the forms of female beauty that would lead to the happiest and most beneficial marriages? Instead of upholding "artificial decency," he argued, men should cultivate "a critical judgment and a pure taste for beauty." Why be a moralistic prude when one could be an empirical observer of the female form?[75]

Like Walker, other scientific thinkers invoked their sexual attraction to white female bodies as evidence that beauty itself was not a socially constructed or culturally specific concept, but rather an objective reality that could be studied, measured, and defined by "the great truths of anatomy and physiology."[76] As the British surgeon and zoologist Thomas Bell put it, "Female beauty differs among the various races of mankind. There is, however, a standard of beauty independent of all idea of that partiality." While admitting that the world's diverse populations would find different bodies alluring, he insisted there was only one "perfect model of beauty": the ancient Grecian female with alabaster skin. Of course, to this Englishman, it also seemed obvious that "the English are the most beautiful people on the globe." Reluctantly admitting that women of African descent were, at times, beautiful, Bell briefly entertained the notion that more people would find Black women attractive if it were not for slavery and the prejudices it engendered. But he quickly abandoned this proposition "in favor of the whites." For him, perfect beauty required aquiline noses, small lips, and blushing cheeks. Leading physiognomists and phrenologists in the United States typically adopted these ideas, treating beauty as both an erotic and a scientific category that simultaneously explained racial and gender hierarchies.[77]

Americans also used physiognomy and phrenology to establish ethnic distinctions at a time when the meaning of "whiteness" was still in flux. Between the 1840s and the 1860s, Irish immigration to the United States increased dramatically, causing anxious white Protestants to brand the Irish

FIGURE 2.5. "Which Is Domestic?" *Human Nature* 9, no. 1 (San Francisco), January 1899. Division of Rare and Manuscript Collections, Cornell University Library. Physiognomy suggested that good wives and mothers would have softer and rounder features, while phrenology claimed that domestic women would have protruding back heads.

FIGURE 2.6. "Contrasted Faces," from Samuel Wells, *New Physiognomy; or, Signs of Character as Manifested through Temperament* (1866; New York: Fowler and Wells, 1871), 537. Courtesy of the American Antiquarian Society.

as insufficiently white and as cunning Catholic invaders who threatened to destroy American cultural heritage. Physiognomists and phrenologists validated these prejudices by giving them the veneer of scientific legitimacy. Through a printed series of "Contrasted Faces," Samuel Wells (husband to phrenologist Charlotte Fowler and supervising editor at the Fowler and Wells Press) juxtaposed the visages of sophisticated individuals with their allegedly less refined counterparts. These images conveyed a clear message: American heritage was British heritage. By comparing sketches of recognizable English beauties with caricatures of Irish women, Wells suggested that Anglo-American women were mentally and physically superior to their Irish counterparts. In a now famous sketch (fig. 2.6), Wells compared Florence Nightingale (a respected English nurse and moral reformer) with "Bridget McBruiser" (a figment of his imagination, intended to stand in for Irish women as a group). US History professors often display these images in the classroom, using them to highlight the extent of anti-Irish prejudice in the nineteenth-century United States. But in the nineteenth century, these images were not simply visual caricatures—they were scientific texts.[78]

By the 1860s, Americans possessed the physiognomic literacy to interpret these images as statements about the character and capacities of Irish immigrants. For readers who needed additional guidance, Wells clarified that McBruiser "lives in the basement mentally as well as bodily." While Nightingale was "bright, intellectual, and spiritual," McBruiser was "opaque, dull, and sensual." Due to her lineage and "personal education," Nightingale was cultured. McBruiser, by contrast, was "rude, rough, unpolished, ignorant, and brutish." If Nightingale's virtue reigned supreme, then McBruiser's "lower or animal passions" triumphed over her moral attributes. While Nightingale made decisions with "reasoning intellect," McBruiser relied on "simple instinct." This meant she only saw value in what could "be eaten or used for the gratification of the bodily appetites or passions." Through such images, Wells rationalized ethnic discrimination while simultaneously articulating a complex (and condescending) defense of universal human equality. The physiognomist never contended that Irish women were incapable of advancement. To the contrary, he insisted that McBruiser might attain "all sorts of virtues and knowledge under the benign influence of long and persistent social, intellectual, and Christian culture." So long as McBruiser abandoned her Catholicism, subjected herself to a rigorous educational curriculum, and carefully adhered to Anglo-American cultural traditions, she could cultivate her mind and improve her body. As Wells put it, "If form corresponds with and indicates character, it must change with the latter, and be, like that, measurably under our control." Another physiognomist phrased it differently: "the mind molds the features." If people could improve their bodies by cultivating their brains, then perhaps there could be hope for Ms. McBruiser after all.[79]

Wells's description of McBruiser exemplified one of the most intriguing inconsistencies of the physiognomical method. On the one hand, physiognomy and phrenology suggested there was a clear hierarchy of humanity, in which some people were inherently superior to others. On the other hand, these sciences insisted on the plasticity of people's bodies and brains, holding out the possibility that all individuals could improve their appearance by refining their minds. Physiognomy and phrenology were not equivalent to biological determinism. To the contrary, they emphasized self-improvement. This meant that despite their retrograde assumptions about "natural" human hierarchies, these disciplines managed to flourish within the reform-minded climate of the antebellum decades.

"THIS LAND OF POLITICAL EQUALITY"

During the mid-nineteenth century, the United States experienced a resurgence of religious enthusiasm. Dozens of new religious denominations emerged, and itinerant ministers began traveling to camp revivals delivering extemporaneous sermons before rapt audiences. Proclaiming that everyone could expunge sin from their hearts, Protestant evangelizers inspired a new class of moral reformers, who pledged to remake society in the image of God. Confident in their ability to craft a better America, they waged war against prostitution, founded temperance societies to curb the overconsumption of alcohol, advocated for universal education, and erected penitentiaries to redeem their "fallen" contemporaries. The most radical among them fought for the "rights of woman" or joined the abolitionist crusade. Meanwhile, some decided society was not worth saving. Instead, they built utopian communities where they combated the uncertainties of the emerging capitalist economy and re-envisioned existing gender and sexual norms.[80]

Undergirding this "Age of Reform" was the belief that the country could—and should—do better. It was an optimistic mindset, rooted in the idea that the United States might be an oasis of egalitarian possibility. More often than not, though, reformers reinforced existing hierarchies in their efforts to improve society. Most were white, middle-class, Anglo-Saxon, and Protestant. They took their own superiority as a given, rarely questioning the assumption that marginalized people would "improve" by becoming more like them. Within this context, physiognomy and phrenology served a dual purpose. By furnishing proof that the elite and middling classes were smarter and better than the souls they sought to save, these disciplines helped bourgeois activists distance themselves from the rabble while simultaneously confirming that everyone was improvable. This gave reformers precisely what they needed: evidence of their own superiority, paired with proof that their political crusades were achievable.[81]

Physiognomy and phrenology ultimately allowed for the coexistence of two seemingly antithetical ideas. They enshrined the notion that all humans were perfectible beings, but they also fostered the belief that individuals were born with a set of unique mental and physical traits which would definitively determine their destiny. In the physiognomic worldview, everyone could pull themselves up by their own bootstraps, beautifying their faces and re-sculpting their stubborn craniums as they acquired more knowledge and sophistication. Reformers latched onto this idea, but it was inherently problematic.

White reformers rarely doubted their own superiority, a fact that becomes visible in the condescending messages they chose to circulate in newspapers like the *National Anti-Slavery Standard*, which argued against slavery and in favor of female suffrage. As the paper proclaimed in 1842, "That the head of the African *has* a less intellectual development, is no proof that it always *will* be so. That the Indian *has* not been civilized, is no evidence that he *cannot* be."[82] The white abolitionist William Goodell made a similar argument, advocating Black suffrage while suggesting that African Americans were physiognomically inferior to their white counterparts. He claimed Black people deserved full legal and political rights because all human beings had been blessed with an "*improveable* mind." For Goodell, the question did not depend "upon the *quantity*" of a person's brain, "but upon the *expansibility* of it." In any case, he believed that education would eventually improve "the forehead of the black man, enlarge his brain, and magnify his 'facial angle.'" Because Black Americans possessed the capacity to improve, he said, they should be eligible for citizenship. Mental enhancements would come first, and would be followed by their physical manifestations. Such statements exposed the bigotry which poisoned even the most progressive physiognomic discourses. Articles like this suggested that people of Indigenous and African descent might someday deserve equality, but only after they had enacted an ambitious cerebral exercise regimen.[83]

Even some of the nation's most radical thinkers used popular sciences to rationalize inequality. The influential abolitionist and women's rights activist Angelina Grimké Weld once took a moment to discuss her family's efforts to informally investigate phrenology's validity. As lovers of the truth, she and her husband wanted to determine whether cranial analysis was a legitimate science or a form of quackery. Thinking himself clever, Theodore "disguised himself as an omnibus driver" and then commissioned a phrenological examination from one of the Fowler brothers. The phrenologist was not fooled. Even though Fowler was unaware of Theodore's true identity, he was apparently "so struck with the supposed fact that an omnibus driver should have such an extraordinary head, that he preserved an account of it." Angelina then presented this tale as proof of phrenology's legitimacy. After all, her husband was a college graduate who made a name for himself as a speaker, writer, and editor. Surely his head would be more impressive than that of a common bus driver! For her, it was important that Fowler saw beyond the disguise, recognizing cranial greatness in her husband.[84]

Fowler's ability to detect Weld's neurological gifts seemed to prove that phrenologists were not reading their own biases onto the skulls of unsuspecting clients, but rather making empirical observations that sometimes militated against their preexisting assumptions. When the Welds latched onto this narrative, they exposed their unspoken (and perhaps subconscious) internalization of class hierarchies. In this scenario, both the phrenologist and the client started with the assumption that educated, white, middle-class Americans would *naturally* have more impressive heads than the working masses. They admitted that individual exceptions might be possible, making allowances for this mythical and mentally distinguished omnibus driver. In the end, though, Fowler did not force his clients to examine their own biases critically. When the "truth" was finally exposed, the omnibus driver turned out to be one of the United States' most illustrious anti-slavery intellectuals. Class hierarchies were righted in the end.

In a variety of ways, physiognomy and phrenology helped people square the nation's egalitarian principles with the inequities of the emerging market economy. Magazines told stories of happy poor people whose heads and faces exposed their internal refinement, allowing prescient observers to see past their outwardly degraded circumstances. In one fictional account published in the *Farmers' Gazette* in 1840, a wealthy man walks into a shop to get his shoe mended. Observing the shopkeeper's face, the author thinks, "What an expansive forehead! What an intelligent countenance! What an expressive eye! There is truth in physiognomy, exclaimed I to myself—that fellow's brains are not made of green peas!" After this initial exultation, the visitor descends into despondency. What a shame for a man with so much intellect to be confined to such a menial position. Tentatively, the story's narrator inquired of the cobbler, "are you happy here?" To his surprise, the cobbler responded that he was entirely satisfied with his lot in life. Why would he need money or possessions when his contentment depended on inner peace? Such saccharine narratives valorized bourgeois work ethics, soothing readers who were unnerved by the upheavals of the market revolution and the economic inequality it engendered. If poor men could be as happy as their wealthy counterparts, then inequality might coexist with social harmony. On the surface, such tales assured people that physiognomic virtue was not limited to the upper classes. In the end, though, they merely promised elite and middling Americans that corporeal analysis would help them discern precisely who among the rabble might be worthy of their esteem.[85]

Protestant reformers were especially anxious to distinguish between the virtuous and the vicious poor. They were determined to spread God's love, but only to those who were eager for salvation. Harriet Wadsworth Winslow, for instance, visited families to discern their suitability for aid. In 1840, the American Tract Society published her letters, which recounted her reform efforts. In one instance, Winslow described her interactions with a poor woman who professed that she had been eagerly reading the Bible. Winslow had her doubts. Using her physiognomic skill set, the reformer declared the woman's "countenance indicated the lowest grade of vice." When the woman finally "confessed she was a sinner," Winslow felt vindicated.[86] The *American Monthly Magazine* recounted a similar tale in 1837, in which the author argued that there were three types of poor people: "the virtuous poor, the vicious poor, and those who are poor from sheer shiftlessness." According to the article, the "miserable monotony of a vicious poverty" led people to develop a "stupid state of heart" that could be "read in the very countenance." Internal viciousness, the author assured readers, would reveal itself in the "wooden features" and "vacant stare" of unrepentant paupers.[87]

Some authors went so far as to argue that social class itself was visible in the lineaments of the human face. The *Christian Disciple and Theological Review* argued that social position affected people's appearances. "Occupations and professions give to those engaged in them a specific, moral, and intellectual physiognomy," the author wrote, arguing that "rank, wealth, and power" produced internal changes which molded the countenance.[88] Daniel Jacques, an influential physiognomist, phrenologist, and popular author, made a similar claim: "each profession and occupation has a tendency to impress its peculiar lines upon the physical system of those habitually exercising it; so that we may generally know a man's trade by the cut of his features." Referring to the United States as the "land of political equality," Jacques claimed that Americans were "constantly rising from lower to higher social grades, acquiring at the same time, measurably at least, the physical traits of the rank which they assume." For Jacques, social status was an anatomical attribute. At the same time, he viewed mental and physical beauty as acquired traits, accessible to all. Arguing that people's faces would become more beautiful as they attained "education and political enfranchisement," he presented a physiognomic vision that controlled for class mobility.[89]

Numerous historians have argued that by the middle decades of the nineteenth century, American scientists and politicians increasingly described social inequality as a natural reality—the inevitable result of anatomical and physiological differences between different groups of people. But these historiographical narratives also obscure a more complicated reality. Americans did, in fact, look for evidence of social hierarchies in the human form, but they did not uniformly believe bodies were permanent and inflexible shells, which invariably reflected people's innate characteristics. Because physiognomy and phrenology were intellectually flexible disciplines, they simultaneously justified existing hierarchies and helped Americans believe that they were living in a meritocratic land of equality and opportunity. Despite providing "evidence" that working people were less beautiful and intelligent than the middle and upper classes, these sciences never precluded the possibility that all Americans might eventually elevate their minds and beautify their bodies. This was a flexible understanding of human nature—one that undercut emerging theories of biological determinism. Within this framework, it was not the innate or essential nature of people's bodies that mattered, but the changeability of the human constitution.

In many ways, physiognomy and phrenology were the perfect sciences for rationalizing the contradictions that lay at the heart of the American Dream. The nation's founders had pledged to create a society where every individual could rise. Natural distinctions of virtue and intellect would trump artificial distinctions of birth and wealth—at least in theory. In reality, the United States was no meritocracy. It was an inequitable society rooted in female subordination, racial discrimination, and economic inequality. Despite the lofty promises of the Declaration of Independence, the American Dream remained largely unattainable for white women, African Americans, immigrants, and working people. The intellectual flexibility of popular sciences merely allowed people to reconcile the nation's egalitarian rhetoric with its palpable inequities.

In the end, it was the optimism of the physiognomic method that made it so appealing—and so dangerous. Because physiognomy and phrenology put the onus of improvement on individual people, rather than on the collective resolution of the structural problems in American society and politics, these sciences sent the message that anyone could beat the odds, improving their lot in life as long as they worked hard enough. This was a hopeful narrative with troubling undertones. If everyone was technically capable of reform, then where did that leave people with retreating foreheads, projecting jaws, and large mouths? Physiognomists and phrenologists suggested that they, too, might achieve mental cultivation and physical beautification. Yet they also assumed that marginalized Americans were indeed deficient, effectively making it impossible for people to value their own brains and bodies without first acknowledging their alleged inferiorities and attempting to fix them. Despite this fact, many working people, people of color, and progressive reformers embraced these sciences and reformulated them in imaginative ways. As chapter 3 reveals, many women in the middling and elite classes—both Black and white—enthusiastically adopted these disciplines, creatively using them to craft alternative beauty standards and argue for the power of the female mind.

3
CHARACTER DETECTIVES

In the winter of 1842, Lucy Chase invited one of the United States' most famous phrenologists into her home. After examining her head, Lorenzo Fowler made his diagnosis. Lucy was not pleased. "He told me that I must not study, that my brain is too large, that my physical strength is much inferior to my mental," she lamented. The phrenologist informed her that she needed to "lay aside entirely my course of study, & try to be a character that it has always been unpleasant to me to contemplate, a very common character." Lucy shuddered to imagine herself as "common." She took pride in her mental acumen, regularly attended scientific lectures, and dreamed of being able to attend college like her brother. Yet now she faced an influential phrenologist who was reprimanding her for an oversized brain. The worst part? Lucy cared about his opinion. She continued to attend Fowler's lectures, eagerly watched as her friends and family members got their heads examined, and once cornered him for advice on how she might achieve a "well balanced head." She also continued to discuss phrenology with her loved ones and even tried "examining heads" at home. Still, the phrenologist's analysis had stung. When a family servant later fell ill with the measles, Lucy assumed her kitchen duties. "I presume Fowler would say that is the place for me," she noted sardonically. Evidently, Lucy had other plans. She refused to let one cranial examination thwart her ambitions, even though she remained an enthusiastic phrenological practitioner.[1]

As Lucy's experience demonstrates, male popular scientists often held discriminatory ideas about the capacities of the female brain. Nor did they stop at circulating these ideas in published works. They also enforced them through direct and personal conversations with American women. Sometimes, women disagreed with the results and privately stewed about the findings. Even so, they were enticed by these sciences and dazzled by their promised ability to unveil the mysteries of human nature. While occasionally harboring doubts about the scientific legitimacy of physiognomy and phrenology, countless Black and white women nonetheless embraced these disciplines, using them to learn more about themselves and to study the hearts, minds, and souls of others. When they did so, they came to their own conclusions about the female mind, using science to positively assess the inner worth of those they most respected.

These women were, for the most part, racially and economically privileged. Yet some labored in textile factories, while others were born into slavery or lived as free Black women in the antebellum North. Many attended female academies and seminaries. Most believed the minds of men and women were equal.[2] This put them at odds with some of the most prominent white male physiognomists and phrenologists, who typically started from two presumptions: first, that white women were mentally inferior to white men; and second, that people of African descent were less beautiful—and less intelligent—than their white counterparts. Given this reality, it might at first seem surprising that so many Black and white women embraced popular sciences. After all, they might have rejected such discriminatory discourses. But they didn't. Instead, countless women participated in scientific debates as autonomous practitioners, refashioning existing theories to align with their own perceptions of proper womanhood. How was this possible? In other words, why did so many women adopt the very discourses that deemed them inferior? The answer lies in *how* they used them.

In physiognomy and phrenology, women found ways to unearth beauty and intelligence in heads and faces that the rest of the world considered plain. They accomplished this feat in three major ways. First, women used popular sciences to draw clear distinctions between people that they saw as "perfect beauties" and those who were physiognomically or phrenologically impressive. A "perfect beauty" had symmetrical features, elegant proportions, and a graceful form. Yet she could also be vain, affected, and enamored of her own personal charms. Because "perfectly beautiful" women were perpetually preoccupied with their own attractions, they typically failed to cultivate the "beauties of the mind." As a result, these women supposedly experienced a pinnacle of adoration and flattery in their teenage years that lasted only until their external attractiveness inevitably faded away. Eventually, they were left with nothing but wrinkly skin and vacuous minds. By contrast, a woman with impressive physiognomic or phrenological developments might lack perfectly regular features or an idyllic form, but she usually had an extraordinary cranium, a high forehead, eyes that conveyed intelligence, cheeks that expressed modesty, and lips that spoke the truth. When they differentiated between perfect and scientific beauty, women outlined supposedly good and bad forms of female attractiveness. This allowed them to identify positive mental, moral, and physical traits in women they admired, while simultaneously adhering to gendered stereotypes that denounced the vanity, affectation, and frivolity of statuesque women with symmetrical features.

Second, women used scientific language to distinguish between the "beauty of features" and the "beauty of expression." Any lifeless statue could have a symmetrical bone structure, they suggested, but only the most refined women would convey traits like intellect or benevolence through their beaming eyes and the workings of their facial musculature. Unlike the white male scientists who argued in favor of women's mental inferiority, women rarely suggested there was one scientific standard of feminine loveliness, nor did they assume that beauty of features always corresponded with a refined character. Instead, they argued that it was just as important to analyze people's movable facial expressions as it was to study the more permanent bones and structures of the skull. In doing so, they insisted on a flexible model of corporeal analysis, not an inflexible interpretation of the skeletal structure.[3]

Finally, by reinterpreting the precepts of physiognomy and phrenology, women suggested that virtue and intelligence were determined not merely by heredity, but also by moral and mental cultivation. Describing both mind and matter as transformable, they suggested that internal improvements could beautify the features of the face, making people more attractive as they perfected their minds. These women studied heads and faces to understand human nature, but for them, bodies did not determine destiny. Seizing the language of popular sciences as their own, they argued for the eminence and improvability of the female mind.

These strategies were subversive in some ways, but not in others. Because beauty itself was a racialized category in early America, women's racial identities invariably shaped their experiences as facial and cranial analyzers. Educated white women, for instance, were willing to expand the boundaries of the beautiful to include intellectual women—whom society often dismissed as ugly and unappealing. Yet they continued to describe physical attractiveness as a valid measure of female worth. White women also failed to argue for a concept of beauty that encompassed their Black counterparts. While white women struggled to get people to value their minds over their bodies, Black women faced a white supremacist culture that refused to see them as beautiful in the first place. For this reason, Black women sometimes internalized the dominant culture's beauty standards or developed innovative and metaphorical ways of talking about the brains, bodies, and souls of Black women. Intriguingly, though, women of color did not universally reject popular sciences. Some embraced phrenology as both an amusing activity and part of their scientific education. Others casually analyzed the countenances and craniums of fictional characters, famous authors, or their white abolitionist allies. Even as Black women grappled with physiognomic and phrenological standards that marked their bodies as inferior, they sometimes opted to adopt popular sciences rather than dismissing them entirely.

Female practitioners surely recognized that physiognomy and phrenology could be problematic disciplines that reinforced racial and gender inequities. Despite this fact, both Black and white women adopted these disciplines, using them to learn about themselves and to study other people's characters. Those who analyzed bodies did so on their own terms and for their own purposes. Flouting established theories when it was convenient, they took elements from physiognomy and phrenology that appealed to them while jettisoning objectionable doctrines. In doing so, women deployed popular sciences in imaginative and idiosyncratic ways. By detecting the physical signs of virtue, modesty, and intellect in individuals they admired—and by noting imperfections in those they did not esteem—they identified female role models, denounced disreputable women, and tried to improve their own characters.

BECOMING A SCIENTIFIC PRACTITIONER

Early American women accessed physiognomy and phrenology in a variety of ways: through print culture, visual images, public lectures, and casual conversations. Some got their heads and faces examined by traveling lecturers. Others copied passages from scientific treatises into their letters, diaries, autograph albums, and commonplace books. Some learned about these disciplines at home or in school. All were enmeshed in a transatlantic cultural and intellectual universe that was inhabited by philosophers of human nature and by poets, novelists, scientists, and physicians. As readers, they engaged with authors who used science to outline desirable traits of femininity. Then, as character detectives, they scrutinized heads and faces to decipher the moral and mental capacities of their contemporaries.[4]

It was common for American women to encounter physiognomy and phrenology at a young age, often from their parents or teachers. In fact, popular scientists explicitly marketed these disciplines to mothers, seeing them as the people most directly responsible for shaping their children's character and cranial development. Some women took this advice to heart. Rebecca

Gratz, a Jewish religious educator and philanthropist, declared that "every parent and teacher ought to be a phrenologist."[5] In a similar way, the abolitionist and early feminist activist Georgiana Bruce Kirby and her friend (and fellow radical) Eliza Farnham physiognomically and phrenologically analyzed their own children. Kirby found her own daughter's "head lacking in breadth, imagination, and cautiousness." Luckily, little Ora was "wide between the eyes," which the physiognomist James W. Redfield associated with musical ability. Kirby also taught about popular sciences in a school she opened for young women in her California home. She asked her students to read George Combe's *Constitution of Man* (1828), even though her nemesis, a persnickety "Methodist Minister," kept warning all the local parents "that none but infidels believe in phrenology." Fearing that teenage girls in California were being forced to wed before they were ready, Kirby used phrenology to teach them about "marriage, maternity, the moral and physical laws affecting children and so forth." In this way, she packaged phrenology as a way for young women to learn about their own bodies.[6]

The United States' most famous female intellectual, Margaret Fuller, also instructed her students in physiognomic and phrenological principles. While teaching in Rhode Island's Greene Street School, Fuller transmitted scientific knowledge to the younger generation by conducting informal physiognomic readings on her own students. In 1838, one of her pupils, Evelina Metcalf, recorded the experience: "After we had done talking about the lesson," she wrote, "Miss Fuller told our characters by the expression of our faces and though all was not quite true yet she is a very good delineator of features." Impressed with her teacher's skills, she declared, "I think she told the one that she knows the least the best." Fuller also invited the famous phrenologist Orson Fowler into her school, where he not only examined Fuller's head but also helped "initiate" her "pupils into the mysteries of the mind." Nor did Fuller stop at analyzing the minds and bodies of her students. She also encouraged pupils to scrutinize the heads and faces of famous authors, politicians, and fictional figures. These character studies complemented the other subjects she taught at the Greene Street School, demonstrating how popular sciences could intersect with and reinforce material from art and literature courses.[7]

Although female academies and seminaries did not offer specialized courses on physiognomy and phrenology, these disciplines could nonetheless be an informal part of the educational curriculum at these institutions. This was particularly true after the 1830s, when practical phrenologists began traveling throughout the United States. These itinerant scientists made stops in northeastern cities, rural southern towns, and even in relatively isolated outposts in the West. They visited the parlors and lyceum halls of elite and middling Americans, but they also read the heads and faces of children, factory workers, and ordinary people. Remembering her time as a "Mill Girl" in the boarding houses of Lowell, Massachusetts, Lucy Larcom noted that "Phrenology was much talked about; and numerous 'professors' of it came around lecturing, and examining heads, and making charts of cranial 'bumps.'" Her friend and fellow mill worker Harriett Robinson likewise remembered how one of the Fowler brothers "went about into all the schools, examining children's heads." Robinson did not have a particularly good interaction with the scientist, who gave her an unfavorable reading. Despite her ambivalent feelings about her own diagnosis, Robinson remembered such events fondly, arguing that they facilitated both social bonding and mental enrichment among the working women in the mills.[8]

Sometimes, though, popular scientists were more intrusive than helpful. When the European phrenologist Johann Gaspar Spurzheim toured through the United States in 1832, he visited both the Hancock Grammar School for (white) Girls and the Smith School for African American children, an institution spearheaded by the Black female educator Sarah Paul. In some ways, this experience must have been enormously gratifying for Paul. After all, Spurzheim was one of the transatlantic world's most famous intellectuals. He did not simply visit Paul's school. He also praised the attainments of her pupils, reportedly saying that Black students would "receive their first education as quick, if not quicker than white." At the same time, this visit must have been enormously frustrating. Paul was a Black woman fiercely dedicated to the intellectual advancement of her students. Spurzheim was a white man who clung to white supremacy, even if he sometimes allowed for the possibility of Black excellence. As he saw it, African American students would do well in youth, but their foreheads were "*in general*, smaller than in the whites" and they had diminished "reflective faculties." This would make them "deficient in the English High School." Ultimately, Spurzheim affirmed that Black pupils were smart and capable, but he also insisted that white students would eventually surpass their African American counterparts. Phrenology, in this case, served as a scientific justification for keeping Black teenagers out of the nation's largely segregated high schools. Even so, Paul invited Spurzheim into her classroom, likely taking pride in his positive evaluations of her students while resenting the artificial limits he placed upon them. Her experience shows how popular science practitioners could infiltrate American educational institutions, enforcing prevailing attitudes about race and gender through personal interactions with both students and teachers.[9]

Even when women were not getting their heads and faces examined in school, they could access popular sciences in more informal ways—through personal relationships and social interactions. Two years after the American edition of Lavater's *Essays on Physiognomy* was released in 1794, a young woman named Charlotte Sheldon and her brother went to visit a family friend named Mrs. Beardsley, where they played piano, "read in Lavater & looked at the Heads," then took a walk before finally heading home. Margaret Cary similarly recalled discussing "Lavater's work on Physiognomy" in a drawing room before a dinner party in April 1843. As these experiences demonstrate, popular sciences easily fit into the rhythms of everyday life. The amateur portraitist Ruth Henshaw Bascom, for instance, once sat down to read a phrenological treatise with a friend on a "very rainy day" in 1833. Then, she "cooked & washed dishes, & all 'that sort of thing.'" Mary Ann Mansfield, for her part, discussed science with her father, who wrote to inform her of "a course of lectures" by the phrenologist George Combe. Because he had already "read so much on phrenology," Mary's father attended just one of the lectures, but he made sure to give his daughter a detailed recounting of it.[10]

Popular sciences could even infiltrate intimate discussions of love and marriage. In 1852, a young white woman named Caroline Hance described a curious visit to a fortune-teller, who mused about the physiognomy and temperament of her future husband. The fortune-teller assured her that she would enrapture a lover with "Black hair, splendid dark eyes, fine forehead, not very tall, fine form, noble in character and disposition, intellect stamped upon every feature, and a peculiar mouth." Hance did not take these pronouncements particularly seriously, saying she "had a good laugh and told her I did not believe a word." She nonetheless copied the fortune-teller's remarks into her journal. On another occasion, Hance recounted going on a walk with a man named Fred, who was studying to be a physician. Fred insisted that his medical training prepared him to "read a young lady's character at first sight," so Hance challenged him to read her countenance. He proceeded to diagnose his female companion with:

> a nervous and sanguine temperament, the former meaning that I would commence several things at once and leave them and that I would prefer hard study to hard work—also that I was passionate for a moment and then it was all over—all of which are very true and I suppose it must be the case that he can read character very readily.

Through this "reading," Fred combined humoral theory, physiognomy, mental philosophy, and gendered stereotypes to suggest that Caroline was an intellectually passionate but fickle person. She agreed, then turned her gaze back on Fred, saying it was "very hard to look at him steadily as he has a deep, full, beseeching eye, one which I admire very much indeed." Hance's scientific knowledge came from a wealth of diverse sources: from the texts she read and the men she conversed with, but also from her interactions with an eccentric fortune-teller who attempted to describe her future husband. As her experience shows, physiognomy and phrenology could be subjects for serious intellectuals, but they could also be lighthearted methods that women used to analyze character in social situations.[11]

Well into the late nineteenth century, popular sciences remained relevant for American women. Almost a hundred years after Lavater first published the *Essays on Physiognomy*, female educators were still using his theories to teach their students how to sketch bodies, convey character, and interpret human nature.[12] Phrenology, too, remained popular in the United States, long after most "serious" intellectuals started to denounce it. When writing her memoir in 1929, Ella Sterling Mighels, daughter of a California miner, fondly remembered her time as a young skull interpreter. "Nowadays many people are mad over psychoanalysis," she acknowledged. "But what would they say if you told them that in 1868 we found great discoveries in pursuing the study of 'Phrenology?'" As a child Ella had devoured books about craniums, and her family bared their skulls for friendly examinations by a local cranial analyzer. By the time she compiled her memoirs, Ella knew that people were denigrating phrenology as a "pseudo-science." Still, she recalled her phrenological adventures fondly, writing that they gave her and her family "a great deal of pleasure." She also hinted that there was likely some truth to the cranial readings. In particular, she recalled an instance where a phrenologist had examined the heads of some of the miners in her neighborhood, declaring that "they had a large amount of philoprogenitiveness" (love of children). Ella was not surprised, since those men had been especially kind to the kids in town. Her experience shows that phrenology made a lasting impact on young women. It also reveals that popular sciences were not confined to cities and farm towns in the Northeast. Like the people who practiced them, they traveled throughout the country.[13]

On at least one occasion, popular sciences even made inroads among enslaved people on southern plantations. Hannah Bond, a woman who escaped slavery and wrote a novel based on her experiences, deployed physiognomic language in *The Bondwoman's Narrative*. Throughout the story, Bond portrays herself as an astute observer of bodies and minds. "Instead of books, I studied faces and character," she declares. Bond uses these skills to observe one of her enslavers, making "a mental inventory of her foibles, and weaknesses, and caprices." She emphasizes the woman's "haughty eyes" and "delicate features with the exception of her lips which were too large, full, and red." This physiognomical clue turns out to foreshadow a crucial plot point: the woman secretly had African ancestry. Eventually the enslaver, herself, would be enslaved. Bond hints at this development by calling her enslaver a "small brown woman" and by emphasizing her large lips (which physiognomists typically associated with people of color). Bond also uses facial analysis to make moral judgments about various characters. When speaking of the story's villain, a slave trader, she notes his "keen black eye, and sharp angular features." She also describes one particularly odious white man by saying that his "forbidding aspect" was "so dark, so sinister and sneering" that she assumed there was "nothing of sunshine to his spirit, nothing of love in his soul." By contrast, she describes a comparatively decent white couple by saying "it is not my intention to draw their portraits . . . after all what language could portray the ineffable expression of a countenance beaming with soul and intelligence?" Such language suffuses Bond's narrative. This raises a question: where did she learn to describe people this way? Although Bond takes great pains to detail how she attained literacy, she is mostly silent on how she acquired a facility with physiognomic language. Her enslaver's library catalog provides some hints, though. In addition to novels by authors who used popular sciences to delineate character, the library housed books specifically dedicated to topics like phrenology, magnetism, and mesmerism. As literary scholars have shown, Bond was likely reading her enslaver's books, since their influence is evident in her own literary style. She probably encountered physiognomic and phrenological language in these works, and then repurposed it to paint unflattering portraits of the white figures who enslaved and exploited her.[14]

Mary Virginia Montgomery, also an enslaved woman, more explicitly detailed her relationship to popular sciences. For the first fourteen years of her life, Montgomery was held as the property of Joseph Davis, brother to the president of the Confederacy. Because Davis fancied himself a benevolent enslaver, he entrusted to Mary's father the management of his business enterprises, and the opening of a plantation store. The venture was remarkably successful. Despite their legal status as property, the Montgomery family was able to replicate many aspects of bourgeois existence, even hiring a white tutor to educate their children. When their neighbors discovered this apparent violation of racial hierarchies, they put a stop to the experiment. But the Montgomerys never abandoned their commitment to their children's education. By the time Mary was twenty-three years old, she had not only achieved her freedom; she had also attained a level of intellectual refinement which rivaled that of middle-class white Northerners.[15]

Montgomery was an avid reader who studied a wide array of subjects, including music, history, literature, geography, and chemistry. Among the topics to which she devoted substantial attention was phrenology. In the 1870s, she subscribed to the *American Phrenological Journal* and purchased her own phrenological bust to practice the science of character discernment. On some nights, she occupied herself with "Phrenological pursuits." On other nights, she read works by luminaries such as Plutarch, conducted "crystallizing experiments" in the field of mineralogy, or pored over Charles Darwin's *Origin of Species*. After being accepted to Oberlin College in 1872, Montgomery said a tearful goodbye to her family and embarked upon the trip to Ohio. Her *Phrenological Journal* kept her company along the way, distracting her from her homesickness (and from the racial discrimination she experienced on the journey). After arriving at Oberlin, she used her scientific knowledge for a different purpose: making friends. "I tried to entertain Phrenologically, and had a lively time," she wrote. "I do not feel homesickness so much now."

Montgomery's experience speaks to the broad reach of phrenology and the diversity of those who subscribed to it. She began her life as an enslaved child. By her early twenties, she was both a college student and a practical phrenologist who subscribed to the nation's most popular phrenological journal, owned her own plaster bust, and analyzed the craniums of other young Black women. Montgomery took phrenology seriously, studying it alongside Charles Darwin's theory of evolution. For her, chemistry, natural history, and phrenology were equally valuable forms of knowledge that sparked both curiosity and amusement while simultaneously contributing to her intellectual enrichment.[16]

Why, though, would Black women embrace phrenology? After all, popular sciences had laid the groundwork for scientific racism, and phrenological journals often published racist depictions of African Americans. The answer lies in the mixed messages articulated by some of the nation's most famous phrenologists. Even though they bolstered white supremacy through their publications, phrenologists were, for the most part, opponents of slavery. This meant they occasionally depicted Black people in positive ways. In 1851, for instance, the *American Phrenological Journal* published a portrait of Sarah Margru Kinson Green (fig. 3.1), who had been a female captive on the *Amistad*. Like Montgomery, Margru eventually went on to attend Oberlin College. The journal described her as having "independence, perseverance, energy, and unusual intellectual powers; remarkable memory, and the faculty of acquiring education." The periodical also proclaimed that Margru's mental assets matched her physical appearance: "The forehead is broad and

SARAH KINSON, OR MARGRU.

FIGURE 3.1. Bust-length portrait of Sarah Margru Kinson Green, *American Phrenological Journal* (New York), July 1, 1850. The Library Company of Philadelphia.

high, and particularly prominent in the center," it stated, "the whole head is large, sustained by a vigorous constitution." Such glowing descriptions were usually reserved for white Americans.[17]

The editors of the *American Phrenological Journal* might have been willing to acknowledge Margru's intelligence, but their depiction of her solidified discriminatory tropes. By contending that Margru was "far superior to Africans generally," the writers described her as an exception that proved their racist rule. Despite being willing to highlight specific individuals as mentally superior beings, they claimed that, as a group, people of African descent were intellectually inferior to Caucasians. Still, it is not hard to see why Mary Montgomery might have been buoyed by such descriptions. In a country plagued by white supremacy and racialized violence, the *American*

Phrenological Journal had printed a tribute to a Black woman who attended college and garnered public praise for the power of her mind. Phrenological works also gave Black and white readers alike the encouraging message that "Self-Culture" could lead to both mental and physical improvement. Finally, Montgomery might have found phrenology appealing because it was one of the disciplines that established her own claim to literary and scientific refinement.

To demonstrate a knowledge of popular sciences—and an ability to use them—was a way for both Black and white women to show that they were informed and respectable readers who had access to all the latest books, magazines, and scientific journals. When they used physiognomic and phrenological language to analyze character, women demonstrated that they were cultivated, introspective, and thoughtful individuals who read widely and then deployed their literary and scientific knowledge in practical ways. Physiognomy and phrenology were accessible to the masses, of course, but it was mostly educated members of the elite and middling classes who subscribed to scientific journals, kept abreast of the latest literary works, and used scientific language to analyze their friends and family. By engaging with these disciplines, Black and white women alike could exhibit their refinement and respectability. For them, repurposing physiognomic and phrenological ideas became a way of showcasing one's literary attainments and demonstrating that one fit snugly within the ranks of the educated middle class.[18]

READING BOOKS TO READ BODIES

As Montgomery's experience shows, it was often through print culture that women accessed and exercised scientific knowledge. Even if they never read treatises on physiognomy and phrenology, they typically encountered examples of facial and cranial analysis in published works and then proceeded to incorporate the language into their own writings and practices. A young woman named Lucretia Fiske Farrington, for instance, kept a notebook where she transcribed extracts from various literary and scientific texts, including physiognomical character sketches.[19] The diary of Hannah Margaret Wharton provides yet another example of how women accessed popular sciences and incorporated them into their lives. Unlike Mary Montgomery and Lucretia Farrington, Wharton did not explicitly reference scientific texts and periodicals in her own manuscripts, yet she consistently studied the faces of others and evaluated her own disposition through an analysis of her appearance. Where, then, was she encountering these ideas? Her journal provides several clues. On multiple occasions, Wharton mentioned the books she was reading. Among these titles were numerous poems by Walter Scott, Hannah More's *Coelebs in Search of a Wife* (1809), William Wirt's *Sketches of the Life and Character of Patrick Henry* (1817), and the published letters of Anna Seward (1811). All these authors used physiognomic language to analyze the character traits of their major protagonists. Hannah Wharton did not stop at reading these texts. She also copied portions of them in her journal, showing how popular sciences could move between published sources and daily practice. By reading texts, she learned to read bodies.[20]

The diary of Charlotte Forten, a middle-class Black woman from Philadelphia, similarly demonstrates how popular sciences circulated between published texts and private manuscripts. Forten not only read *Jane Eyre* (1847)—a novel brimming with phrenological language—she also devoured a biography of Charlotte Brontë and then used physiognomical rhetoric to analyze the author's frontispiece portrait, saying that Brontë had "a noble face—which the light of the soul beautifies." On another occasion, Forten read Robert Turnbull's *Genius of Scotland* (1809), which she found "very interesting." Although she was generally pleased by the book, she was especially intrigued by the physiognomic "sketches of the fitted men who are the glory of Scotland."[21] At one point, the author had analyzed a sculpture of the poet Robert Burns, saying, "The forehead is particularly fine—open, massive, and high, with an air of lofty repose." Turnbull nonetheless claimed that "the likeness" was "defective." The artist evidently failed to capture the "intelligence and good humor" of Burns's mouth, portraying it as "unpoetical and vulgar." When Forten commented on such sketches, she signaled that she expected a certain level of physiognomical specificity from the texts she read.[22]

Of course, women were not always happy with the portraits they analyzed, and they complained when they encountered poor likenesses. Anna Gale's diary provides an example of this phenomenon. After reading the memoirs of the Empress Josephine in 1838, she provided a detailed analysis of Napoleon's consort. Based on her own understanding of the woman's inner nature, she relayed what she believed the Empress looked like:

> In Josephine's—character, I discovered, energy, benevolence, firmness, strong maternal affection, united to a most delicate, and refined taste, and a graceful ease of manners, which charmed, and won the admiration of those around her. In her countenance, I could see beaming from her large, soft, blue eyes, an expression of mild benignity.

When Anna wrote that she "could see" Josephine's beaming and benignant eyes, she was not speaking literally. She was engaging in a form of reverse physiognomy. First, Gale read Josephine's memoirs and determined her prevailing character traits. Then, she ventured an imaginative sketch of Josephine's appearance. After discovering a portrait that she had previously overlooked, Gale became frustrated (fig. 3.2). She had pictured Josephine as a stunning and impressive figure, but the image failed to match up to the vision she theorized. "Forming such an idea of her character, and beauty," wrote Gale, "what was my disappointment, when turning to the commencement of the book, I discovered an engraving of her, which before had escaped my notice. It was not only far from being beautiful," she complained, "but it was even coarse, and commonplace; there was no expression in the face, no grace in the form." When imagination confronted reality, Gale found her physiognomic dexterity lacking.[23]

By the early nineteenth century, the act of systematically interpreting visages had become an important part of the process of properly digesting a published work. As one periodical author argued in 1807, "An ingenious author has observed, that a reader seldom pursues a book with pleasure, till he has a tolerable notion of the physiognomy of the author."[24] Samuel Wells, the influential physiognomist and phrenologist, took this argument even further, arguing that people were themselves readable texts. "Every man is a book," he argued; "those who know how, can read him." Through their interpretations of famous people's heads and faces, published authors gave middle-class women a template for analyzing the portraits of the authors they were reading. In instances where an image was unavailable, readers analyzed the ekphrastic clues provided by writers or developed their own textual sketches.

For nineteenth-century women, popular sciences could be forms of empirical observation, practices which heightened the pleasure of reading, or social activities that enlivened dinner parties. Some attended lectures on physiognomy or phrenology, first absorbing the knowledge and then deploying the skills they learned to analyze the heads and faces of the speakers who captivated them. Others used these disciplines to complete assignments for their teachers in academies and seminaries. Still others read phrenological treatises or discussed these works in conversations with friends. And some women subconsciously imbibed scientific language from their favorite poets and novelists. Above all else, physiognomy and phrenology functioned as shared languages of character detection that came as second nature to anyone steeped in the transatlantic intellectual milieu of the early nineteenth century.

A few questions nonetheless remain: Once these women gained access to

FIGURE 3.2. Portrait of Josephine, Napoleon's Consort, in John S. Memes, *Memoirs of the Empress Josephine* (New York: Harper & Brothers, 1837). Courtesy of the American Antiquarian Society. This is likely the image to which Gale was referring.

scientific knowledge, what did they do with it? How, for instance, did they use physiognomy and phrenology to understand the female body and the female mind? And why did both Black and white women embrace popular sciences when most white male practitioners were using these disciplines to foster white supremacy and denigrate the minds of all women? As the rest of this chapter shows, women adopted and reformulated the language of physiognomy and phrenology, using these sciences to envisage a positive image of female intellect and character.

NOT A PERFECT BEAUTY

Margaret Fuller was an ugly woman. At least that was what Ralph Waldo Emerson proclaimed when he noted his friend's "extreme plainness," the "nasal tone" of her voice, and her disagreeable "trick of incessantly opening and shutting her eyelids." In Emerson's telling, Fuller's features "repelled" him. He magnanimously insisted that he managed to get past this "disagreeable first impression," eventually learning to value her mind over her beauty. But Emerson's critique stuck. Well into the twentieth century, his words have been quoted in encyclopedias and repeated by scholars and popular authors alike. In the past and the present, Fuller's critics and admirers have agreed: she was plain in body but impressive in mind.[25]

Throughout her lifetime, Fuller confronted a gauntlet of gendered expectations. Women were supposed to be beautiful, and people seemed to agree that she was not. Her intelligence, at least, seemed undeniable. To reconcile these mixed assessments, Fuller's contemporaries turned to science. The transcendentalist poet Sarah Helen Whitman, for instance, used phrenology to present a "more tender and sympathetic appreciation of [Margaret Fuller's] character and career." Whitman knew that Fuller's contemporaries tended to describe her as ugly, vain, and pompous. One evening, though, Fuller subjected her "haughty head" to the "sentient fingers" of John Neal, a practical phrenologist and advocate of female education. Neal's "masterly analysis" emphasized Fuller's "complexities and contradictions," as well as the struggles she faced as an intellectual woman in a male-dominated world. His phrenological reading provided Fuller's admirers with an alternative beauty standard—one that de-emphasized her purportedly "irregular" features while strategically emphasizing the physical markers of her intelligence.[26]

Many of Fuller's contemporaries used a similar strategy, acknowledging Fuller's alleged unattractiveness while arguing that her brilliance was visually perceptible in her outstanding physiognomical and phrenological profile. As *Godey's Lady's Book* declared in 1876, "Margaret Fuller had a splendid head; but her features were irregular, and she was anything but handsome." In a similar way, Caroline Healey Dall (a fellow female intellectual) emphasized Fuller's "sharp features" as well as her "thin" and "ungraceful" mouth. She nonetheless said that her forehead was a "good height," and that her eyes were "very vivid." Publicly quoting from a private letter, the feminist activist Anna Brackett likewise declared that Fuller "was not handsome, yet her very speaking and deep eyes and her intellectual forehead attracted attention always." Edgar Allan Poe also described Fuller's appearance in ambivalent ways. Yet even he conceded that her eyes were "full of fire" and that her "mouth when in repose indicates profound sensibility." Poe did not find Fuller beautiful, but he was impressed by her "capacious forehead" (which was, after all, the ultimate symbol of mental excellence). In all these instances, physiognomy and phrenology provided people with something that nonscientific discussions of Fuller's beauty never could: evidence that she was intellectually exceptional.[27]

Popular sciences also seem to have shaped visual depictions of Fuller. When compared with her daguerreotype portrait (fig. 3.3), one portrait accentuated the anatomical markers of Fuller's superior brain by exaggerating the size of her cranium, the breadth of her forehead, and the prominence of her eyes (fig. 3.4). The engraving erected at Fuller's gravesite more closely replicates the features in her daguerreotype. It portrays her in profile, a position that would have allowed viewers to scrutinize her facial angle (fig. 3.5). Perhaps the artist intended for people to analyze this marbleized tribute scientifically. Perhaps not. In any case, at least one of Fuller's admirers did just that. After Fuller's death, the abolitionist Sallie Holley discovered the "exquisite chiselling" on a tour through Mount Auburn Cemetery. She was deeply affected. Marveling at Fuller's "most admirable" head, "and on her face, full of character, generous enthusiasm, and high aspiration," Holley wondered how anyone had ever described the woman as "plain-looking." Fuller might not have exemplified existing beauty standards, but she *did* meet the physiognomic and phrenological standards for internal excellence. As Holley saw it, Fuller's beautiful mind expressed itself in the "fine and noble development" of her cranium. If not a stereotypical beauty, she was a stunning physical specimen.[28]

This subtle distinction—between a perfectly beautiful and a physiognomically or phrenologically impressive countenance—was central to how nineteenth-century women evaluated each other. In both published works and private manuscripts, Americans distinguished between those who were merely beautiful (an empty accomplishment) and those with heads and faces that conveyed their inner worth. In 1829, for instance, the *Ladies' Literary Portfolio* drew a physiognomical character sketch of a fictional female paragon, Miss Neville. According to the description, Neville "was not what the gay world would call a perfect beauty." Yet there was "a bewitching expression in her countenance, which rendered her peculiarly interesting. Her eyes were a dark blue, and rather piercing in their look; her nose somewhat aquiline; her lips thin, and well formed." Although she lacked the markers of classical beauty, Neville had the physiognomic indications of a "superior" mind and soul. If not a faultless belle, she had a "peculiarly interesting" countenance.[29]

At first glance, articles like this conveyed simple messages: Beauty is not important. Heroines can be homely. Character and intelligence matter more than appearance. On closer inspection, though, such pieces made scientifically inflected claims about female minds and bodies. If beauty was not important, then authors wouldn't have spent so much time describing women's appearances. The truth was that beauty *was* important. Even so, women's magazines decried society's obsession with "perfect beauty" and instead encouraged readers to search for more *legitimate* indicators of female attractiveness. Through physiognomy and phrenology, they suggested, people could develop a better system for detecting brainpower in the bodies of intellectual women. This would allow observers to see past exterior superfluities and instead discern the internal dispositions of their female contemporaries.

To get this message across, authors expended much intellectual energy trying to distinguish between "good" and "bad" kinds of beauty. The most common strategy was to contrast two types of women. On one side was the beautiful and fashionable—but frivolous and coquettish—belle. On the other side was the cultivated woman. The cultivated woman perhaps had irregular or unsymmetrical features, but she also displayed an "interesting," "engaging," or "intelligent" physiognomy. Emma C. Embury, a regular contributor to the *New York Mirror* and *Ladies' Magazine*, articulated this idea most clearly in a fictional story about two women: Aunt Mabel and Aunt Silly. Both were old maids. Yet the two were opposites when it came to moral and mental cultivation. Aunt Silly, as her name suggested, valued nothing but fashion, frippery, and her own beauty. Aunt Mabel, on the other hand, was a woman of good sense, intellectual sophistication, and moral quality.[30]

As the story makes clear, "perfect beauty" was not the same as a pleasing or intelligent physiognomy. "Mabel Morrison could never have been styled a beauty," wrote the author, but "she certainly possessed of that which is far rarer than beauty—I mean loveliness." The author then distinguished

FIGURE 3.3. Daguerreotype of Margaret Fuller, taken by John Plumbe (1846). National Portrait Gallery, Smithsonian Institution. This is the only known photograph of Fuller. It emphasizes her identity as an intellectual, portraying her in the act of reading a book.

FIGURE 3.4. Portrait of Margaret Fuller (ca. 1840–1880). Prints and Photographs Division, Library of Congress, LC-USZ62-47039. Images such as this one portrayed Fuller with a large cranium and expansive forehead. Nineteenth-century Americans believed such traits signified an individual's capacity for profound thought.

FIGURE 3.5. Margaret Fuller's grave at Mount Auburn Cemetery in Cambridge, Massachusetts. Public domain. The engraving on this "handsome monumental stone" impressed the abolitionist Sallie Holley so much that she physiognomically analyzed the "fine and noble development" of Fuller's cranium in a letter to her friend.

between the two sisters in old age. Aunt Silly had once been stunning, but her beauty had faded. She now had so many wrinkles and so few teeth that her face resembled a "musty parchment." Aunt Mabel, by contrast, was never as classically beautiful as Aunt Silly in her prime. But her countenance had always been the mirror of her meritorious mind. This meant that Aunt Mabel's face remained attractive throughout her entire life:

Few are so unskilled in physiognomy as to require to be told that the most beautiful faces are not always the loveliest. A mouth may be as perfect as if formed by the chisel of a Phidias, and yet, if unadorned by the smile of good humor it will never be lovely—an eye may be as brilliant as the diamond, yet if it lack the *inward* light of intellect, or if it be overhung by the scowling brow of habitual anger, it will never awaken the feeling of tenderness. A face may possess a combination of features, which, according to the rules of art, constitute the perfection of beauty, but it may be utterly deficient in loveliness; and a face utterly destitute of regular beauty may, if intelligent and amiable, be exceedingly *lovely*.

Such excerpts suggested that women should aspire to be beautiful. They simply had to cultivate the "beauties of the mind," rather than the "beauties of the person." If plain in body, they should be attractive in mind. Mental excellence would then reflect itself in the form.[31]

Just as Embury distinguished between Aunt Mabel and Aunt Silly, nineteenth-century women used physiognomic and phrenological language to identify virtuous and intelligent women and then contrast them with their less impressive counterparts. Harriet Story White Paige, for instance, compared the faces of Jane Seymour and Caroline Norton, granddaughters of the "celebrated Sheridan." While saying that both women excelled in "personal beauty," Paige thought Norton's face was "more intellectual, less regular, and of a decidedly higher order than Seymour's." Norton had "a large mouth, betokening energy, and decision," while Seymour had only "regular loveliness." In other words, Seymour was a perfect beauty with symmetrical features, but Norton's face conveyed intellect, energy, and decisiveness.[32]

Margaret Bayard Smith engaged in a similar process when comparing the minds and appearances of Thomas Jefferson's daughters in 1802. When describing Maria Eppes, she wrote that her subject was "beautiful," with a face demonstrating "simplicity and timidity personified." Even though Martha Randolph was "rather homely" in comparison to her sister, Smith found her the more "interesting" of the two. Smith insisted Randolph was the superior sister, because she had a "countenance beaming with intelligence, benevolence and sensibility." Her face conveyed merit, claimed Smith, and "her conversation fulfills all her countenance promises."[33] Harriet White Paige and Margaret Bayard Smith believed that appearances correlated with inner nature, but they also distinguished between the "perfect beauty" of "regular features" and the beauty that resulted from an "intellectual" or "dignified" countenance.

It was this strategy that Fuller's admirers would use when analyzing her head and face. While agreeing that she was not traditionally beautiful, they suggested that her genius and good character were visible in her head and face. Fuller's students got particularly defensive when people accused their instructor of being unattractive. In their view, Fuller's intellect brought "dignity to her appearance and such fine expression to her countenance that her companions, especially the younger members of the class, went away impressed with 'her beautiful looks,' and would on no account allow people to call her plain." The popular British author Robert Chambers made a similar point, arguing that Fuller's countenance, when "lighted up with feeling and intellect, dissolved its plainness, if not deformity, in beauty of expression." Even as they acknowledged that she did not have stereotypically beautiful features, Fuller's admirers insisted that her superior mind imbued her face with a pleasing cast. Popular sciences provided the linguistic toolkit that helped them do so.[34]

It should not be surprising, then, that Fuller herself used scientific language to evaluate other women. After meeting the female author George Sand, she gushed about the woman's appearance, intelligence, and character. Fuller swore that Sand's features were not "vulgar," as some caricatures had suggested. Instead, Sand's countenance blended both masculine and feminine traits: "the upper part of the forehead and eyes are beautiful," Fuller wrote, while the lower part of her face was "strong and masculine, expressive of a hardy temperament and strong passions, but not in the least coarse." With physiognomical aplomb, Fuller rejected a simplistic question: *Was George Sand beautiful*? Instead, she shifted the terms of the conversation, emphasizing the "expression of goodness, nobleness, and power" in her idol's face. Rather than make Sand's beauty the primary topic, Fuller used the language of popular sciences to highlight the "masculine" strength, "hardy temperament," and noble mind of a fellow female intellectual. Although Fuller believed that such traits were visible on Sand's countenance, she did not try to pretend that Sand was perfectly beautiful in the eyes of men. In this case, physiognomy and phrenology gave Fuller a template for analyzing Sand's appearance while simultaneously highlighting the importance of her powerful brain.[35]

By differentiating between "regular beauty" and physiognomical beauty, women made two seemingly antithetical arguments. On one level, they suggested that attractiveness did not matter. At the same time, they confirmed that appearances were important. They reconciled these claims by insisting that so long as one looked for the *proper* features in the body—attributes like intelligence, generosity, modesty, virtue, and good sense—then it was entirely acceptable to analyze appearances. This allowed women to insist that they valued inner characteristics over personal beauty. Even so, women rarely stopped at examining the manners or conversation of the women they encountered. Imagining that character traits could express themselves as physical realities, women searched for hidden messages in people's heads and faces. Sometimes, though, they disavowed the importance of features entirely, focusing instead on facial *expressions*. By emphasizing the movable—and more controllable—elements of people's faces, women tinkered with the rules of physiognomy and phrenology to suggest that both intellect and virtue could infuse a lasting charm over the countenance, illuminating even the homeliest faces.

DISCERNING CHARACTER

Among the most dedicated and creative facial analyzers in the early decades of the nineteenth century was a wealthy young woman of the Boston elite named Anna Cabot Lowell. Between 1818 and her death in 1894, Lowell kept detailed diaries which together comprise more than twenty boxes of manuscripts. These journals detail her daily experiences, religious opinions, and intellectual development, as well as her interactions with the fashionable world of Boston society. As the journals reveal, Anna Cabot Lowell assiduously analyzed her own character and regularly tried to discern the moral and mental attributes of her friends, family members, and acquaintances.

When analyzing others, Lowell distinguished between facial features and expressions—a technique that helped her find laudable internal traits in women with less-than-pleasing external characteristics. In one instance, she described a young woman named Elvira Degin, writing that she was "certainly plain, yet she has occasionally such a sweet expression & her countenance is so faithful a mirror of what passes in her mind, that I look at her with far more pleasure than I do many beauties." Similarly, on another occasion, Anna spoke of her friend Sarah Sullivan, saying it was a "privilege" to be in her company. Even though Sullivan was "without any pretensions to beauty," Lowell wrote that her "traits of character are so depicted on her countenance, that I take far more pleasure in looking at her than I do in contemplating many handsome persons." On yet another instance, she evaluated her friend Emmeline Austin, saying, "Without being handsome, she has a delicacy of features & a sweetness of expression which has all the charm of beauty." Lowell acknowledged that women like Sullivan, Degin, and Austin were not attractive—at least not according to existing societal standards. She nonetheless admired them. In Lowell's mind, intelligence and virtue electrified their faces and beautified their bodies.[36]

By focusing on the expressive nature of faces, rather than their permanent traits, women such as Anna Lowell found beauty in those who might have appeared homely at first glance. Determined to find "beauty" in the faces of morally and mentally distinguished women, female physiognomists distinguished between the "beauty of features" and the "beauty of expression." While they denigrated "coquettes" and "celebrated beauties" for their vanity and affectation, they typically found moral and mental beauty in the faces of the people they esteemed, even when those individuals had "irregular features." As Susan Fenimore Cooper once stated, "It will often happen that the most intelligent countenance is connected with ill-formed features, that the best expression of kindly feeling, or generous spirit, beams over the homely face."[37] When people made such arguments, they were not merely arguing that society should value women for their minds, rather than their beauty. They were instead suggesting that inner beauty could transform a person's outer appearance, infusing it with an alluring, if intangible, appeal.

The importance of expressive beauty was reinforced in female academies and seminaries. At the Annual Examination of the Canton Female Seminary in 1843, one speaker argued that the "symmetry of mental faculties is of higher importance than that of bodily form, and the beauty of moral character is far more excellent than that of the countenance." Even so, he insisted that a "richly cultivated and refined mind is capable of diffusing a lasting charm, over the features of any countenance."[38] In a similar way, a speaker at the "Opening of the New-York High-School for Females" proclaimed that "intelligence" could "irradiate the form in which it dwelt."[39] Through such strategies, nineteenth-century Americans described intellect and character as almost ethereal entities, capable of conjuring the illusion of attractiveness where none had previously existed. It was not that people thought female beauty was inconsequential. They simply argued that expression mattered more than features.

This was not physiognomy proper. In fact, Lavater had explicitly distinguished between physiognomy (the analysis of the *features*) and pathognomy (the analysis of the *expressions*). Yet Lavater himself acknowledged that the boundary between the two disciplines was often messy. Like many of the women who adapted physiognomic principles for their own purposes, Lavater contended that a "beautiful mind" could sometimes express itself through a form "which is the reverse of corporeally beautiful." Despite this fact, he insisted that physiognomy was a real science, while pathognomy was not. While he found value in pathognomical observation, he insisted that real physiognomists also had to analyze the skeletal structure.[40]

Disregarding Lavater's strict distinction between physiognomy and pathognomy, nineteenth-century Americans often interpreted the permanent bones and structures of the face while simultaneously focusing on what one Massachusetts minister described as the movable "muscles of mental expression."[41] Rachel Van Dyke, for instance, focused equal attention on facial features and expressions. In 1810, she studied the countenance of a man named Mr. Davis, an attractive flautist who had enraptured much of her town through his charm and musical talent. Despite the enthusiasm of her friends, Van Dyke was unimpressed:

> He is handsome enough, has good regular features, but I don't see much expression in his countenance. It is a certain expression in the eyes, a certain cut of the mouth that always strikes me. I am a great physiognomist. It is an amusement for me to observe the various countenances of persons who pass the windows. Methinks I can tell accurately their dispositions and characters.

Like Lavater, Rachel Van Dyke drew a distinction between facial features and facial expressions. But unlike him, she valued the latter over the former. Although Mr. Davis apparently had "good regular features," she found there was something missing in his face. His lack of expression troubled her, leading her to believe that he was not as wise, talented, or attractive as others had imagined. Rachel Van Dyke nonetheless styled herself a "great physiognomist." She did not seem to notice that her own method was at odds with Lavater's supposedly more "scientific" system, nor was she particularly troubled by the discrepancy.[42]

The influential educator Emma Willard similarly blurred the boundaries between physiognomy and pathognomy. In a letter to her sister, Willard analyzed the countenance of the Scottish intellectual Dr. Chalmers, whom she met on her trip abroad. In her analysis, Willard distinguished between superficial beauty and physiognomical attractiveness. Even though she found Chalmers to be "inelegant in exterior," she was impressed by his animated expressions. "Had I marked his physiognomy, merely in a quiescent state," she wrote, "I am not certain I should have detected the hidden fire within; but from the play of his features in speech, I could clearly discern the marks of his genius and benevolence." Had Willard been following Lavater's rules of physiognomical discernment, she would have focused on Chalmer's face in its motionless state. The *Essays on Physiognomy* clearly defined physiognomy as the "observation of character in a state of tranquility, or rest," warning that paying too much attention to the muscular movements of someone's visage could be distracting. But Willard did not rigorously adhere to Lavater's techniques. She cared far more about analyzing Chalmer's countenance in action than about studying it in its "quiescent" state. Through his facial *expressions*—not his *features*—she was able to discern his "genius and benevolence."[43]

In another instance, Willard used a combination of physiognomy and pathognomy to condemn a British schoolteacher. Identifying the woman only as "Miss A," Willard described the displeasing traits she had discerned in the woman's character upon meeting her. Miss A was attractive and had "regular" features, but her "physiognomy and manners" put Willard on edge. The schoolteacher tried to maintain a calm demeanor, but "her countenance assumed an air of spite and vexation." Through conversation, Willard then confirmed what her physiognomical instincts had already hinted: the woman was a hostile and ungenerous individual who proved reluctant to support a fellow female educator. In some ways, Willard was using physiognomy just as Lavater had intended. Miss A was trying to conceal her true disposition, and Willard was using scientific observation to unearth her dissimulation. Yet Willard invested most of her energy in tracing the woman's *expressions*. Through this method, she triumphantly unmasked Miss A as a jealous woman who was willing to lie about her boarding school and the accomplishments of her students.[44]

Later that day, Willard encountered a far more amenable governess at a different institution. If Miss A had been off-putting and cold, then Miss Y proved to be honest, generous, and endearing. Willard wrote that the woman's "countenance and the play of her features in speaking, impressed me with an idea that she possessed strength of mind and character, with kindness of heart." When comparing these two educators, Willard focused primarily on how the two women's faces appeared when they were engaged in conversation. Lavater might have referred to this as a "pathognomical" investigation. Willard nonetheless insisted she was reading the "physiognomy" of others.[45]

Willard was also happy to blend physiognomy and pathognomy with phrenology. While in Europe, she socialized with the famous phrenologist George Combe. She was highly pleased by the encounter, calling Combe an "eminent philosopher." Willard also used a combination of physiognomy and phrenology to analyze the craniums of eminent men. The Baron Cuvier, she argued, had a "large and strongly marked head" which she described as "sublime." Willard then "compared the physiognomy of Cuvier, with that of Lafayette," the famous general who helped Americans secure independence from Britain. While both men had "noble countenances," she found more "mental strength" in Cuvier (the scientist) and more "benevolence" in Lafayette (the military hero). Willard effortlessly traversed the boundaries between pathognomy, physiognomy, and phrenology. For her—as for other women—these were not mutually exclusive forms of scientific observation, but symbiotic tools of character detection.[46]

Women were certainly interested in the skeletal conformations of people's countenances and craniums, but they often emphasized the importance of facial expressions over immovable features. This strategy allowed them to distinguish more effectively between people with honorable characters and those with vicious inclinations. It also helped them find positive traits in the countenances of otherwise unattractive individuals. Sometimes they took this argument to its logical extreme, arguing that moral and mental cultivation could physically transform the face, leading to a more pleasing appearance.

INTELLECT BEAUTIFIES

In 1823, the Bostonian poet Sarah Wentworth Morton published a collection titled *My Mind and Its Thoughts*, which included a chapter on "Physiognomy." Although Morton engaged with Lavater, she was not terribly impressed. Referring to the *Essays on Physiognomy* as a "specious" set of guidelines, she criticized the "various twistings" of Lavaterian logic. Yet even as Morton criticized the famous physiognomist, she never condemned his central premise. She, too, believed external traits conveyed internal merit, merely critiquing his tendency to treat physiognomy as an infallible scientific system with universally applicable rules. For her, it was entirely possible to study a person's countenance without "critical regard to the complex laws of Lavater." Facial analysis was useful, she proposed, but it was a far more complicated— and more fluid—process than Lavater's "altogether systematic and artificial" maxims suggested.[47]

One of Morton's critiques centered on Lavater's obsession with the permanent and unmovable features of the human countenance. Morton was more concerned with expressions than with features, and she believed inner transformations could alter people's appearances. In her own remarks, she insisted "that the moral habits, the disposition, the understanding, and the passions, give *expression*, and in effect, stamp *character* on the features, without changing the tints, or altering the strong lineaments of original nature." In other words, the face did, indeed, convey inner characteristics, but the bone structure was not as revelatory as Lavater had proclaimed. As people cultivated certain traits, these moral and mental elements would "stamp *character* on the features," even if they did not fundamentally change the skeletal conformation of one's countenance. The general appearance of the face was more important than individual features, and faces could change as one's inner qualities dictated.[48]

Many of Morton's female contemporaries were less invested in studying the unmovable elements of people's skeletal structures than in tracking how people's heads and faces transformed over time. If a polished mind could imbue an otherwise homely countenance with physiognomical loveliness, then it stood to reason that people's bodies themselves could also transform. Education, in other words, might transform women's physical features, making them more beautiful as they improved upon their reasoning capacities. In the novel *Northwood* (1827), for instance, the famous author Sarah Josepha Hale described a woman's intellectual maturation, writing, "Her mind had been developed and disciplined, and its pure free light seemed to irradiate her face with intelligence, and gave a lustre to her eyes."[49] In a similar way, the female educator Almira Phelps suggested that women's status as the intellectual equals of men required a new conceptualization of female beauty. "Women are now looked upon as rational beings," she proclaimed, arguing that beauty "must give precedence to the plainest features irradiated with intelligence and good sense." Later in the book, Phelps returned to this theme: "As to personal beauty," she wrote, "it depends so much upon the expression of mental qualities" that if women wanted to appear beautiful, they had to first embark upon an internal makeover, manufacturing attractiveness from the inside out.[50]

Nor were educated women and female authors exceptional in arguing that intellectual cultivation magnified women's beauty. Male advocates of women's higher education made similar claims. In 1792, one author contended that "beauty is little more than the emanation of intellectual excellence" and that the passions had a "mechanical effect upon the aspect." "Beauty, therefore, depends principally upon the mind, and consequently may be influenced by education."[51] Similarly, in *Advice to Young Ladies on the Improvement of the Mind* (1808), Thomas Broadhurst argued that "a well-cultivated mind... imparts to every feature of the moral character an indescribable charm," saying it was visible "in the very looks and language of the countenance."[52] In an address to students at a female academy in 1841, the speaker Samuel Galloway similarly urged women to cultivate their minds and improve their beauty. He asked, "Is she not lovelier... whose eye kindles with the mingled fire of elevated thought and pure feeling, and whose countenance is invested with those radiant lines of thought, which like stars on the broad canopy of heaven, tell of a bright spirit within?" For Galloway, engagement with "true science" and the cultivation of "elevated thought" could beautify young women's faces.[53]

In both published works and private manuscripts, Americans claimed that moral and mental development could enrich women's personal charms. Writing to her family in 1844, the avid physiognomic and phrenological practitioner Rebecca Gratz mused about the importance of female beauty. She acknowledged that individuals often waxed poetic about the beauties of the mind. In reality, though, most women were desperate to be traditionally beautiful. Emphasizing the importance of internal cultivation, Gratz proclaimed that "an expression of goodness is so necessary to beauty, that I believe it is independent of features in a great degree." For this reason, she believed that a "good mother, may always beautify her daughters by bringing them up well."[54] In a similar way, the physician, scientist, and female suffragist Mary Putnam claimed that "at twenty-five, beauty is of less consequence than at sixteen, because the character has become more formed and positive, and moulds the body surprisingly." Character, she hinted, could physically imprint itself upon one's lineaments over time.[55]

Even as some anatomists and physiologists claimed that all individuals inherited their minds and bodies and could do little to alter them, women claimed that mental refinement could beautify the body. In doing so, they echoed the language of physiognomists and phrenologists who argued that the "mind molds the features." Bathsheba Crane, for instance, published a volume of reminiscences in 1880, which included copies of the letters she had written between the 1840s and the 1870s. In one letter, she declared "the face, like a dial, is a revelation of character." She then launched into a discussion of how the mind fashioned the form, proclaiming that "every human being carries his life in his face." Quoting Lord Lytton, she wrote, "It is said, 'men and women make their own ugliness.'" Suggesting that mental and physical characteristics were mutable, she continued:

> On our features the fine chisels of thought and emotion are forever at work. The passions of the soul—love, hate, revenge, jealousy, tenderness, and sorrow—steal into the lines of the face, are stamped in the deep iris of the prophetic eyes, and breathe in the magic power and pathos of the wonderful voice. Where a fine organization and deep sensibility accompany the practice of intellectual pursuits, the soul revels in light and shadow upon the face.... A soul full of sunshine will light up the face and give it a charm mere beauty could never impart.

Crane did not come up with this quotation on her own. It had been widely reprinted throughout the nineteenth century. Perhaps she found it in *The*

Argosy, The Ladies Repository, or *The Casket of Literary Gems*. Or perhaps she encountered it in *New Physiognomy* (1866), a book written by the phrenologist Samuel Wells. Throughout the book, Wells claimed that moral and mental improvement led to physical beautification. As he put it, "We are free to choose what course we will pursue, and our bodies, our brains, and our features readily adapt themselves and clearly indicate the lives we lead and the characters we form." It is entirely possible that Crane copied this excerpt from Wells's book. In any case, she used her reminiscences to reiterate the same point: the body reflected character, yes, but the body was not a hard and inflexible shell that displayed one's hereditary traits. It was an impressionable and dynamic entity, continuously changing according to the dictates of one's internal faculties and personal decisions.[56]

Anna Cabot Lowell articulated this idea on several occasions, but particularly when analyzing the moral, mental, and physical changes she observed in her cousin, Mary Lowell. Recalling their last meeting, she remembered being unimpressed by Mary's character and appearance. Fortunately, though, "a considerable change took place in her" by the fall of 1825. To illustrate the shift, Lowell melded a description of Mary's outward transformation with a discussion of her inner refinement:

> In the first place, she grew up from a little, puny thing, with pale cheeks, ~~though bright~~ eyes, to be a tall, fine-looking young lady, with eyes beaming with intelligence, & cheeks whose "mantling blood in ready play, rival the blush of rising day": indeed, she has become really beautiful.

In attempting to describe Mary's moral metamorphosis, Anna began with a delineation of how her physical features had changed since their last meeting. Though Mary had previously demonstrated a "cultivated" mind, her eyes were now "beaming with intelligence." Though she had previously been a "puny thing" with pale cheeks, she was now "a tall, fine-looking young lady." By enriching her mind, she had beautified her body.[57]

At first glance, this entry reads like a simple descriptions of Mary's external alterations. But through this physical description, Anna was describing a change in Mary's character. Previously, she had barely been comfortable admitting her cousin's eyes were "bright" (as indicated by her efforts to cross out the word). When she described Mary in 1825, by contrast, she employed the phrase "eyes beaming with intelligence." In doing so, Anna tapped into a larger cultural tradition that conflated eyes and intelligence. An 1831 periodical, for example, described the writer Margaret Derenzy as a woman, "tall and graceful," with "fine black eyes, beaming with intelligence." Her eyes, the author said, "threw a halo of intellectual expression over [her] countenance." Another article described one fictional figure as a paragon of pious femininity whose "expressive blue eyes, beaming with intelligence, declared the mildness of her disposition." And, in the *Memoirs of Sir Joshua Reynolds* (1817), Leonardo da Vinci is described as having had "a form perfect in proportion" and "eyes beaming with intelligence and fire." By 1839, the *American Phrenological Journal* assured readers that "If the countenance beam with intelligence and goodness, there is a predominance of the moral and intellectual regions of the brain." Such a fascination with "beaming" faces reflected a broader cultural attachment to physiognomy, which viewed the eyes as the "window of the soul."[58]

As women described the faces of their female contemporaries, they used similar phrases. In the autograph album of a Philadelphia woman named Elizabeth Clemson, one of her friends wrote a poem about Elizabeth's "beaming refinement of mind," which "spread o'er the whole of [her] face." Margaret Bayard Smith had also described her meeting with Thomas Jefferson's daughter by saying Martha Randolph had a "countenance beaming with intelligence." Margaret Fuller's students used similar language within the classroom. Mary Ware Allen, for instance, described a fictional heroine by writing, "Intellect beamed from her large, brilliant eyes." And when Anna Gale imagined the countenance of the Empress Josephine, she declared she "could see beaming from her large, soft, blue eyes, an expression of mild benignity." Similarly, Catharine Maria Sedgwick wrote to her niece Katharine about the actress Ellen Tree, saying that she had "a face beaming with expression" and a countenance that reflected her frank and natural disposition. When viewed in this context, it seems clear that Anna Lowell was not merely describing the appearance of her younger cousin when she used this phrase. She was also participating in a scientifically informed conversation about the connections between facial beauty and moral and intellectual excellence.[59]

Of course, physiognomists and phrenologists were not the only people who spoke of beaming eyes. What made such language popular science? Were all discussions of heads and faces scientific? Not exactly. Long before Lavater published the *Essays on Physiognomy*, poets and novelists had written about beaming eyes and countenances. But such phrases usually functioned primarily as physical descriptors, rather than code words that helped readers discern character. Still, it is important to remember that physiognomy was a long-standing cultural tradition. It did not suddenly emerge in the late eighteenth century. As far back as the sixteenth century, authors referred to the eyes as the "windows of the soul." In the 1770s, though, Lavater was able to capitalize on this widespread belief, popularize it, and encourage people to think of facial analysis as a true *science*, rather than an art. In doing so, he imbued existing aesthetic and literary traditions with new meanings.⁶⁰

By the early decades of the nineteenth century, Americans were not merely providing physical descriptions in their letters and diaries. They were *using* physical descriptions to convey deeper information about people's inner dispositions. Beaming eyes, on their own, might be beautiful, alluring, or captivating. But it was something different to argue that someone's eyes were "beaming with intelligence," or to suggest that their countenance beamed with "an expression of mild benignity." When Anna Cabot Lowell used such language, she demonstrated her facility with the scientific language of the day. Even if she did not think of herself as a trained physiognomist, her papers show that she was aware of physiognomy and phrenology. In one of her notebooks, she included an "Extract from the private Journal of Lavater." She also mentioned going to a social gathering where the famous phrenologist "Dr. Spurzheim was the theme of conversation here as everywhere else." Within this context, it should not be surprising that she regularly analyzed faces to interpret people's character.⁶¹

Lowell also indicated her entanglement in the literary and scientific culture of the day when she described Mary's cheeks, "whose 'mantling blood in ready play, rival the blush of rising day.'" Here, she quoted directly from the same Walter Scott poem that Hannah Wharton had copied into her journal in 1813. By quoting Scott, both young women indicated their familiarity with physiognomic principles.⁶² Lowell's discussion of blushing also echoed the racialized assumptions of the scientific literature of the period. In his *Notes on the State of Virginia* (1787), Thomas Jefferson notoriously claimed white bodies had a "superior beauty," which signaled their racial superiority. The British naturalist Charles White made a similar argument in 1799, this time explicitly drawing a connection between cranial developments and intelligence. Suggesting that Europeans were smarter and more beautiful than other human beings, White asked, "Where shall we find, unless in the European that nobly arched head, containing such a quantity of brain . . . ?" He followed these rhapsodic musings with a racist tribute to white beauty, saying that "the beautiful women of Europe" had a "general elegance of features and complexion," which expressed "the amiable and softer passions in the countenance." For white naturalists, blushing cheeks became the "emblem of modesty, of delicate feelings, and of sense." Lowell did not question (or even address) this assumption when she copied this passage into her journal. Instead, she emphasized the expressive nature of white women's faces, suggesting that inner goodness could "beam" through otherwise unimpressive external traits.⁶³

When Black women engaged with such concepts, they waded into a sea of racist cultural and scientific beauty standards. Continually confronted with prejudicial ideas, Black women sometimes internalized them. There were times, for instance, when Charlotte Forten almost seemed to accept the notion that Blackness and beauty might be antithetical. After the Civil War, she moved to South Carolina to teach formerly enslaved people. While there, she described a "little mulatto child" with "large beautiful black eyes and lovely long lashes." She then wrote, "The mother is a good-looking woman, but quite black," a quick aside that revealed her tendency to associate beauty with lighter skin. As the scholar Brenda Stevenson has argued, "Charlotte certainly adopted white standards for beauty that caused her to think of herself as unattractive." In 1858, for instance, she subjected herself to "a thorough self-examination," which precipitated "a mingled feeling of sorrow, shame, and self contempt." Comparing herself unfavorably with her peers, Forten lamented, "Not only am I without the gifts of Nature, wit, beauty and talent; without the accomplishments which nearly every one of my age, whom I know, possesses; but I am not even *intelligent*." By contrast, she referred to Maria Chapman, a white abolitionist, as "the most beautiful woman I ever saw." Another time, Forten described a conversation with John Greenleaf Whittier's sister, who showed her a "picture of an Italian girl" and said that there was a "striking likeness" between the two women. "Everybody else agreed that there was a resemblance. But I utterly failed to see it: I thought the Italian girl very pretty, and I know myself to be the very opposite." On another occasion, Forten visited with a white woman with "long, light hair, and beautiful blue eyes." She described her as "a little poetess,—a sweet, gentle creature. I have fallen quite in love with her." During the visit, the women discussed Delia Bacon, an author who garnered attention by contending that many of Shakespeare's plays were not written by him alone but by a group of his contemporaries. Forten disagreed. To support her point, she invoked phrenology: "I *know* that Shakespeare, and *he* only wrote those immortal plays. To look at that noble unrivalled head alone, would convince me of that." As this example reveals, Forten was clearly familiar with popular sciences. Throughout her journals, she analyzed the heads and faces of both real and fictional figures. Curiously, though, she rarely used these techniques to talk about herself, or to evaluate other Black women.[64]

While Forten sometimes called her friends and family members "pretty" and "intelligent," she mostly reserved explicitly physiognomic and phrenological descriptions for famous white people. She rhapsodized about William Lloyd Garrison's "noble face," and Nathaniel Hawthorne's "splendid head." After meeting Lucretia Mott, she proclaimed that Mott was "saintly in look as in spirit, for a beautiful soul shines through her beautiful face." She also gushed about the "noble, spiritual face" and "glorious eyes" of John Greenleaf Whittier, one of her intellectual idols. Intriguingly, though, Forten primarily used physiognomy and phrenology to identify moral and mental excellence in her white counterparts.[65]

Perhaps Forten was hesitant to analyze the bodies of other Black women because she had encountered so many racialized depictions of beauty in the novels, newspapers, and magazines she read. Forten's literary diet included a heavy dose of anti-slavery works. Even in these books, though, she confronted prejudicial ideas about the Black female body. In the early 1850s, she read both Harriet Beecher Stowe's *Uncle Tom's Cabin* (1852) and Mary Hayden Pike's *Ida May* (1854). Both novels were immensely popular, both were written by white women opposed to slavery, and both contained some tepidly positive depictions of Black female beauty. Yet both portrayed white and mixed-race people as more attractive than their darker-skinned counterparts. When introducing a "dark mulatto" named Venus, Pike described her as having small eyes and "broken teeth," writing, "her negro features gave her a stupid and morose appearance." By contrast, when depicting the book's heroine—a white child who was illegally sold into slavery—the book said she was not only "beautiful," but also "so delicate in appearance, it hardly seems that she can be a servant." Through this anecdote, Pike implied that only those of "white parentage" could be truly beautiful. Then, she egregiously used a woman's white body as physical evidence that she was unsuited to slavery.[66]

Pike herself was clearly steeped in the physiognomic milieu of the era, because she used many of the same strategies that female authors typically deployed when comparing morally and mentally distinguished women with their "perfectly beautiful" counterparts. When discussing a female enslaver—who was admittedly *not* a heroine in this anti-slavery work—Pike portrayed the woman as having a "stately head," "classic features," and "alabaster" skin. According to the author, "the delicate bloom which tinged her cheek and mantled into brightness on her curved lips" made her alluring. Even so, her beauty was deceiving. The "contour of her face and head" might have been "perfect," but she was not a true physiognomic beauty. Her spiteful and superficial nature made her morally suspect. But even as Pike critiqued this anti-heroine, she described her as "radiantly and peerlessly beautiful." By contrast, she typically depicted the Black people in her story as physically flawed in some way.[67]

When Forten read such novels, she confronted the idea that white bodies could be beautiful, even when they belonged to women with reprehensible characters, while "negro features" gave one "a stupid and morose appearance." Is it truly surprising, then, that she was hesitant to apply physiognomic rhetoric to her Black friends and family members? Even in the works of anti-slavery authors, prevailing beauty standards were bound up with white supremacy. At the same time, Black women did not unequivocally reject popular sciences. Although Forten never recorded detailed physical descriptions of Black women, she did use physiognomic and phrenological language to evaluate white intellectuals and activists. Mary Montgomery went even further, carting her phrenological journals with her to Oberlin and entertaining friends by examining their heads and faces. In such instances, physiognomy and phrenology became individualized practices where women had the option to read bodies as they saw fit. Despite being deeply problematic disciplines, popular sciences gave both Black and white women the latitude to find moral and mental value in the features of the people they most admired.

Physiognomy and phrenology ultimately helped women cultivate uniquely gendered knowledge systems and allowed them to uncover moral and intellectual value in others. When viewed against the writings of some of the white male thinkers who used science to insist on the inferiority of female minds, women's uses of physiognomy and phrenology were both creative and subversive. Most of them either ignored or explicitly rejected essentialist discourses that emphasized the ineradicable differences between the sexes. Yet they, too, reinforced the problematic notion that women's bodies could be "read" for signs of inner capacity. White women, especially, helped solidify—or at least failed to question—the racialized beauty standards that excluded Black women.

In the end, both Black and white women were working within—and not outside of—a broader gender system that venerated white female beauty. They maneuvered around its edges, but most of the time they reiterated both the import of whiteness and the significance of female attractiveness. Still, countless women embraced physiognomy and phrenology rather than explicitly rejecting these sciences as harmful and discriminatory. By the 1840s and 1850s, many women's rights activists had enthusiastically adopted these sciences, using them to further their own social and political goals. As they saw it, physiognomists and phrenologists were not enemies but allies. In response, some gender conservatives attempted to discredit both popular sciences and the female activists who embraced them. For men who were threatened by women's political engagement, physiognomy, phrenology, and feminism seemed like entangled ideologies that might undermine the status quo.

4

THE MANLY BROW MOVEMENT

In the middle decades of the nineteenth century, Americans began sniping at each other over a curious topic: female foreheads. Women had apparently started parting their hair in the center, combing it down on each side, and plastering their locks tightly against their skulls. To make matters worse, some had shaved the top of their hairlines, all in an attempt to make their foreheads appear larger and more prominent. It was an ugly and unappealing trend—a fashion craze exemplifying everything that was wrong with nineteenth-century gender relations. At least that's how some people saw it. Others argued that large foreheads were not only attractive but also a sign of mental distinction. If a woman had been blessed with a high brow, why shouldn't she display it? After all, what could be more appealing than a belle whose prodigious forehead revealed her beautiful mind? The debate raged for more than three decades, with editors reprinting dozens of missives denouncing the "strong-minded women" who immodestly bared their brows. But why? Were women actually engaging in such a practice? If so, what made it so disturbing to male observers? And why should historians care? As it turns out, these questions get to the heart of some of the gendered political debates that roiled the United States during the middle decades of the nineteenth century.[1]

At first glance, the culture wars over women's foreheads might seem like frivolous exchanges over female hairstyles. In reality, they were scientific debates about the nature, sphere, and political destiny of women. By the mid-nineteenth century, Americans mostly agreed that well-developed foreheads were anatomical symbols of intelligence. Still, they found it difficult to agree on something else. Were high and broad foreheads beautiful in women or not? Some commentators answered that question with a decisive "no." Contending that large brows gave women "a masculine and defying look," they fulminated against the "bluestockings," "bloomers," and "female brawlers" who deliberately exposed their foreheads for public view. Others suggested that high brows were indeed appealing, precisely *because* they signified mental distinction. Believing in the intellectual equality of the sexes, those individuals saw women's foreheads as physiognomic proof of their impressive cerebral capabilities.[2]

Through these public disputes, conversations about women's beauty became scientific deliberations about the social and political status of female citizens. Debates reached a fever pitch in the 1850s, a time when the women's rights movement was gaining both strength and visibility. Women's benevolent activity had been widely accepted since the time of the American Revolution, but in the 1830s, female activists pioneered novel forms of political engagement and began advocating for a more sweeping reorganization of society. As the primary drivers of moral reform crusades, women launched assaults on the nation's drinking culture and sought to push temperance laws through state legislatures. Some embarked upon campaigns to eliminate prostitution and expose the behavior of lascivious men. Thousands more joined the anti-slavery movement.[3] Many also criticized the legal doctrine of coverture, which treated women as if they were subsumed by their husbands' identities and unable to claim legal, political, or economic rights of their own. Explicitly challenging men's economic authority, female activists and their male allies began advocating for women's property rights. By the time Elizabeth Cady Stanton and Lucretia Mott convened the Seneca Falls Convention in 1848, some of the nation's most progressive thinkers were already demanding female suffrage. A broad-based movement had coalesced by the 1850s, with activists organizing national conventions where they publicly advocated for "the rights of woman." These reformers practiced what Anne Boylan has called a "new style of female politics." Pursuing ambitious and expansive reform agendas, they tackled policy matters that many Americans considered to be outside the purview of proper femininity. In response, they found themselves vilified in society's more respectable circles.[4]

It was within this context that newspapers and magazines began railing about female hairstyles (and the foreheads so immodestly accentuated by them). Threatened by the rise of women's activism, gender conservatives sought to discredit their enemies by denigrating their appearance. In women's high brows they saw a hideous fashion trend, but they also saw an ominous challenge to existing gender conventions. Scholars have long recognized that antebellum Americans relied on the power and prestige of science to rationalize and justify political inequalities. As they have rightly pointed out, individuals invoked disciplines like physiognomy and phrenology to argue that men and women were mentally and physiologically distinct beings, destined by "nature" to occupy "separate spheres" within society. And yet, the conflict over female foreheads exposes a far more complicated story. As the nation became embroiled in acrimonious exchanges about the "rights of woman" between the 1830s and the 1860s, fights about fashion became quarrels about political inclusion, and popular science functioned as a weapon that opposing groups could deploy in debates about gendered power.[5]

There were certainly people who invoked popular sciences to rationalize and justify inequality. More often than not, though, cranial analyzers aligned themselves with—and not against—female intellectuals. In this battle of the brows, feminists and phrenologists found themselves on the same team, joining up to critique the gender conservatives. For this reason, many women's rights activists imagined phrenology as an exciting new science that might establish the intellectual equality of the sexes and facilitate their crusade for gender equality. As combatants on all sides acknowledged, discussions about women's bodies were debates about women's brains—and debates about women's brains were disputes about the role women were going to play within the broader society.[6]

"A FINE HIGH FOREHEAD"

By the middle decades of the nineteenth century, physiognomy and phrenology had come to shape how women saw both themselves and others. They also influenced female fashions. In 1817, the professional stylist John B. M. D. Lafoy wrote a hairdressing manual in which he insisted his colleagues should consult the "science of physiognomy" when styling a woman's coiffure. He argued that "not even a hair should be adjusted" before analyzing the facial features, as well as "the soul itself!" As he saw it, one could not dress a woman's hair without studying "the unerring principles of the celebrated Lavater."[7] Periodicals similarly suggested that women might consult popular sciences when fashioning their appearance. Reprinting an article from the *London Monthly Magazine* in 1825, both the *New-England Galaxy* and the *Athenaeum* urged women to consult the science of cranial analysis: "To apply phrenology to hair-dressing, may appear fantastical and ludicrous," the author sheepishly admitted, "yet we will trust our demonstrations to the trials of any one who chooses to make them." Though regretting their lack of instructional drawings, the magazine editors trusted readers would use their own "taste and knowledge" to craft scientifically informed hairstyles.[8]

Recognizing the cultural prevalence of popular sciences, women in the Oneida utopian community seized upon the opportunity to experiment with an innovative coiffure. They claimed that long hair was "burdensome and distasteful" because it hindered their ability to work effectively. According to the group's Annual Report, "any fashion which required women to devote considerable time to hair-dressing, is a degradation and a nuisance." Even though they desired to cut their hair short, Oneida's women faced a roadblock. In the Bible, the apostle Paul had suggested that long hair was essential in women because it preserved female modesty. In a utilitarian maneuver to get around the problem, Oneida residents invoked science. Phrenologists had purportedly proven that the organ of "Amativeness" was located in the back of the skull, at the nape of the neck. Since this was the faculty that signified one's libido, it followed that those who embraced the typical style of the 1840s—by sweeping their long hair into a chignon at the back of their head—were actually exposing their sexual propensities to the world. The only rational—and modest—solution would be to wear the hair short, with the back of the head covered by their flowing locks. "Accordingly, some of the bolder women cut off their hair, and started the fashion, with soon prevailed throughout the Association."[9]

While the Oneida women focused their attention on the back of the cranium, other Americans fixated on the female forehead. The famous sculptor Hiram Powers, for instance, relied on phrenology when deciding how to arrange the coiffures of his patrons in his marble busts. In an 1856 exchange with his client, Mary Sargent Duncan, Powers wrote: "I will arrange the hair if you will trust me in the capacity of a hair dresser—and I flatter myself—that although your head may not present such enormous organs of self-esteem, philoprogenitiveness, etc., yet it will lose nothing of its intellectual developments under my hands." Powers accomplished this goal by sculpting her hair in a manner that heightened her brow.[10]

Some people went to extreme lengths in their quest to acquire high foreheads. Long before the Swedish intellectual Fredrika Bremer became a globally recognized author and women's rights activist, she was a small child with a big problem: a low brow. When she was young, her mother repeatedly criticized her "uncommonly low forehead." Feeling the sting of this insult, Bremer decided to take matters into her own hands. In an effort to heighten her brow, she started "cutting away the hair at the roots all round the forehead." At first, Bremer panicked. What had she done? But her mother's response was unexpectedly positive. Remarking that her daughter "looked unusually well," she added, "Your forehead is, after all, not so very low." Bremer was ecstatic. The next time unwelcome bristles began poking through her scalp, she acquired "a pair of tweezers" and plucked the strands along her hairline, "root and all." Eventually the hair stopped growing back, leaving her with "a fine high forehead."[11]

Her transformation was apparently quite successful. When the American novelist Catharine Maria Sedgwick first met Bremer, she described her as "plain." On closer inspection, though, she described her as physiognomically beautiful. "I like her more and more," wrote Sedgwick, "and, as the soul comes out and overspreads the features with its beaming and beautiful light, I am ashamed to have called her 'plain.'" By claiming that Bremer's moral and mental excellence added charm to her face, Sedgwick built on themes that she had encountered in print culture. Ten years earlier, a similar anecdote had appeared in *Miss Leslie's Behaviour Book* (1839). In this guidebook for young women, Eliza Leslie admitted that Bremer had several "personal defects." At the same time, Leslie pointed to her "broad and intellectual forehead" as a signifier of her genius. "Physiognomists say that the eye denotes the mind, and the mouth indicates the heart," she wrote. "Truth it is, that with a good heart and a good mind, no woman can be ugly." Although Bremer lacked the "regular beauty" of a stereotypical belle, she had a "fine high forehead" that showcased her powerful mind. By pointing this out, both Sedgwick and Leslie judged Fredrika Bremer according to *physiognomical* standards rather than conventional notions of perfect female beauty.[12]

Bremer's case is particularly intriguing because writers—including her sister—seemed convinced that her high brow was not natural but rather the result of a calculated beauty regimen. This exposed the potential pitfalls—but also the exciting possibilities—of the physiognomic and phrenological method. A high forehead was supposed to be a natural sign of mental refinement, something far more desirable than the simpering smile or fashionable adornments of a socialite. Intellectual women had initially found popular sciences so appealing precisely for this reason: these disciplines privileged an alternative beauty standard that valued science and health over vanity and style. In theory, physiognomy and phrenology provided women with a way of seeing past superficialities. The problem? Scientific observers could not always detect the maneuverings of clever self-fashioners, and by the mid-nineteenth century, high brows had *themselves* become a fashion trend. Within this context, a woman who went out of her way to emphasize her expansive forehead could come under fire for her vanity.

Where, then, did this leave the unfortunate souls who inherited low brows? Their first option was to style their coiffures strategically. According to antebellum beauty books and magazines, women could enhance their beauty by subtly showcasing their brows. *The Toilette of Health, Beauty, and Fashion*

(1833), for instance, dedicated an entire chapter to the female forehead. In a direct engagement with the science of phrenology, the author claimed that "a high forehead, full and broad" illustrated "a predomination of the intellectual faculties." The writer was particularly piqued by the prevailing custom of wearing a vast profusion of curls around the forehead. This habit not only detracted from beauty; it also destroyed the physiognomical effect. Instead, he urged women to exhibit their foreheads and dress the hair with simplicity and taste. Dr. John Bell's manual, *Health and Beauty* (1838), similarly instructed women to part their hair in the middle, comb it down closely upon the upper part of the cranium, and expose the forehead. He insisted that this arrangement would imbue the countenance with "an intellectual air" because the forehead was "itself an index of the mind."[13]

Portraits from the period suggest that many women took this fashion advice seriously. There was a marked transformation in women's hairstyles during the second quarter of the nineteenth century. In the 1810s and 1820s, women often ornamented their brows with curls and wavy wisps of hair. By the 1840s, most women were parting their hair in the center, pulling it away from their foreheads and meticulously securing it against their skulls. Of course, there were multiple forces influencing female fashions during this period. As visual and material culture scholars have shown, the severe gothic styles of the late 1830s and 1840s demonstrate a resurgence of early modern fashions. Intriguingly, though, this shift in women's hairstyles coincided with phrenology's peak popularity in the United States. Perhaps women were consciously adhering to scientific ideals. Perhaps they were simply following the fashionable dictates of the day. In any case, their coiffures tended to align with the instructions in physiognomically and phrenologically informed beauty manuals.[14]

Women's magazines also show hints that physiognomy and phrenology shaped how women viewed their own bodies. Editors not only urged women to take up popular sciences, but also gave readers advice on how to attain physiognomical standards. *Godey's Lady's Book*, for instance, quoted the famous physiognomist Johann Caspar Lavater, and instructed women to use scientific facial analysis when contemplating their coiffures. Because everyone's skull was unique, the magazine declared that "Phrenology should be called in" when arranging the hair. "Woman has very much the advantage over man in this respect," the magazine announced. "She can make her head show, phrenologically, for pretty much what she pleases." Through articles like this, women learned that they could use clever hairstyling to disguise unappealing "bumps" and emphasize more desirable propensities.[15]

When hairstyles did not do the trick, people could adopt Fredrika Bremer's approach, snipping the hair at the top of their foreheads. If newspapers are

FIGURE 4.1. Thomas Sully, *Mary Sicard David* (1813). Cleveland Museum of Art. Public domain.

any indication, some women took the risk. In 1840, one periodical mocked the "ladies of Baltimore," who had supposedly "started a *beautiful* and *intellectual* fashion—that of shaving their foreheads in order to make more prominent and visible the intellectual regions." Sneering at the phenomenon, the author declared, "if shaving a part of the forehead makes a young lady look *handsome* and intellectual, how much more handsome would she look, and how much more intellectual would she be, if the whole head were shaved."

With palpable sarcasm, the newspaper hinted that these women were engaging in a foolish behavior, precisely because they wanted to look wise.[16]

If scissors, tweezers, and strategic coiffures failed, Americans had another option: depilatories. These products usually came in powdered form. When mixed with water, they could be plastered upon the skin, where they would weaken the follicles and strategically singe "superfluous hair" from the body. Advertisers insisted their products were safe, speedy, and effective. They lied. Depilatories often contained quicklime, arsenic, and other harmful chemicals. As a result, critics warned readers against using these products,

FIGURE 4.2. Portrait of an Unidentified Woman (ca. 1825). Courtesy of the American Antiquarian Society.

FIGURE 4.3. George Chinnery, *Harriet Low* (1833). Peabody Essex Museum. Public domain. Intriguingly, Low sat for this portrait around the same time that she met a woman who "had her hair dressed in the new style, as plain as possible." Less than two weeks later, she conversed with "a famous phrenologist." Then six months later, she started reading Spurzheim's works on phrenology, which, as she put it, helped her "begin to look at people's heads very scientifically." See Katharine Hillard, ed., *My Mother's Journal: A Young Lady's Diary of Five Years Spent in Manila, Macao, and the Cape of Good Hope from 1829–1834* (Boston: George H. Ellis, 1900), 231–32, 276, 284.

especially on their faces. Newspapers and magazines nonetheless began marketing them in the 1830s.[17]

The country's most influential depilatory dealer was a man named Joseph W. Trust, who started advertising a product called "Poudre Subtile" in 1841. Adopting the alter ego of "Dr. Felix Gouraud," Trust promised his

FIGURE 4.4. Joseph Goodhue Chandler, *Amoret Gillett Austin* (1846). Courtesy of Susan Garton. This portrait exemplifies the new hairstyles of the late 1830s and 1840s. The hair is parted in the center and fastened tightly to the skull. Physiognomically and phrenologically informed beauty manuals advocated for such styles because they showcased the forehead and cranial developments.

product would remove "the hair concealing a broad and elevated forehead." His advertisements directly invoked popular sciences, citing the physiognomist Johann Caspar Lavater and declaring that there was much "truth" in phrenology. "Dr. Gouraud" marketed himself as an "ingenious chemist," promising his products would "eradicate every fibre of superfluous hair wherever applied, and display the hidden beauties and intellectual developments of either male or female foreheads." Several ads specifically targeted female customers, declaring the company's products could remove undesirable moustaches, uproot the hair from unsightly moles, or eliminate hair from otherwise "beauteous" and "intellectual" brows. One frequently reprinted advertisement addressed itself directly "TO THE LADIES," proclaiming that "a broad and elevated forehead" was both an "element of personal beauty" and a "mark of intellect." The ad assured potential customers they might *already* be the lucky possessors of high foreheads, but that their otherwise imposing

FIGURE 4.5. Portrait of Abby Kelley Foster (1846). Courtesy of the American Antiquarian Society.

brows were currently "obscured by the encroachments of a too luxurious growth of hair." Luckily, customers could easily solve this problem. For just $1 a bottle (about $30 in today's currency), Gouraud's "Poudre Subtile" would remove superfluous hair from people's foreheads so they could unveil their intellectual developments for the world. Gouraud's company reprinted hundreds of these ads throughout the nineteenth century. Despite facing competition from "Chastellar's Hair Exterminator" in the 1860s, the family business flourished, remaining in operation until at least the 1930s—long after Gouraud himself had died.[18]

Those who lacked access to these alleged miracle cures could also find recipes for DIY depilatories in middle-class beauty manuals. One book recommended a recipe made from the "juice of the milk-thistle mixed with oil," and suggested "the gum of the cherry tree" might stop hair from growing in the first place. It also provided a detailed recipe for a "stronger depilatory," made with vinegar, crushed ant eggs, and arsenic. The author warned readers they should not inhale the harmful concoction, but he had no qualms about telling women to smear it upon their skin, as long as they speedily wiped it off with a linen cloth. The surgeon and ethnologist Daniel Brinton also gave women a recipe for a homemade depilatory in his scientific beauty manual. Since the original mixture contained arsenic, he recommended an alternative that substituted quicklime and "sulphurit of calcium." Brinton nonetheless informed his readers there was no chemical substitute for "mechanical" depilatories. The best results could be obtained by "simply pulling the hairs out 'by the roots.' This is an alarming sound, and suggests torture," he admitted. "But there is no occasion for terror. If properly performed, the operation is painless." In any case, beauty required sacrifice.[19]

By mid-century, depilatories were attracting the ire of popular writers. Lydia Maria Child lamented the widespread influence of "Dr. Gouraud" and his cosmetics in 1845, and *Godey's Lady's Book* warned women against using depilatories in 1853. "Nothing is more vulgar than the attempt to heighten the forehead by removing the hair at the parting," the magazine declared.[20] Even phrenologists decried depilatories. Maintaining that an intellectual forehead resulted from the shape of the brain and the conformation of the skull—and *not* the height of the hairline—they mocked the "simple, silly creatures" who thought they could cheat science. No, they sneered, one could not "enlarge the understanding" by plucking out the hairs above the brow.[21]

Disgruntled authors nonetheless blamed two causes for the rise of women's unbecoming hairstyles: phrenology and female intellectuals. Several US magazines, for instance, reprinted this reprimand from a London periodical: "Some naturally pretty women, following the lead of the strong-minded high-templed sisterhood, are in the habit of sweeping their hair at a very ugly angle off the brow, so as to show a tower of forehead, and as they suppose, produce an overawing impression. This is a sad mistake." With this comment, editors critiqued the "strong-minded" women who tried to artificially enhance the height of their brows.[22] In a similar way, another author railed against a fictional "Blue-Stocking" who was so committed to the "truth of phrenology" that she "cut away" the curls around her face in order to expose her high forehead. The article called the woman short and fat, emphasizing her "shrill" voice and peculiar appearance. To highlight "her own possession of intellect," the bluestocking had swept her hair away from her brow. But this woman was no real intellectual. She was merely a "deceived creature" trying to look smarter than she was.[23]

Other writers made similar arguments, using popular sciences to distinguish between the "bluestocking" and the "true woman." They typically agreed that intelligence was a positive trait in the female sex. They nonetheless insisted that modesty was a paramount characteristic. Excessively cerebral—and overtly political—women were vain and misguided. True women, by contrast, were humble, attractive, and agreeable. These were the women who loved God and family, the ones who put the interests of others ahead of their own. One article, for instance, extolled the virtues of "Cousin Nell," a fictional heroine with a "low, sweet forehead." Cousin Nell "was no ultra reformer; no 'Woman's Rights' person! she was only a young and gentle woman—true to her womanly nature—loving all created things, and her Creator more than all." Regrettably, not all women were like Cousin Nell. Others were arrogant and ostentatious bluestockings, or (even worse!) women's rights activists. Americans denounced such women for their vanity and political activism. Nestled within these missives were criticisms of strong-minded women and their supposed fondness for phrenology. If Cousin Nell had happily displayed her "low, sweet forehead," then bluestockings exhibited a suspicious hubris through their desire to expose their intellectual brows.[24]

At issue here was female modesty. By the mid-nineteenth century, few Americans insisted that women should be ignorant simpletons. In fact, most people argued that female education was essential, since it would help the next generation of mothers raise virtuous and productive citizens. Intelligence was acceptable. Pedantry was not. In a similar way, Americans expected women to be attractive, even as they denounced those who absent-mindedly adhered to the dictates of fashion. Beauty was essential. Frippery was objectionable. The connecting thread between these arguments was the discourse of female modesty, which demanded that women cultivate beautiful minds and bodies while being endearingly oblivious to their own charms.

When women artificially heightened their brows to expose their phrenological developments, they violated the demands of this discourse. It was not just that they eschewed natural beauty for a strategically fashioned coiffure; it was that they engaged in this practice with the explicit aim of displaying their mental power. An intelligent countenance was well and good, but a woman who styled her hair to achieve a scholarly appearance? That was vanity personified. What could be showier than a lofty brow? Despite paying lip service to the importance of female education, writers typically demanded that women stay quiet about their capacities, revealing their neurological gifts only when it benefited men. Since female activists violated this expectation by attracting public attention, they engendered backlash from gender conservatives in both Europe and the United States.

Like popular sciences themselves, discourses about the foreheads of female intellectuals were transatlantic in nature. In 1846, the British caricaturist George Cruikshank released a print titled "A Woman of Mind" (fig. 4.6). In this sketch, an ill-featured woman ignores her husband and children in favor of literary pursuits. The family's home is chaotic, with torn wallpaper, broken windows, and crooked picture frames. In the background, an older daughter attempts to soothe a screeching toddler while her younger sister weeps beside her. In the center, a disgruntled husband holds an angry baby, flummoxed by the disordered state of his domestic abode. Meanwhile, the literary woman sits at a desk covered with books, evidently unbothered by the mayhem that surrounds her. She wears reading glasses, and her hair is fastened tightly to her skull. Her brow is almost comically large. Engrossed by intellectual work, she dismisses her needy family with a flick of her wrist. Clearly, this is a household in a crisis spawned by a woman's dereliction of her domestic duties.[25]

Cruikshank published this image alongside a poem by Henry Mayhew, the influential editor of *Punch* magazine. Mayhew began with a critique of unapologetically intellectual women and then proceeded to ridicule phrenology:

> My wife is a woman of mind,
> And Deville, who examined her bumps,
> Vow'd that never were found in a woman,
> Such large intellectual lumps.
> 'Ideality' big as an egg,
> With 'Causality'—great—was combined;
> He charg'd me ten shillings, and said,
> Sir, your wife is a woman of mind.

FIGURE 4.6. George Cruikshank, "A Woman of Mind," printed in *The Comic Almanack for 1847* (London: David Bogue, Fleet Street, 1846). The British Museum. The satirical print portrays a bluestocking (identifiable by her large blue spectacles) with her hair pulled back from a large forehead. She is ignoring her domestic responsibilities in favor of literary pursuits.

Mayhew's talk of "lumps" and "bumps" was strategic. Because phrenology's critics denounced the discipline as "bumpology," Mayhew deployed this phrasing to lampoon the entrepreneurial figures who commodified cranial analysis (in this case, the British practitioner James De Ville). After sneering at the fictional bluestocking's "large intellectual lumps," Mayhew censured her husband for paying "ten shillings" for a worthless examination. Establishing a link between fashion, phrenology, and female arrogance, Mayhew crafted a female villain who adorned her face with "horrid blue spectacles" and swept her hair back to expose a voluminous brow. As he saw it, this was nothing less than female vanity disguised as modesty:

> She's too clever to care how she looks,
> And will horrid blue spectacles wear,
> Not because she supposes they give her
> A fine intellectual air;
> No! she pays no regard to appearance,
> And combs all her front hair behind,
> Not because she is proud of her forehead,
> But because she's a woman of mind.

The poem, in other words, argued that female intellectuals made absurd fashion choices because they hoped others would recognize in their foreheads the phrenological signs of intellect. This poem then inspired the composer Jonathan Blewitt to adapt it into an "Eccentric Song," which he sang "in his own racy style," causing "quite a sensation" throughout Britain. A cartoon mocking a high-browed woman thus became a spectacle for lowbrow audiences.[26]

By mocking the "woman of mind," Cruikshank, Mayhew, and Blewitt did more than criticize phrenology. They also insinuated something else: bluestockings made bad mothers. As historians of gender and sexuality have long contended, nineteenth-century scientific and popular writers often intimated that women's intellectual developments might inhibit their motherly instincts. Yet Mayhew's poem also raised an issue that scholars have less often addressed: the connection between gender politics and popular sciences. By styling her hair to showcase her brow—and by commissioning a cranial examination—Cruikshank's bluestocking flaunted her "fine intellectual air" while pretending she was "too clever to care how she looks." Meanwhile, she failed to make her husband's meals, refused to sew his clothing, and showed little interest in soothing her distressed children. Preoccupied by her desire to enrich her mind and advertise her intelligence, she let her household disintegrate.[27]

Taken together, this cartoon, poem, and song raise some important questions: Why did these artists portray their anti-heroine as a phrenological enthusiast? Were mentally distinguished women, in fact, enamored of cranial analysis? If so, why? If not, why did so many authors and artists make associations between popular sciences and female intellectuals? Phrenologists, after all, regularly insisted that male brains were bigger, better, and more powerful than those of their female counterparts. Why would any self-respecting woman embrace such a discriminatory discipline? And why were critics like Cruikshank, Mayhew, and Blewitt so threatened by such a phenomenon? The answers to those questions reveal the connections between science, fashion, and women's activism between the 1830s and 1860s.

PHRENOLOGY AND FEMINISM

When authors, editors, and artists drew connections between phrenology and female activism, they were, indeed, onto something. During the middle decades of the nineteenth century, countless intellectual women embraced popular sciences—and they were particularly enthusiastic about phrenology. They did so despite the fact that a great many physicians, scientists, and popular writers were using disciplines like physiognomy and phrenology to defend female subordination. If men were uniquely suited for public activism and political power, popular writers suggested, then women were destined for marriage, motherhood, and domesticity. America's most famous phrenological family, the Fowlers, published particularly egregious missives on the "nature," "role," and "sphere" of "woman." They gave public lectures on the glories of female domesticity and ran dozens of articles with variations of the same title: "Woman—Her Character, Influence, Sphere, and Consequent Duties and Education." In almost all these pieces, they declared that women, by their very nature, were suited for one aim and one aim alone: motherhood.[28]

To the modern ear, the Fowlers come off as unrepentant misogynists who believed women should be confined to the home and to their "one great destiny" as wives and mothers. But this wasn't entirely true. In the pages of their books, almanacs, magazines, and pamphlets, the Fowlers provided Americans with a series of multifaceted, complex, and often incongruous visions of ideal womanhood. They valorized female domesticity and perpetuated the cult of true womanhood, but they also printed editorials advocating female "emancipation" and explicitly declaring their support for women's rights conventions and the "noble-hearted women" who organized them. The Fowlers vociferously campaigned for female education, and published physiognomical profiles gushing about the superior brains of prominent activists like Elizabeth Cady Stanton, Susan B. Anthony, Lucretia Mott, and Amelia Bloomer. In turn, many of these women came to view phrenology as a rational science that might further their own personal and political goals.[29]

Phrenology appealed to women for a multitude of reasons. As Carla Bittel has shown, phrenologists claimed that men were mentally superior beings, but they also encouraged women to cultivate their intellectual faculties. They were quite clear that female brains were smaller and less developed than male brains. Even so, phrenologists insisted that all people could improve, and this included women. What's more, phrenologists lauded women for their supposed moral superiority. This logic seemed to provide scientific evidence for bourgeois gender conventions, which stressed the moral superiority of women. In other words, phrenological science assured women that they *should* be cultivating traditionally feminine characteristics, allowing middle-class women to take comfort in their existing gender identities. At the same time, phrenology confirmed that male and female brains were more similar than different, and that women had all the same basic faculties, organs, and propensities as men. This meant that if individual women wanted to cultivate traits like boldness, strength, and intellectual profundity, then they could do so without undermining their femininity.[30]

Phrenologists also marketed their science directly to women, encouraging them to be practical thinkers who empirically investigated the complexities of human nature. In doing so, they treated women as scientific practitioners and serious intellectuals. When Lorenzo Fowler examined the head of a woman named Lucy Rider, he started by telling her she was "naturally dependent," with a "naturally strong sympathy with children." At the same time, he praised her "good memory" and "elevated mind," saying she had "an uncommon curiosity to see and know; to learn and gain information." He explicitly encouraged Rider to think of herself as a scientific practitioner. After reading her cranium, he posited that she would be "quite fond of Astronomy and of the Sciences generally." Even more directly, he urged Rider to take up the study of skulls. "You have the faculties to advance quite rapidly as a scholar, would be much interested in the study of Phrenology," he wrote. On the surface, this is at odds with Fowler's public insistence that women's reflective capabilities lagged behind those of men. On closer inspection, though, his conclusions are hardly surprising. Phrenology presumed that all human beings had the same basic organs and faculties—just in different proportions—making the minds of women and men different in degree rather than in kind. Plus, Lorenzo was used to interacting with female scientists. He lived with one. His wife, Lydia Folger Fowler, was a physiologist and phrenologist who became the second woman to earn a medical degree in the United States (behind Elizabeth Blackwell). She also played an active role in the temperance movement and the women's rights crusade. Lorenzo evidently saw value in his wife's identity as a female physician and activist, and he seems to have encouraged other women to follow a similar path.[31]

Heeding this clarion call, women, like men, became enthusiastic popular scientists. They, too, attended public lectures and submitted their skulls for examination. They also internalized many physiognomical precepts, using them to evaluate others and to interpret their own characters and capacities. Some women even became traveling lecturers in their own right. Despite this fact, scholars have mostly emphasized how popular sciences laid the framework for gender discrimination. Only recently have historians begun to emphasize that some of the nation's most enthusiastic phrenological practitioners were not defenders of the status quo but abolitionists and women's rights activists.[32]

Popular sciences intrigued these women because they provided a scientific framework for discerning physiognomical indicators of moral and intellectual excellence in their own bodies—and in the bodies of their female friends and allies—even when the rest of the world deemed them unattractive. Margaret Fuller, remember, was fond of physiognomy and phrenology—and why wouldn't she be? Facial and cranial analysis helped the world identify the external signifiers of her superior mind. They also gave her tangible, anatomical evidence that she was constitutionally suited to be one of the United States' leading intellectuals.

Like Margaret Fuller, many of the nation's most influential female reformers enthusiastically adopted physiognomy and phrenology. The abolitionist Lucretia Mott, for example, believed in the promise of scientific character analysis and used phrenology to make sense of her own commitment to racial justice and gender equity. Although phrenologists regularly argued for the mental inferiority of both women and people of color, Mott used the science to analyze the personalities of her political allies and craft a neurological explanation for her own commitment to abolitionism. While she was not impressed by "practical phrenologists" like the Fowlers, she greatly admired George Combe, seeing him as one of the world's leading intellectuals.[33] It is also possible that Mott found value in phrenology because phrenologists found value in her. Orson Fowler, for instance, argued that women excelled in perception while men excelled in reflection. He claimed that women's foreheads were generally narrow, yet conceded that "occasionally women have high, wide, bold foreheads, like Lucretia Mott."[34]

Mott's abolitionist ally Abby Kelley similarly saw how phrenology could be both personally and politically useful. After reading some of Lorenzo Fowler's books, she described him not as a stodgy conservative but as a progressive reformer who, like her, believed in human agency and emphasized the improvability of both individuals and society. "I am quite captivated with his straight forwardness and blunt radicalism," she wrote to her future husband. For Kelley and her friends, physiognomy and phrenology were not silly pseudosciences or discriminatory discourses. Quite the contrary. They were disciplines that contributed to women's mental enrichment. In one letter, Kelley's friend Anna Breed expressed her frustration at the lack of intellectual stimulation that was available to her in Lynn, Massachusetts. She was enormously grateful when she finally had a social call with a group of like-minded intellectuals, calling it "quite a treat." Breed made it clear the group had not discussed frivolous matters: "What did we talk? not about *pretty babies*, nor *new gowns* and *caps*. possible! and in Lynn, too? Our principal subjects were Phrenology, Physiology, Physiognomy, climate of the south—residence there—J. G. Whittier and his poetry." For Breed, talk of babies and dresses was for ordinary women. Cultivated women cared about science and poetry. Kelley agreed. Her letters abound with phrenological language. She and her fellow abolitionists used popular sciences to analyze their own brains and bodies, as well as their political leanings. Kelley also used science to explain why she continued to lecture in one particular town, rather than moving on. She felt she had not yet convinced the population of the merits of abolitionism. "Phrenologists say I have a great development of concentrativeness," she wrote, "which, I suppose, gives me the desire to see work *well done up* wherever I go, before leaving." Her friend William Bassett, by contrast, lamented that he had a "small" organ of "Concentrativeness," which he used to explain the "confused *medley* which I have *jammed* into this letter." His topics? Temperance, religion, physiology, phrenology, and the success of abolitionism in Massachusetts.[35]

By the mid-nineteenth century, activists were steeped in the culture of popular science. They advocated for phrenology in their newspapers, wrote about it in their letters, and talked about it at social occasions. Amelia Bloomer, for instance, used her women's rights newspaper, *The Lily*, to advertise phrenological magazines, declaring, "No family can be without these most useful journals without depriving its members of much useful and valuable information." She also published a testimonial from a woman who got her skull examined, and once criticized the "old fogies" who spouted antiquated ideas that had been forged "before the days of brains and phrenology." Bloomer claimed women would not be truly enlightened until they understood "the sciences of Phrenology and Psychology." Her paper suggested women's rights activists and phrenologists were joined in a struggle against the same crusty conservatism, fighting for change in a world that refused to take them seriously.[36]

Elizabeth Cady Stanton likewise viewed her introduction to phrenology as a discovery that eschewed the superstitions of religion for the rational enlightenment of scientific investigation. After reading phrenological treatises for the first time, she felt she had finally come into the "clear sunlight of truth." She subsequently used phrenology to analyze people's heads and faces, guide her child-rearing practices, and help her select a family cook. The latter experience made her question her skills as a facial and cranial analyzer, as Rose turned out to be less than reliable despite her "large head, with great bumps of caution and order." Stanton was initially impressed by "her eyes," which "were large and soft and far apart," since physiognomic theory suggested that eyes too close together signaled a propensity for thievery. She was thus overjoyed by her good fortune—at least initially. "In selecting her, scientifically, I had told my husband, in triumph, several times what a treasure I had found." Unfortunately, Rose turned out to have a drinking problem, which led

Stanton's husband to remark "that, in selecting the next cook, I would better not trust to science." Despite such mishaps, Stanton viewed phrenology as a logical discipline that complemented her other reform efforts. She not only forged personal friendships and political alliances with female phrenologists such as Lydia Folger Fowler and Charlotte Fowler Wells; she also described phrenology as one of the "isms" that made her a "radical"—a discipline that facilitated women's rights by demonstrating the power of female minds and characters. As she saw it, her commitment to cranial analysis gave "soul and zest" to her life and distinguished her "from the common herd."[37]

For female activists, phrenology was not merely a discriminatory science that established the superiority of white men's minds. It was a liberating tool that would enlighten women about their own mental and physical abilities. Sure, some of the most prominent phrenologists argued that men were more likely to have larger and more powerful brains. But they also argued that women had all the same faculties as men. This meant that any individual woman might exhibit strong reflective powers, the capacity for profound thought, and the ability to engage in sustained deliberation, even if these traits were not universally common among *all* women.

Phrenology also delivered encouraging evidence to white female intellectuals who yearned for tangible proof of their own mental distinction. In 1853, Stanton subjected her cranium to the probing fingers of Lorenzo Fowler. He informed her that her brain was "above the common size," saying her "intellectual organs" were especially large. He told her she was "naturally kind, humane, generous," even if she was also "capable of being very sarcastic." Fowler assured Stanton she was an original thinker, "afraid of nothing," and brave enough to speak out about topics that others found controversial. That same year, Susan B. Anthony commissioned an exam from Nelson Sizer, a manager at Fowler and Wells. She learned that she had a "terse, sharp, spicy and clear" mind, and that she could "sustain mental effort with less exhaustion than most persons." Physiognomists and phrenologists had long argued that men were thinkers, while women were feelers. They were nonetheless willing to recognize exceptions. While examining Anthony's head, Sizer said her "mind more naturally runs in the channel of intellect than of feeling." Beyond simply telling her that she was intelligent, he assured her that she was better than other women. Phrenology, in such cases, provided activists with a comforting revelation: they were smarter, bolder, and more distinguished than their contemporaries.[38]

For women who prided themselves on their intellectual acumen, cranial analysis seemed to provide scientific proof of what they had suspected all along: They had powerful minds and well-organized brains. They were leaders, not followers. They were thoughtful reinterpreters of the existing social and political order. They were not simply the intellectual equals of men; they were superior to the general population. This might also explain how white activists reconciled their commitment to anti-slavery activism with their belief in the mental superiority of white Americans. Phrenology suggested white Americans were generally superior to their Black counterparts (although it sometimes made allowances for exceptional African Americans, just as it made exceptions for intellectually distinguished white women).[39]

Understandably, Black women had a far more complicated relationship with phrenology than did their white female counterparts. On the one hand, they were clearly aware of physiognomical and phrenological principles and occasionally used them to analyze the heads and faces of their contemporaries. On the other hand, they did not publicly advocate for phrenology in the same way that many Black men and white abolitionists did. Perhaps this is because the most famous phrenologists fostered both racist and sexist stereotypes in their public writings—or because they tended to ignore Black women entirely. By the 1840s and 1850s, the nation's phrenological establishment was, for the most part, opposed to slavery. Yet they were far more outspoken about issues like temperance, capital punishment, and women's rights than about racism and abolitionism. They also regularly published articles about the supposed superiority of Caucasian brains and skulls. Phrenologists tended to balance their discriminatory pronouncements about the female brain with glowing physiognomical profiles of white female activists. But no such printed pantheon existed for women of color.

With the exception of one laudatory phrenological portrait of Sarah Margru Kinson Green, one of the captives on the *Amistad*, the *American Phrenological Journal* rarely elevated Black women as models of moral and mental distinction. Hesitant to alienate southern subscribers, the magazine did not even print an analysis of Frederick Douglass—the most famous Black man in America—until after the Civil War. When they finally did so, they waffled between encomium and insult, balancing each compliment with a snide remark. The editors praised Douglass's large brain, but wrote that "his forehead does not indicate the philosopher." They discovered dignity in his cranium, but also found a large organ of "Destructiveness." Then, they clarified that he did not have a "repulsive face," as if this would otherwise have been a perfectly logical assumption. They also included a footnote acknowledging they had received numerous requests to publish "a portrait and sketch of this distinguished personage," but that they had delayed due to lack of a "perfect photograph." The veracity of this claim is questionable at best. Douglass was the most photographed nineteenth-century American. Could it *really* have been that hard to find a good portrait of him? The editors also made it clear they did not intend to "take sides with either of the political parties," a disclaimer they usually did not include when providing physiognomical portraits of other political figures. Evidently, they knew some part of their audience would be unnerved by an effusive review of a Black man's countenance and cranium. While they seemed willing to alienate readers who opposed women's rights, they mostly bowed to the wishes of white supremacists.[40]

Perhaps this explains why Sojourner Truth never got her phrenological profile published, as many of her white female counterparts did. In 1867, she got her head examined by Nelson Sizer. When doing so, "She took the risk of being subjected to racism masquerading as science." Yet the examination itself was rather positive. Sizer was part of the leadership team at Fowler and Wells, but he was also a man of abolitionist sympathies. His reading of Truth is refreshingly free of the usual racist stereotypes. He diagnosed her with a "moral firmness of character" and a love of truth. Calling her "upright" and a "good reader of character," Sizer also praised her "confidence" and "self-reliance." Unlike the *American Phrenological Journal*'s analysis of Douglass, this reading did not contain subtle digs at Truth's countenance, cranium, or character. Even so, it emphasized her moral fortitude over her mental acumen. Sizer told Truth she was "ingenious," but by this he meant she could "learn to use tools, and be a good mechanic." He also told her she had an "excellent memory" for "faces," "places," and "facts," but he did not describe her as an intellectual powerhouse. As a point of comparison, he *did* describe Susan B. Anthony that way. Still, Sizer's analysis of Truth was undeniably flattering. Despite this fact, Truth never had it published. Perhaps the *American Phrenological Journal* was not interested in broadcasting the face of a Black female activist. Perhaps Truth herself was ambivalent about phrenology. Or perhaps she simply knew it would be more helpful to circulate—and profit from—her own likeness than to rely on a white man's dissection of it. Still, it is entirely possible that phrenology could have appealed to Truth for the same reasons it appealed to other abolitionists. Phrenology delivered a flexible interpretation of the human brain. If all people had an idiosyncratic profile of moral and mental talents, then anyone could be exceptional, regardless of race or gender.[41]

In the end, though, phrenology left both Black and white women with mixed messages. Were they equal to men? Or fundamentally different? Could female brains be both equal to and distinct from male brains? Was the brain itself gendered? If so, what did that mean for social and political relations? And what about women of color? Were African Americans the intellectual equals of whites? Or were intelligent Black thinkers merely exceptions that validated the rule of white supremacy? Different people answered those questions in different ways, and influential popular scientists fell all over the political spectrum. Although some gender conservatives invoked science when arguing for women's mental inferiority, many white activists and abolitionists put their faith in phrenology, seeing it as a discipline that demonstrated the intellectual equality of the sexes. Recognizing this reality, some of their critics took a different approach. As the women's rights movement gathered steam in the early 1850s, they began describing phrenology and female activism as mutually supportive and equally threatening challenges to the existing social and political order.

THE "UNSIGHTLY SPREAD OF SUPERFICIAL INTELLECT"

The very same year that Elizabeth Cady Stanton and Susan B. Anthony got their heads examined, *Putnam's Monthly Magazine* published a critique of women's hairstyles. Despite being skeptical of phrenology's legitimacy, the author acknowledged its massive influence on the public imagination. Americans, he claimed, were supremely "proud" of their foreheads because they were "higher and wider" than those of other people. Despite these natural endowments, some people sought to heighten their brows by artificial means. "Now ever since phrenology began to finger our craniums, our vanity has been very busy in smoothing the way for its titillating advances," the author complained. Desperate to display their high brows, Americans were failing to enrich their minds. Instead, they used "brushes and depilatories" to create an artificial impression of mental excellence. He feared that intellectual women were especially egotistical. "A high, expanded, arched forehead may be excellent in man, as indicative of intellectual force, the power of knowledge; but it is a positive blemish in a female." Quoting another author, he argued "that a large, bare forehead gives a woman a masculine and defiant look." He was particularly exasperated with the female intellectual, who "plasters and presses and glues, and posts, like a bill-sticker, her front hair on either side of her forehead." Urging women to ignore "that ugly Frenchwoman, Madame La Mode," he told them to let their clustering curls and wavy locks flow freely.[42]

This article was a denunciation of female vanity, but it was also more than that. Why should it matter if women wanted to pull their hair away from their forehead and cement it inelegantly against their skulls? It mattered because people in the nineteenth century attached political significance to women's skulls. Since they believed high brows indicated intellectual eminence, women who bared their foreheads were violating a central tenet of true womanhood: female modesty. If these women were willing to alter their appearance, they might be willing to challenge other gender conventions, too. Connecting phrenology and the burgeoning women's rights movement, *Putnam's* stated the problem explicitly:

> We recommend our beauties to cultivate the low forehead, and advise our mannish women of the Woman's Rights Convention, to transplant the hair from their heads to their chins, and with bold fronts and strong beards, make good their claims to man's privileges and his wardrobe, to his boots and his walks in life.

In other words, *Putnam's* made it clear that women's decision to exhibit their high foreheads was not merely an aesthetic one. It was a political choice—one that gender conservatives found both frustrating and threatening.

After *Putnam's* scolded bluestockings for styling their hair in ways that emphasized the power of their brains, Limerick, Maine's *Morning Star* published a brusque response. The newspaper argued that intelligence was a positive trait in women—one that heightened their charms (and their brows). "Is there no beauty in intelligence?" the paper asked. Answering its own question, the *Morning Star* professed that all estimable women would have "a capacious and active intellect" and criticized the "foolish sentiment" that "a low forehead is the climax of beauty in woman." Finally, the author wondered why *Putnam's* had been so preoccupied with female beauty and so dismissive of female intellect: "Away, away with that sickly notion which would make dolls of the female world, the mere play things of man!—They have work to do, responsibilities to bear, errors to correct, souls to save, instruction to give, as well as men, and they should be educated accordingly." Through these words, Limerick's *Morning Star* provided a rousing defense of female education, paired with a simultaneous vindication of phrenology and women's political activism.[43]

The public debate became especially acrimonious after the *Boston Post* published an article about female foreheads in the spring of 1855. The piece inspired dozens of responses and reprintings from various newspapers and magazines between the 1850s and the 1870s—in both the United States and Europe. Most authors repeated some version of this refrain:

> The notion that high foreheads, in women as well as men, are indispensable to beauty, came into vogue with phrenology, and is going out with the decline of that pretentious and plausible "science." Not long ago more than one "fine lady" shaved her head to give it an "intellectual" appearance, and the custom of combing the hair back from the forehead probably originated in the same mistaken ambition.[44]

If these authors were to be believed, high foreheads made women look "bold" and "masculine," and were therefore antithetical to both "modesty" and "gentleness." Writers proclaimed that unless women wanted to look like "shrews," they should cultivate the "low forehead." Ostensibly, these articles were clever mockeries of intellectual women and their misguided faith in a dubious pseudoscience. Yet they also betrayed a deeper anxiety: perhaps phrenology truly did reveal something about the human brain. Despite mocking the legitimacy of the science—and claiming that it was on the decline when it was still massively popular—these authors clearly internalized some of its major contentions. If high and broad foreheads signaled mental profundity, then surely these traits were suited to men, not women.

It is telling that censures of women's lofty brows were usually paired with an explicit commitment to gender hierarchies. Genio C. Scott, a fashion illustrator and publisher from New York, combined his attack on phrenology with a critique of the dress reform movement that was being spearheaded by feminist activists like Amelia Bloomer, Elizabeth Cady Stanton, and Susan B. Anthony. He lambasted the "strong-minded Bloomers" who not only tried to wear the pants in their relationships, but were also "in the habit of shaving their foreheads for the purpose of getting up an artificial intellectuality." There is no evidence that female activists were indeed shaving their foreheads. But Scott was right to link high-browed women with the dress reform movement. In fact, many of the same activists who embraced phrenology also advocated for the Bloomer costume, which consisted of loose pants underneath a calf-length skirt. This included Amelia Bloomer herself, the phrenological enthusiast and feminist activist who popularized the outfit. Dress reformers argued that women's health was far more important than superficial beauty. They decried fashionable corsets, which crushed women's rib cages and made physical exertion difficult. They also railed against the long, impractical skirts that made women's daily errands and household chores into a constant struggle. Leading phrenologists made many of the same claims, urging women to focus on their bodily health rather than shallow fashions. Both phrenologists and dress reformers argued that societal conventions prohibited women from perfecting their bodily and mental health. Orson Fowler even embarked upon a crusade against tightlacing. When Genio Scott criticized "strong-minded Bloomers," then, he wasn't being irrational. He was attacking a group of female activists who were, in fact, challenging existing fashions and simultaneously flirting with phrenology.

Scott also expressed anxiety about the potential dissolution of class hierarchies. It was one thing if elite and middle-class bluestockings wanted to heighten their foreheads through a "most unsightly" fashion trend. "But a bad example is sure to find plenty of imitators; for we have noticed of late that several maid-servants (of a strong-minded turn of mind, we presume) have actually been resorting to the same barbarous practice." His advice? Men should "lock up their razors" and put a stop to this "unsightly spread of superficial intellect, before it has fairly turned the heads of all our cooks and nursery-maids." For him, phrenological hairstyling was a troubling development. It not only made women less beautiful; it also emboldened them politically. Such fashions, he believed, were irritating when adopted by women in the upper and middling classes, but they became dangerous when embraced by working women.[45]

In the minds of gender conservatives, women's decision to highlight their high brows was more than a fashion problem. It was a disconcerting challenge to patriarchal hierarchies, triggered by a cabal of immodest bluestockings. In response, the phrenologists rallied to women's defense, arguing that a beautiful skull signaled a beautiful mind, and encouraging women to be proud of their cranial endowments. To phrenologists, it seemed a great crime to tell women that low brows were beautiful, for "if a low, small forehead" was "essential" to "female beauty, it is so because it shows a lack of wide-reaching and deep-searching intelligence!" Unwilling to consign women to perpetual mental mediocrity, the Fowler family insisted that men who preferred short brows were specifically seeking out stupid women. As they saw it, a disingenuous exaltation of female beauty was no substitute for true respect.[46] William Stillman and John Duran agreed. As editors of *The Crayon*, a magazine dedicated to art history and the graphic arts, they claimed that a high forehead was not a defect but rather "an addition to the beauty of its owner." Why? Soaring brows indicated intelligence, and mental refinement made women more—not less—attractive.[47]

As they watched these debates unfold in public, it would have been obvious to female intellectuals that the phrenologists were not their enemies. Even after the nation started to reject phrenology as a dubious pseudoscience, some suffragists refused to give it up. When activists found themselves in need of a publisher for the first volume of *The History of Women's Suffrage* in 1881, they naturally turned to the Fowler family. The women then dedicated their first volume to luminaries like Mary Wollstonecraft, Margaret Fuller, and Sarah and Angelina Grimké, but also to Dr. Lydia Folger Fowler, the female physician and phrenologist who had held several leadership positions within the women's rights movement. *The History of Women's Suffrage* also contained a tribute to Lydia's sister-in-law and fellow phrenologist, Charlotte Fowler Wells, as well as numerous references to popular sciences. Phrenology, though hopelessly racist and misogynistic by modern standards, quite

FIGURE 4.7. "The Manly Brow Movement," *Harper's Bazaar* 2, no. 52 (New York), December 25, 1869. Courtesy of the American Antiquarian Society.

clearly appealed to early feminists and abolitionists. Within this context, it makes sense that gender conservatives directed their ire against both phrenology and the allegedly immodest women who subscribed to it.

By 1869, elite scientists were trying to distance themselves from phrenology, and the discipline was declining in popularity. Americans remained physiognomically literate, though, and they continued to draw connections between phrenology and female intellectuals. On Christmas Day, the fashion magazine *Harper's Bazaar* published a satirical critique of "The Manly Brow Movement" that mocked the women who tried to look intelligent by sweeping their hair away from their high foreheads (fig. 4.7). By this point, phrenology had already lost much of its purchase in the United States. Even so, *Harper's* editors knew their readers would know how to read this physiognomical imagery.

The short piece included three images. "A" and "B" were sketches of women who dressed their hair in the fashionable style of the late 1860s. Image "C" was different. It portrayed "Miss Oldboy," a woman who did "not see why she should wish to hide an Intellectual Forehead." A pithy caption referred to the first two women as "horrid Frights" (despite their relatively standard appearance) and expressed a desire to "see a little more of" the "Manly Brow" in sketch "C."[48] The sarcasm was palpable, and the message was clear: women who tried too hard to look like intellectuals were ugly, misguided, and masculine. Even if they *thought* they were attractive, they were ultimately just shrews who assaulted men's aesthetic sensibilities by trying to broadcast their brainpower. By this point, Americans would have been well aware that fashion statements were political statements, and that the public battles over female brows were not flippant conversations about women's aesthetic choices, but scientific meditations on the capacities of the female brain.

When people fought over women's high foreheads and curious coiffures, they were asking bigger, thornier questions about the social and political position of women. Were women the intellectual equals of men? Did they deserve political rights? Would a rigorous education inevitably masculinize a feminine belle? And should women be able to dedicate themselves to literary pursuits, even if it came at the expense of their attractiveness to men? For nineteenth-century Americans, these were not just social or political questions but scientific ones that could be answered through phrenological means. When they took up phrenology, women's rights activists showed that popular sciences could facilitate rather than stymie reform. Gender conservatives similarly linked phrenology with female activism, but they did so to discredit both intellectual women and the scientific frameworks that those women used to make sense of their bodies, identities, and political priorities. But these were not the only groups to draw connections between science and politics. As chapters 5 and 6 will demonstrate, prison reformers and Black abolitionists also embraced physiognomy and phrenology, using these disciplines to reimagine American society.

5
CRIMINAL MINDS

In a letter to her lover in 1845, Margaret Fuller described a visit to a group of women who had recently been released from prison. She took special interest in one, zeroing in on her face. "Her eyes were brown and very soft," Fuller wrote, "around the mouth signs of great sensibility." Despite her beauty and refined delicacy, the woman appeared to be suffering from consumption. This caused Fuller to lament the allegedly degraded position of incarcerated women. She nonetheless saw them as reformable, saying, "I like them better than most women I meet, because, if any good is left, it is so genuine, and they make no false pretensions." After meeting with the prisoners, she made a public appeal for funds to provide the women temporary asylum and find them useful employment.[1]

Sara Jane Lippincott—a nineteenth-century writer, abolitionist, and women's rights activist—developed a different interpretation of the incarcerated people she encountered in her travels. Like Fuller, she used physiognomic language to interpret their characters. But unlike Fuller, she viewed them as "hard, Heaven-forgotten looking creatures." When visiting the Newgate Prison on a trip to London, she said her "very soul shuddered and sickened at the sight of beings seemingly so helpless, hopeless, and redemptionless." For Lippincott, the prisoners' faces reflected their wickedness: "I think I never saw human eyes which had so lost every ray of the primal soul light, seeming to give out only a deathly, pestilential gleam from moral vileness and corruption—faces into which all evil passions were so stamped as by the iron hoof of Satan himself." In her mind, evil could be ineradicably imprinted on people's countenances, and the minds and bodies of incarcerated people reflected their inherent depravity.[2]

The two women's divergent uses of popular science shed light on a set of larger debates that preoccupied middle-class reformers in the United States.

Were prisoners redeemable? Or were they a unique class of corrupt and incorrigible beings? Was vice itself an innate propensity? Or were most social problems precipitated by environmental forces beyond people's control? Antebellum reformers developed a diverse assortment of answers to these questions. Yet many of them were united in one assumption: countenances and craniums reflected character. While Margaret Fuller discovered sensibility in the mouth of an incarcerated woman, Sara Lippincott found "evil passions" in the eyes of Newgate's prisoners. Both women, despite their differences, relied on science to understand vice and to craft strategies for combating it.[3]

Throughout the antebellum decades, a wide variety of bourgeois reformers evaluated the craniums, countenances, and characters of the prisoners and sex workers they sought to "save." Through the language of physiognomy and phrenology, they attempted to reconcile two apparently incongruous understandings of human nature. On the one hand, facial and cranial analysis promised that reformers might peer into people's souls, giving them a way to see people's innate propensities in the bones and sinews of their bodies. On the other hand, reformers typically viewed the mind and body as malleable entities, subject to change and capable of cultivation. They believed, in other words, that environmental forces—paired with people's personal choices—could transform the mind and alter the body. Within this context, science had a prophetic ability to predict people's future behavior, but it was not necessarily deterministic. Bodies did not dictate destiny, but they might provide hints about people's prospects and possibilities.[4]

Because physiognomy and phrenology did not doom any individual to perpetual degradation, moral reformers found these disciplines useful in their political crusades. Activists had to believe that all people could attain redemption. Otherwise, what was the point of their life's work? Luckily, the tenets of popular sciences were flexible enough to account for the fact that all humans were perfectible beings. Ultimately, though, physiognomy and phrenology provided reformers with a way of reifying existing hierarchies by rooting them more securely in the human body. As many reformers saw it, most incarcerated people had bad heads, vicious countenances, retreating brows, vacant eyes, and hardened features. Predictably, they detected the visual signs of remorsefulness only in the lineaments of those they deemed capable of redemption. All others, they suggested, deserved their degraded position because of their bad choices and inherited propensities.

"THE IMPRESS OF VILLANY"

As early as the 1790s, American authors began using facial analysis to suggest that criminality manifested itself in the human countenance. Lavater had included brief remarks on the subject in the *Essays on Physiognomy* (1775–1778). "Whoever has frequently viewed the human countenance in houses of correction and jails," he argued, "will often scarcely believe his eyes, will shudder at the stigmas with which vice brands her slaves."[5] American authors then took up this argument, claiming criminality was visible in the visage. In 1796, for instance, the *Weekly Museum* published a short story titled "Physiognomy" in which a character managed to save his own life by detecting "the formation of some shocking design" in the visage of a homicidal visitor.[6] Similarly, in an oration delivered to a group of freemasons in 1798, the speaker R. W. James Mann described a "murderous culprit" who was unable to escape "retributive justice" because his face disclosed the "conspicuous mark" of criminality. In one of the story's footnotes, the author wrote: "Whoever attentively reads Lavater on physiognomy, will be convinced, that man carries the traits of his immoral part, upon his countenance." This tale assured anxious listeners that criminals could not hide their true natures. They would always be identifiable, for their physiognomies would give them away.[7]

One story in *The Literary Mirror* similarly featured a narrator who believed criminality exhibited itself in facial features. Although he found the *Essays on Physiognomy* to be helpful, he did not think the volumes spent enough time talking about the countenances of those who flouted the law. As a result, he spent two years collecting "the silhouette of every offender that came under my jurisdiction, carefully delineated," and then compiled "an appendix to the Lavaterian codex." This man believed the proper identification of criminals constituted the most useful branch of the human sciences, for it was "far more important to distinguish at first sight the house-breaker, the highwayman, the adulterer, or the murderer . . . than to analize [sic] the poetical, heaven-directed ethereal soul."[8] Decades later, the New York physician J. F. Daniel Lobstein would make a similar argument in a treatise on medical physiognomy. He claimed that the "habit of crime" was visible in the eyes, and then followed this assertion with a set of physiognomical instructions for determining a person's guilt or innocence in the courtroom.[9] In these instances, popular and scientific writers alike suggested there was something unique about criminal faces. If only people could learn to read offenders' appearances properly, they might be able to prevent their diabolical designs and mitigate social suffering.[10]

By the middle decades of the nineteenth century, phrenology added further detail to these physiognomical understandings of crime. Although phrenology assumed that all human brains contained the same set of organs, it also suggested that all individuals would have unique brains that were differently developed in distinct areas. Good wives and mothers would have a large bump at the back of their skull to indicate their organ of "philoprogenitiveness." Philosophers and mathematicians would have high and broad foreheads to indicate their powerful perceptive and reflective faculties. The heads of those who committed crimes, by contrast, would be widest above the ears, due to a bulging organ of "destructiveness." As Courtney Thompson has illustrated, phrenologists initially referred to this area as the "organ of murder," assuming it would be larger in those who broke the law. Leading phrenologists also divided the brain up into three major parts: the intellectual *faculties*, the moral *sentiments*, and the animal *propensities*. Incarcerated people, they suggested, had defective "organizations" in which the propensities triumphed over the faculties and sentiments. Because of phrenology, phrases like "the propensity to steal" and "a propensity to murder" became part of a transatlantic "scientific vernacular" that people used to talk about crime.[11]

When discussing vice and the groups of people who fell victim to it, nineteenth-century reformers used science to draw complex—and often contradictory—conclusions about human nature. Arguing that virtue was both a state of mind and a physically discernible characteristic, they claimed that a life of crime or dissipation could "stamp" certain physical features upon offenders. The Massachusetts clergyman Cyrus Bartol, for instance, imagined that vice disfigured people's appearances. As a Unitarian minister, he visited sick parishioners and provided charity for the poor. One day in 1834, he met with a young woman who had been "seduced" and abandoned "by one who promised marriage." Bartol pitied her, but also claimed the woman's "lineaments indicate the *material* for a fine strong character," pointing out the "remains of beauty in her countenance." He regretted that her misfortunes had wreaked havoc upon both her inner and outer beauty. Comparing her atrophied countenance to the "ruins of the most Splendid Temple," he lamented, "How sad the spectacle of a mind whose native capacities are thus crushed!" These were not throwaway comments. Throughout his personal writings, Bartol engaged directly with physiognomy and phrenology. He described the latter discipline as useful but imperfect: "a track to truth" if not an infallible science. Even so, he admitted that Johann Gaspar Spurzheim had been an uncanny judge of human character. How could this be explained? Bartol decided that Spurzheim likely supplemented his phrenological knowledge with his own skill in physiognomy, using his "keen and observant eye" to "pierc[e] through the expressions of countenance to the mind beneath." Although Bartol trusted physiognomy more than phrenology, he thought that valuable information could be gleaned from cranial analysis. Still, he resisted phrenology's deterministic impulses. In his worldview, minds and bodies could change over time, responding to the vagaries of free will, education, and personal experience. It was this belief that convinced him that his female parishioner's "seduction" had initiated a process of moral decay, which transformed her body into "ruins." Remnants of her virtuous character were still discernible in the "lineaments" of her countenance, he argued, but the evidence of her former virtue was barely visible.[12]

Just as they sought to discern the physical symptoms of vice in people's heads and faces, moral reformers also relied on science to make artificial distinctions between individuals who might be capable of reformation and those who were not. In 1844, the Prison Discipline Society compared the countenances of two male offenders with the visage of an unfortunate young mother who had been caught up in the criminal escapades of an unsavory associate. "The impress of villany [sic] was deep on the countenances of the white men," the report noted. "One was a thief, the other a counterfeiter." By contrast, when the authors then described the female prisoner, they wrote that she "showed in her countenance no marks of peculiar depravity. Her crime was apparently being associated with a Black man, who stole from, and set fire to, a building in which no one lived." Condemned by the crimes of someone else, the woman now faced death, and her innocent baby confronted the prospect of life without a mother. At first glance, the racialized implications of this story seem clear: a white woman was supposedly led astray by a Black man. But as scholars of phrenology have shown, scientific understandings of crime did not initially view people of color as inherently more likely to engage in felonious activities. At least in the early nineteenth century, the stereotypical criminal tended to be a white man with large organs of "destructiveness" and the "impress of villany" upon his countenance. Even as physiognomy and phrenology contributed to typecasting, they suggested that those who committed crimes could be mentally and physically diverse.[13]

As a way of showing that incarcerated people did not constitute a monolithic class of deplorables, reformers often highlighted the stories—and appearances—of previously virtuous but ill-fated offenders. John Luckey, an administrator at the Sing Sing Prison outside New York City, described one particularly penitent prisoner as having a "manly and dignified bearing, intellectual countenance, and serious deportment."[14] In a similar way, the Reverend Ansel D. Eddy published a religious pamphlet analyzing the appearance of Jacob Hodges, a reformed man. After meeting Hodges for the first time, Eddy remembered that he "had a full view of his broad African face, every line of which spoke the language of a mind and heart of no ordinary character. There was a subdued, tender, yet cheerful aspect to his countenance, as if fully conscious of what he had been, yet blessed with the conviction of a new heart, and in hope of a better state yet to come." Hodges's appearance apparently identified him as a living, breathing example of God's forgiving grace.[15]

Similar descriptions appeared in popular magazines. In 1839, Emma Embury published an article in *Godey's Lady's Book* where she analyzed the head and face of the fictional Newton Ainslie, a previously respectable man who had accidentally maimed a young boy and landed himself in jail. Although the prisoner's body was both "shrunken" and "attenuated," he apparently exhibited "a head so fully developed, it would have thrown a phrenologist into ecstasies." Ainslie's "forehead was high and broad, his eyes piercing and intelligent, his features delicately formed." The man's appearance indicated that he had to be more than a "common felon." In addition to showing "remarkable intelligence," he was "a good classical scholar" and mathematical genius. Ainslie had been a good—but ill-tempered—man who gave in to his basest passions and destroyed his own life. Embury published this story as a cautionary tale: Anyone could succumb to the temptations of vice without proper self-regulation. Even the smartest individuals were not exempt. Her story also articulated another theme: people could improve, but if they were not careful, they might instead change for the worse.[16]

If reformers recognized penitence in the faces of model prisoners, they also discerned traces of virtue in the faces of individuals who had given themselves over to vice. The New York Female Moral Reform Society, for example, regularly visited sex workers in hopes of "redeeming" them. In the Society's magazine, *The Advocate of Moral Reform*, one reformer told of an encounter with a woman "of the town" on her approaching death. The woman allegedly did not think herself worthy of redemption. The reformer was nonetheless determined to fight for the final salvation of her soul and set about scrutinizing her body for signs that she was capable of reform. The initial diagnosis seemed promising. "Her personal appearance very much surprised me; and her intense anguish of spirit awakened all my sympathy," the author recalled. "She was evidently no ordinary woman. Her stature was tall, her figure, though large, was elegant, and of beautiful symmetry. Her eye was dark, full, and piercing; her forehead high, and her whole countenance strong marked, and indicative of a high order of intellect." These physiognomical signals showed she had once held the capacity for virtue and intelligence. But to the reformer's vexation, the woman proved resistant to salvation. As someone who was suffering from painful physical ailments, the sex worker surely resented being subjected to the glares and condescending speeches of an officious moral reformer. The reformer, for her part, was clearly frustrated by her own inability to lead the young woman to God, insisting that she died a miserable death while surrounded by her brothel "associates." This was not a hopeful tale of redemption. Like Emma Embury's story, it warned readers that even the most sophisticated individuals were not immune to vicious inclinations.[17]

The same magazine published an analogous account in 1852, which followed a member of the New York Female Moral Reform Society as she visited the syphilitic ward of a women's hospital. Voyeuristically, the reformer guided her readers through the hall, describing the women she encountered along her path. For the author, it seemed obvious that these women were suffering from the corporeal ravages of a devastating disease. But she was ultimately less interested in the women's bodily suffering than in their sexual misconduct, alleging that their mental and physical ailments could be traced back to the folly of "ungoverned passion." She argued most of the patients were "brutal, sodden, animal in their expression," claiming there were some "from whose features vice and brutality have worn every trace even of womanhood." If she admitted that sickness was the immediate cause of their physical degradation, she nonetheless claimed it was the women's moral turpitude that had precipitated their corporeal decay. In doing so, the reformer melded older understandings of beauty and vice with newer, more physiognomical ideas. As far back as the 1730s, the artist William Hogarth had depicted prostitution as a trade that destroyed women's bodies. In *The Harlot's Progress* (1732), for instance, he visually conveys the story of a woman who becomes a sex worker and then contracts syphilis. The disease not only marks her face with open sores, covered by patches; it eventually takes her life. Such stories of syphilis leading to physical deterioration were common in the eighteenth century. By the mid-nineteenth century, though, authors knew how to meld such narratives with scientific language. Descriptions of foreheads, countenances, and craniums became moralistic lessons about how vice itself could disfigure people's physiognomies.[18]

To the moral reformer who toured a syphilitic ward in 1852, it seemed logical that she would be able to read more than illness on the faces of those she visited. On these women's countenances, she detected something "keener than the pang from tortured nerve and burning tissue": the physical sign of debilitating guilt. One patient particularly caught the reformer's eye, causing her to declare, "it was evident, at a glance, she must have been of a very different class from her companions." Although the woman exhibited the physical symptoms of syphilis, her "crime had not worn away noble features." Her forehead remained "full and finely arching, the head high, complexion soft and delicate, unnatural intensity of expression, and the lips firm and fixed, all the lines of her face fine and noble, and every feature expressing the proud and refined woman." In this instance, the reformer acknowledged that disease could destroy the body, but she also suggested that this woman's distinguished cranium, full forehead, and "noble features" indicated her unrealized potential. In instances like these, reformers did not use physiognomic and phrenological language to argue that sex workers had innate criminal propensities. Instead, they used science to uncover traces of virtue in the faces of "fallen" women. Then, they relayed these women's stories as cautionary tales. Vice could annihilate anyone, they warned. Every woman must be vigilant.[19]

In addition to being practitioners of facial and cranial analysis, reformers sometimes discovered that they, too, were the objects of scientific scrutiny. In the *Illustrated Phrenological Almanac for 1850*, for instance, Lorenzo Fowler juxtaposed two portraits. One depicted the "remarkable head" of Mary Runkle, a suspected serial killer who faced the executioner's noose in 1847. The other portrait was of Margaret Prior, one of the founders of the American Female Moral Reform Society. In the first image, of Runkle's head, Fowler detected a lack of "refinement," "delicacy," "intellectuality," and "warmth." He argued that she was deficient in the organ of "Benevolence." By contrast, Prior's portrait supposedly revealed a "full and expanded frontal lobe" and a head of great "height," which signaled "an elevated tone of mind and feeling." If Runkle's head demonstrated her lack of compassion, then Prior's revealed precisely the opposite: a tendency "to assuage the pangs of suffering humanity." In this instance, phrenology did double duty. First, it identified and denounced a woman with alleged criminal propensities. Second, it assured Americans that the examination of heads and faces would allow the general populace to distinguish between wrongdoers and reformers. Such juxtapositions would have given female activists confirmation that their very bodies and brains suited them for benevolent work. They also would have provided reformers with a way of self-righteously distinguishing between themselves and the objects of their reform efforts.

Ultimately, organizations like the New York Female Moral Reform Society sought to alleviate the physical suffering of marginalized women, but their main goal was to obliterate the moral corruption that they saw at the heart of society. When they sent visitors to brothels, poor homes, workhouses, and hospital wards, their primary aim was not to palliate the bodily pain of the women they harassed. Instead, reformers intended to use these harrowing accounts of women in distress as warnings to those who still retained their virtue. Their accounts were usually predicated on three assumptions: human nature was perfectible, vice was avoidable, and redemption was possible. Certain people might inherit vicious propensities or a disordered mental organization, but that did not mean they were destined to be immoral forever. By contrast, even initially respectable individuals might suddenly find themselves on a precipitous descent into the pits of vice. These ideas reflected a new understanding of crime: a theory of wrongdoing that rooted bad behavior in the moral failings of individuals. It was a philosophy that had been decades in the making by the mid-nineteenth century.

THE RISE OF THE PENITENTIARY

Beginning in the late eighteenth century, both Americans and Europeans started reconceptualizing criminal behavior and searching for ways to grapple with it more effectively. For much of the colonial period, Americans punished lawbreakers through spectacles of public punishment. They hanged offenders, whipped them in the town square, cropped their ears, or trapped them in stockades during public rituals of suffering and humiliation. In the decades following the American Revolution, however, Americans became uncomfortable with such visible displays of state-sanctioned cruelty. Throughout the nation, reformers began arguing that those who broke the law might instead be removed from society and reformed. Ideally, offenders would occupy prisons, workhouses, and asylums. During this period of temporary sequestration, they would labor to improve themselves, attain salvation, and develop the moral character necessary for eventual republican citizenship.[20]

By the early decades of the nineteenth century, the strategic containment of convicted citizens had largely replaced the spectacle of public punishment. This transformation was inspired by the humanitarian impulses of the Enlightenment and the egalitarian spirit of the American Revolution, but it was also entangled with the Second Great Awakening's emphasis on human perfectibility. Protestant reformers, in particular, argued that criminals needed—and were capable of—redemption. Together with Quaker activists, they crafted a new system based on the notion that crime was best prevented through moral reform, rather than physical punishment. Even so, punitive measures never disappeared from the criminal justice system. Corporal punishments merely moved inside the prison, where the probing eyes of a voyeuristic public could no longer witness them. Rather than seeking to eliminate pain entirely, reformers argued for the power of "redemptive suffering" behind institutional walls. Their new establishments were called penitentiaries—a name reflecting this new preoccupation with personal penitence.[21]

To rehabilitate the unredeemed, reform-minded citizens tried to guide them on the path toward salvation. In addition to visiting the homes of the poor, invading brothels and public houses, and building "Houses of Refuge," elite and middling Americans trekked to prisons to visit prisoners, sex workers, and allegedly "debauched" or "fallen" individuals. Penitentiaries likewise hired prison chaplains and "moral instructors" to guide inmates toward redemption. In principle, these reformers were committed to the notion that all people could change. But they also sought to distance themselves from those they endeavored to improve. If reformers saw certain individuals as redeemable, they described others as inherently hopeless. In doing so, they contributed to the idea that a degraded criminal class existed within society: a group of people who were, by their very natures, different from respectable citizens. For this reason, they tried to develop metrics for distinguishing between the potentially virtuous and the permanently vicious. While deeming the first group worthy of rehabilitation and aid, they dismissed the second group as people who deserved only condemnation.

Of course, the success of these crusades depended on the very people subjected to containment. Effective prison management, in other words, relied on inmates' willingness to voluntarily enact the rituals of redemptive suffering. It required prisoners to first acknowledge their alleged misdeeds and then make conscious efforts to address their previous wrongs. Prisoners did not always cooperate, which frustrated zealous reformers and caused them to complain about supposedly deceptive inmates. Physiognomy and phrenology emerged as two paired strategies for countering dissimulation. By reading the heads and faces of prisoners, reformers sought to discern deeper truths about the people they thought they were saving.

"PERHAPS HIS FACE TELLS A LYE"

In 1826, Levi Lincoln Jr., the governor of Massachusetts, bemoaned the state of penal institutions in his state. "There is much reason to believe," he wrote, "that, as a Penitentiary, the system is utterly ineffectual to purposes of reform or amendment." Although convicts were enriching the state through forced labor, it appeared they were not becoming more virtuous. To make matters worse, it seemed that confinement was only further corrupting inmates. The governor complained about how four to sixteen prisoners were forced to sleep in the same cell at night. Locked together in these "committee rooms of mischief," they supposedly took advantage of close quarters to engage in precisely the types of behavior that reformers were trying to prevent. To combat this tendency, the Massachusetts State Prison turned to the expertise of its prison chaplain, Jared Curtis. Between 1829 and 1831, Curtis recorded detailed memoranda about inmates at the Massachusetts State Prison. In these assessments, he tried to gauge prisoners' reformability by evaluating their appearances.[22]

Curtis knew first impressions were not always what they seemed. Still, he found it useful to analyze prisoners' faces, even when he acknowledged he might later be proven wrong. On one occasion, he described a man named Thomas Jennings as "Not very frank & communicative," writing, "Should think him a shrew'd crafty fellow—not much feeling—& cut out for a rogue." But Curtis also acknowledged that he had come to this interpretation merely by examining Jennings's physical features. "Perhaps, however, I judge him too hard," he wrote. After all, the man seemed to be reading the Bible. Was he truly irredeemable? Surely not! Still, Curtis found it hard to shake his initial assumptions, admitting he did "not like his appearance." On one level, the prison chaplain was hopeful that individuals might exceed his expectations. But he also assumed that others would disappoint him. One day, he appraised a man named Josiah Harris, writing that he "appears like a sincere man—frank, ingenious & humble—Conduct good." Characteristically cautious, Curtis added, "Still he may be very deceitful though he is far from having that appearance." With such individuals, only time would tell.[23]

On other occasions, Curtis expressed more confidence about his own skills of character detection, favoring his own physiognomical interpretations over the stories prisoners told him. John Reed, for instance, insisted he had been temperate, but Curtis claimed that "his face tells a different story." The man looked like "rather a hard character," wrote Curtis, but "perhaps his face tells a lye." On another instance, he dismissed a man named Charles Watson as a "wild fellow" with "a bad face." Even more forcefully, he described George White as having "a very bad face—& from his life, it does not belie him." White insisted he was innocent of his most recent offense, but he was also a perpetual offender. Because of his record and his visage, Curtis scorned him as a "a finish'd scholar in his line of business": theft.[24]

Curtis was especially skeptical of Irish prisoners, demonstrating his prejudices as a Protestant prison chaplain. After meeting with a man named

Thomas Goffs, he wrote: "Appears frank & clever—like many Irish men—but is, most likely, like many Irish men, better outwardly, than at heart—Drink—drink—drink—is the ruin." On another occasion, Curtis evaluated an Irish inmate named Thomas Baron, saying he "appears mild and pleasant and well dispos'd." Curtis seemed hopeful for the man's good behavior, particularly because he had been diligently reading the Bible in his prison cell. "But an Irishman is not always to be seen through with a glance," he warned. He would have to wait to see how the man performed. In all these encounters, Curtis searched for a sense of certainty. He desperately wanted the comfort which came from knowing whether a prisoner would reform or not. At the same time, he knew that certainty was elusive. Curtis nonetheless analyzed faces to validate his own assumptions and develop institutional strategies for reform.[25]

These instances of face-to-face evaluation were not unique to the Massachusetts State Prison. Scholars have long known about Eliza Farnham's efforts to use phrenology to diagnose and treat female prisoners during her time as the Matron of the Female Ward at Sing Sing State Prison. During the 1840s, Farnham became notorious for her radical efforts to reimagine prison discipline. She allowed inmates to talk to each other, a sweeping change for an institution that previously relied on a system of mandatory silence and forced labor. In addition, she provided inmates with novels and phrenological manuals, established a nursery for those who gave birth while incarcerated, and furnished the women with a decorative parlor, complete with a piano and music books. As an evangelist for the science of phrenology, Farnham claimed that incarcerated people were not entirely responsible for their actions and should be treated rather than punished. Farnham pointed out that the working classes often lacked the opportunities for educational cultivation that were available to more privileged Americans. This, she believed, prevented them from strengthening their more refined faculties, and allowed the more vicious elements of their character to reign supreme. As she saw it, penitentiaries could counter people's undesirable propensities by developing their virtue, and phrenology could be used to develop a more rational, more scientific, and more humane system of rehabilitation.[26]

Farnham worked closely with the phrenologist Lorenzo Fowler. She first required numerous Sing Sing prisoners to sit for daguerreotypes while Fowler provided phrenological readings. Then, in 1846, she compiled an American edition of the British reformer Marmaduke Sampson's book, *Rationale of Crime, and Its Appropriate Treatment*. In the book's appendix, she included excerpts from Fowler's phrenological descriptions, alongside portraits of the Sing Sing inmates.[27]

With public exploits such as this, Farnham acquired enemies quickly. John Luckey, who served as prison chaplain during most of her tenure, resented her obsessive reliance on phrenology and her rejection of religion. By the late 1840s, Farnham had attracted an avalanche of unfavorable media attention, and her attempts to profoundly reimagine the Sing Sing State Prison eventually got her fired. This has led scholars to emphasize Farnham's supposedly unique affection for the popular psychological sciences of the day, as well as her revolutionary effort to remake the prison according to phrenological dictates. Farnham, to be sure, was far more radical than most of her contemporaries. She was also exceptional in her effort to elevate phrenology above all other types of reform. Other reformers pushed Bibles onto incarcerated people, but it seems that only Farnham furnished prisoners with phrenological texts, encouraging them to cultivate their minds and embark upon a journey of craniological self-knowledge. In analyzing the heads and faces of prisoners, though, she was far from unique. In fact, her methods reflected practices that were happening at other penal institutions throughout the country.[28]

"I DISCOVERED WHAT HE WISHED TO CONCEAL"

Seven years before Farnham was appointed the Female Matron at Sing Sing in 1844, Philadelphia's Eastern State Penitentiary hired the Reverend Thomas Larcombe to serve as the prison's "moral instructor." In this capacity, Larcombe regularly visited prisoners to assess their mental, moral, and physical characteristics. Every time he entered an inmate's cell, he would jot down short entries where he analyzed their countenances, appraised their religiosity, and made conjectures about their potential for future redemption. Like Farnham, Larcombe relied on physiognomy and phrenology to guide him in these pursuits.

Because he was always searching for ways to separate the improvable from the irredeemable, Larcombe tried to distinguish between the individuals who were smart and clever and those who were both intelligent *and* virtuous. He knew he had to be careful with the first group of inmates, fearing they might use their purportedly calculating natures to manipulate prison workers. The second group, however, he deemed capable of redemption. Larcombe developed a shorthand for discussing prisoners' reformability. When certain inmates were discharged, he assumed they would quickly resume their felonious activities. He marked those entries with a short addendum at the bottom of the page: "No hope" or "n. h." In other instances, Larcombe was more optimistic, predicting that there was "considerable hope" that an individual would reform.²⁹

Although he went to great lengths to determine the sincerity of the prison's inmates, Larcombe rarely trusted the words of the incarcerated. He even questioned the evaluations of the prison workers. On one occasion, he recorded an entry for a man named Daniel Davis, noting, "Keeper thinks him sincere." Initially, he thought the man might embrace religion and atone for his crimes, but after several visits he decided Davis was "Too old to learn." In a similar way, he once described a prisoner named William Johnson as having "some appearance of tenderness, affirms innocence, had a good character." After another meeting, however, Larcombe crossed out the phrase "had a good character," and instead wrote "NO HOPE" in the margins.³⁰

Larcombe was even harsher with other prisoners, using phrases like "Ingenious, cunning, skeptical & hopeless," "harden'd & reckless," "incredibly vicious," or "Hardened & hopeless to the last." After several encounters with George Hark, a man convicted of burglary, Larcombe described him as "palpably deceptive." Although Hark initially made "pretences to religious experience," he had "since confessed he was only hypocritically pretending and seems filled with a diabolical spirit." This prisoner's behavior earned him an evaluation of "n. h." from the moral instructor. Larcombe projected the illusion of control by taking detailed notes, but he was deeply disconcerted by people like George Hark. They threatened to make fools of reformers. Hark, for instance, knew the script he was expected to enact and performed his artificial religiosity with great aplomb. In the end, neither Larcombe nor his reform-minded contemporaries could effectively distinguish between allegedly "hopeless" individuals and those who showed "hope of M[oral]. Ref[orm]." They might have been able to imprison people's bodies by containing them within the penitentiary, but they could not control their brains and souls. Nor could they trust their own observations.³¹

Reformers were nonetheless sanguine about their ability to separate the disingenuous from their undesigning counterparts. To identify dissemblers, they deployed popular sciences, resting their hopes in the idea that external features proclaimed internal dispositions. Thomas Larcombe, for instance, once congratulated himself for discerning a lack of "tenderness" in one prisoner's appearance. When recounting his encounter with Lewis Williams, Larcombe employed a similar strategy, noting that he "looks and acts in the silent doggest manner of an old convict." As for John Dickerson, Larcombe apparently "Never saw a more ruffianlike expression." Then, when evaluating David Baggs, Larcombe regretted the man had "no proper sense of his moral condition," arguing that he had the "peculiarly strong look of a Sharper." This was a term nineteenth-century Americans used to describe particularly clever or conniving tricksters who conned people out of their possessions. Even though Baggs made "strong promises" about his desire to improve his character, Larcombe did not believe him. Sardonically noting that the man was "no doubt Sincere," the moral instructor declared that his "appearance is much against him." Rather than trust Baggs's words, he analyzed his features, expressing skepticism that the man could reform.[32]

On other occasions, Larcombe more explicitly invoked the language of popular sciences. He suggested, for instance, that William Thompson had "strong Phrenological manifestations of a thief." In a similar way, Larcombe visited the cell of a man named John May, concluding he "looks guilty & like one whose head denotes a thief." Then, when describing William White, he said the man was "an old convict no doubt" because he was "Secretive" and had "the appearance of a thief. (eyes close)." More dramatically, he referred to Brian Monahan as "a low browed Stupid, animal," with an inherent "Propensity" to criminality. He likewise insisted William Jones was a man with "Dishonesty & low trickery impress'd upon his face." Perhaps Larcombe poked and prodded these men's skulls as a true phrenological investigation would have required. This process would have allowed him to enact his own privilege and power through a violation of prisoners' bodily integrity. But perhaps he simply stared at inmates, using his physiognomical "skills" to interpret their inner characters. In any case, Larcombe believed the men's craniums and countenances conveyed deeper truths about their criminal proclivities.[33]

Unsurprisingly, inmates took exception to the prying eyes of their observers. In 1836, the Annual Report of the Board of Managers of the Prison Discipline Society admitted that every prisoner suffered from "a daily exposure to the gaze of idle curiosity, cast upon him by the thousands, who flock in to witness the state of degradation to which his crimes have reduced him." Criticizing the disciplinary system practiced at Auburn State Prison in New York, the authors complained that the "young offender" was constantly subjected to the "gaze of men, women, and children, who line his pathway, and scrutinize his countenance." This sort of surveillance would eventually obliterate a prisoner's "sense of shame, generate bitter and revengeful feelings, and set him more at war with society than before," they argued. By contrast, the authors thought Philadelphia's system of solitary confinement was more constructive. It seems they were not overly concerned with the privacy of prisoners, though, for they never questioned the pernicious effects of subjecting Philadelphia's prisoners to the unrelenting gaze of meddlesome men like Thomas Larcombe.[34]

Larcombe's notebook itself provides evidence that certain prisoners resisted scrutiny, and that they were resentful of his efforts to draw conclusions about their character through invasive observations. William Thompson, for instance, made efforts to disguise his own countenance and instead turn his eyes upon prison reformers and administrators: "Seems closely to watch, the person who speaks to him as if to ascertain his object & character & to not commit himself," Larcombe recorded. In a similar way, George Ryno begrudged his meetings with the moral instructor. It was "Evidently a painful struggle to the prisoner to meet with me," Larcombe wrote. During each interview, Ryno tried "to make his face as Brass." The moral instructor then dismissed his resistance strategy as just "another hardning [sic] process." Larcombe was clearly aware that incarcerated subjects begrudged his scrutiny. He noted, for instance, how one man demonstrated "evident uneasiness under the gaze of the person speaking." A different man, he complained, "Seems ashamed to look at me," while another, a man named Jesse Quantrill, "hates to see me." Quantrill's antipathy toward the moral instructor was justified. After all, Larcombe had asserted that the man's visage looked "like a basilisk" (a giant deadly serpent), and said "the Countenance of this prisoner, indicates, passion the most baleful & deadly." Although Quantrill was eventually discharged, Larcombe insisted he remained "*as wicked as ever.*"[35]

In addition to the paid employees who invaded prisoners' cells to facilitate their own reform agendas, there were also ordinary citizens who took it upon themselves to meet with the "fallen" and help them attain salvation. William Parker Foulke, for instance, was a lawyer, abolitionist, and philanthropist who joined the Philadelphia Society for Alleviating the Miseries in Public Prisons in 1845. In this capacity, he regularly visited the Eastern State Penitentiary. Foulke was an avid reformer and an evangelist for the "Pennsylvania system" of "cellular isolation." Incarcerated people needed solitary confinement, he insisted, lest they pick up bad habits from their fellow prisoners. Foulke nonetheless believed that individuals might benefit from conversations with allegedly virtuous reformers. Viewing himself as such a heroic figure, he regularly visited inmates and conversed with them on a variety of topics. These discussions usually centered on religion, but he also encouraged his interlocutors to tell him about their personal lives and contribute their own thoughts on prison reform.[36]

Foulke wanted to distinguish between people who seemed reformable and their more hardened counterparts. To accomplish this aim, he studied their heads and faces. On one instance in 1846, Foulke visited Prisoner No. 1623, whom he described as "a man whose countenance indicates firmness, & yet want of moral principle," admitting, "I did not like him in first view."

But Foulke continued visiting the man, and over time he began to think differently. On one visit, he said the prisoner was "much changed. Instead of the sullen, stubborn looks with which he used to meet me, he seemed cheerful—showed me his book & asked for a better one—which I provided." According to the prison keeper, the man was behaving superbly. He had been reading religious books and practicing arithmetic—a positive change for someone who had previously threatened to murder a prison officer! Yet this man's story also exposed the limits of reformers' work as character detectives. Foulke assumed the man had been cheerfully adjusting to prison life, but the following August he hanged himself in his cell.[37]

In general, Foulke's notebooks disclose a basic inability to understand the priorities of incarcerated people, paired with a spirit of willful ignorance. Above all else, Foulke longed to believe that penitentiary reforms were having their desired effect. For this reason, he regularly asked the occupants of Eastern State Penitentiary if their experience had been beneficial. Sometimes, inmates brazenly rejected Foulke's idealized narratives. On other occasions, they humored him, telling him what he wanted to hear. With boyish enthusiasm, Foulke recorded those exchanges, assuring himself that the "separate system" had been splendidly effective and then exulting about how even its victims recognized its many benefits. Prisoner No. 1831 apparently said the Eastern State Penitentiary was more effective than the New York House of Refuge, Blackwell's Island, or Moyamensing: institutions where individuals were apparently more likely to cultivate "bad associations." Even so, the man criticized the Philadelphia system and pointed out that incarcerated people always found ways to communicate with each other, regardless of restrictions. Prisoner No. 1454 similarly discussed the detrimental "effect of our system on his mind," saying "it was like the effect of keeping a limb out of use—that it benumbed it." But Foulke pressed on, eager for confirmation that solitary confinement compared favorably with systems at other prisons. The prisoners undoubtedly knew what Foulke wanted to hear. Sometimes, they indulged him and assured him the Eastern State Penitentiary was the best system in the nation.[38]

Because Foulke was personally invested in the notion that prisoners were being truthful in their praise, he relied on physiognomy and phrenology to validate their candor. Seeing what he wanted to see, Foulke described one individual as "a young man of good head & countenance; active mind—very intelligent—frank & manly in his bearing." When speaking of another man, he wrote that he was "quiet of manner," exhibiting "sincerity of *expression* in his eye, & a freedom from cant & strong professions which won my confidence." Thrilled to have concrete examples of his own success, Foulke became dependent on the positive affirmations of the people he was supposed to be reforming. His (in)ability to see sincerity in the visages of incarcerated people only convinced him of his own effectiveness.

On other occasions, Foulke ignored evidence that solitary confinement was harming the mental health of prisoners, even when his own standards of physiognomical evidence seemed to suggest otherwise. Through his regular visits and his correspondence with Dr. Given, the prison physician, Foulke became aware that mental illness was troublingly prevalent within penitentiary walls. Although he was inclined to trust the doctors, he was also determined to examine the "physical aspect of *actual* cases" by visiting the afflicted individuals himself. By this point, the Eastern State Penitentiary was facing serious reproaches from medical professionals, popular writers, and reformers within and beyond the country. Critics argued that by isolating prisoners from human contact, prison administrators were subjecting them to a particularly cruel form of mental degradation—a regime of torture which destroyed inmates' sanity and sense of self. Charles Dickens, for instance, famously commented on the prevalence of mental illness among those subjected to solitary confinement:

> On the haggard face of every man among these prisoners, the same expression sat. I know not what to liken it to. It had something of that strained attention which we see upon the faces of the blind and deaf, mingled with a kind of horror, as though they had all been secretly terrified. In every little chamber that I entered, and at every grate through which I looked, I seemed to see the same appalling countenance. It lives in my memory, with the fascination of a remarkable picture. Parade before my eyes, a hundred men, with one among them newly released from this solitary suffering, and I would point him out.[39]

When critics attacked Eastern State Penitentiary's "separate system," they claimed it imposed cerebral agony upon prisoners, eventually wearing them down until they were incapable of living beyond the penitentiary walls. For Dickens, Philadelphia's inmates were men who had been "buried alive ... dead to everything but torturing anxieties and horrible despair."[40]

Foulke was aware of—and threatened by—these criticisms. He also was familiar with the medical literature on physiognomy, phrenology, and mental illness. In the early decades of the nineteenth century, American and European medical professionals regularly argued that mental illness could manifest itself in the countenance. Prison reformers followed these debates and sometimes incorporated medical physiognomy into their activism. The Prison Discipline Society, for instance, published a report in 1830 stating:

"With the insane, it is emphatically true, that the dark shadows of the mind are visibly projected upon the face." They also suggested that those who were recovering would exhibit signs of their newfound health through physical changes in the "outward aspect" of their countenances.[41]

In a strategic reformulation of physiognomic ideas, Foulke used facial analysis to assure himself the situation at the Eastern State Penitentiary was not nearly as bad as it seemed. After seeing a young African American prisoner, Foulke ventured an interpretation of the man's mental state: "His eyes had a dull, heavy look; he had a sprightly smile—His air was rather bashful," he recorded. According to the penitentiary's physician, the man was suffering from dementia. But the physiognomical evidence did not convince Foulke, and he relied on racial prejudices to bury his concerns. The man's appearance resembled "what one often sees in blacks of sound mind," he insisted. Still, he worried the man's time in prison might have worsened his mental condition, noting, "If the dull look has been occasioned by the imprisonment, it deserves notice." On another occasion, he inspected a young white prisoner's physiognomy for signs of mental malady. Disturbingly, the man showed symptoms of delusion, even though "Dr. Given says that when he came into prison, he exhibited no indication of weakness." Even Foulke had to admit that he "Looks feeble in body & there is an expression of face indicative of mentle [sic] feebleness." It seemed imprisonment itself had been the cause of his current mental state.[42]

Part of the reason Foulke made these visits in the first place was because Dr. Given had asked him to do so. The physician worried the penitentiary's regime of solitary confinement was negatively affecting prisoners' minds. Foulke was not so sure. Although he was open to small reforms, he was unwilling to give up on his commitment to the "separate system." Still, Foulke continued meeting with the prison's physicians, and after speaking with another doctor, he acknowledged the signs of mental illness were, indeed, "alarming." He nonetheless insisted the physicians remained supportive of solitary confinement. In such instances, Foulke took the advice of medical professionals, but he also used facial and cranial analysis to draw his own conclusions about the brains and bodies of incarcerated people.[43]

Like Foulke, other Philadelphia reformers cautiously used popular sciences to combat criticisms of solitary confinement. In 1848, the *Pennsylvania Journal of Prison Discipline and Philanthropy* published a discussion of how the "separate system" was affecting the minds and bodies of incarcerated people. The periodical admitted there were certain "peculiarities" in the countenances of "men who have been kept in comparative seclusion," but the authors did not agree that solitary confinement destroyed people's mental faculties. In a similar way, Foulke confessed that he discerned mental maladies in the visages of certain inmates. But he preferred to posit other reasons for these disquieting psychological symptoms. He believed, for instance, that most of the problems had been caused by the prisoners' supposedly unnatural attachment to masturbation, rather than their enforced isolation from human contact. Despite relying on facial analysis, he used it to confirm "facts" he already believed to be true.[44]

In the end, reformers such as Jared Curtis, Eliza Farnham, Thomas Larcombe, and William Parker Foulke used physiognomy and phrenology as tools to mediate their relationships with prisoners and support their preferred reform agendas. By scrutinizing the facial and cranial features of incarcerated individuals, reformers tried to accomplish a few goals: ascertain if inmates were reformable, differentiate the liars from the truth tellers, and detect those who might have been born with innate criminal propensities. Their scientific conclusions, however, were always shaped by preconceived ideas. These reformers all believed, for instance, that individuals were redeemable. Otherwise, why waste so much time meeting with them privately, bringing them books, and tracking their progress? Every now and then, they identified a promising individual with a "good head and countenance." Yet reformers were also eager to distinguish between themselves and the incarcerated people they intended to remake in their own image. In instances where their efforts proved fruitless, physiognomy and phrenology provided convenient justifications. Certain prisoners, they hinted, might not be capable of reform, particularly if they had the "strong Phrenological manifestations of a thief" or signs of "Dishonesty and low trickery impress'd" upon their visages. By insisting they could unmask prisoners' duplicity, reformers both emphasized and exaggerated their own power. Through the empirical power of science, they assured themselves they would be able to separate the "good" prisoners from the "bad"—a task that was far easier in theory than in practice.

In 1873, a local New York magazine told a happy tale of a reformed female felon. Elizabeth D. had once been an evildoer, but she attained grace, escaping "a life of infamy and a drunkard's grave." Her face had apparently improved alongside her moral character. According to the reformer who encountered her later in life, "nothing in her queenly air, fine features, and intellectual countenance, would have suggested the life she led. Truly all things are possible with God." By the late nineteenth century, optimistic narratives like this one would become more and more infrequent, as both Americans and their European counterparts increasingly conceived of criminality as an ineradicable identity that resulted from innate and intrinsic deficiencies. In the first half of the nineteenth century, however, many Americans embraced a more flexible version of physiognomic determinism—one that aligned with their complex and contradictory ideas about crime and personal responsibility.[45]

Antebellum reformers were committed to the idea that all individuals were perfectible beings, believing that drastic social change was not only possible but desirable. This mindset served as the intellectual foundation for abolitionism, the growth of temperance crusades, the rise of utopian communities, and the various outpourings of religious enthusiasm that characterized the Second Great Awakening. It was this same attitude that undergirded penitentiary reforms and inspired the establishment of mental asylums, houses of refuge, charitable organizations, and moral reform societies.

With the aid of popular sciences such as physiognomy and phrenology, reformers had been able to argue that *some* people were hardened, recidivistic criminals, while *others* were penitent and reformable wretches. This duality required an understanding of human nature that was both deterministic and flexible. On the one hand, elite and middle-class white Americans wanted to distance themselves from the prisoners, poor people, and sex workers who served as the targets of their reform efforts. On the other hand, the entire premise of their political agenda rested on the notion that all humans were capable of reform. Middle-class activists wanted to believe that the people they sought to save were capable of inner transformation. But they also needed an intellectual philosophy that accounted for those who seemed incapable of reform. Physiognomy and phrenology provided both.

Reformers regularly argued that vice imprinted itself on the human form. At times, they even suggested that particular individuals were born with an innate predisposition to commit crime. But they rarely argued that vice itself was inevitable. Instead, they suggested that if everyone took proper precautions—by carefully cultivating virtuous traits over vicious ones—then it was possible to avoid a calamitous descent into degeneracy. Even after an individual had transgressed social norms, there might still be hope for redemption. Chaplains and moral reformers thus encouraged prisoners to read their Bibles, strive for personal penitence, and follow the instructions of their self-imagined saviors. Sometimes, they went so far as to argue that positive moral changes would imprint themselves on the countenance, beautifying the face as the mind improved. In doing so, they forged a middle path between environmentalist understandings of human nature and the starker forms of biological determinism that would triumph in the latter half of the nineteenth century. This was a clever—if convenient—adaptation of popular sciences, demonstrating the intellectual flexibility and creativity that these disciplines generated. As chapter 6 will demonstrate, prison reformers were not the only people who managed to imaginatively interpret—and sometimes purposefully misinterpret—physiognomic and phrenological doctrines. Confronting the rise of scientific racism in the antebellum decades, African Americans, too, turned to popular sciences. In their hands, physiognomy and phrenology became weapons against the white scientists who questioned their humanity.

6

FACING RACE

In 1849, Frederick Douglass published a scathing critique of white painters in the *North Star*: "Negroes can never have impartial portraits, at the hands of white artists," he wrote. "It seems to us next to impossible for white men to take likenesses of black men, without most grossly exaggerating their distinctive features." For Douglass, the reason for this impartiality was obvious. White observers declared their incapacity to distinguish between the faces of Black individuals, proclaiming instead that they all looked alike. Ignoring the great "variety of form and feature" among Black Americans, even the most sympathetic white artists imposed a preconceived understanding of Black features upon people of color. By invariably sketching them with "high cheek bones, distended nostril, depressed nose, thick lips, and retreating foreheads," white portraitists crafted visual caricatures of African Americans that emphasized their supposed "ignorance, degradation, and imbecility."[1]

Scholars have cited this quotation repeatedly, using it to explain why Douglass was so invested in sitting for his own photograph. Only photographs could be true-to-nature portrayals, Douglass often suggested, for only photographs could portray African Americans as they truly existed, rather than how they looked in the minds of prejudicial painters. Yet most scholars do not interrogate one of the primary reasons Douglass thought pictures were so consequential in the first place: like most nineteenth-century Americans, he interpreted portraits using the popular sciences of physiognomy and phrenology. Like many of his contemporaries, Douglass believed it might be possible to detect people's character through a reflective process of rational discovery. When he complained about the artists and naturalists who distorted Black features, he was invoking a broader transatlantic discourse and revising a scientific tradition that white thinkers had been using to denigrate Black minds and bodies since the late eighteenth century. For decades, white Americans had used visual and scientific depictions of Black features as evidence that proved the alleged reality of Black degradation. Even when they rejected physiognomy and phrenology as "hard" sciences, white people used them to rationalize and defend the political disenfranchisement of people of color.[2]

Physiognomy and phrenology were malleable discourses, though, and this allowed both white and Black Americans to interpret them in unique and oppositional ways. Although white artists and scientists relied on physiognomic evidence to argue that African Americans were mentally and physically inferior beings, Black Americans co-opted the very discourses that had been central to the rise of scientific racism, crafting an alternative method of corporeal analysis to argue for equality. When wielded by Black hands, physiognomy and phrenology did not solidify white supremacy; they vindicated the mental capacities of African Americans.[3]

Of course, Black thinkers recognized that they were dealing with problematic discourses. If they embraced physiognomy and phrenology wholeheartedly, then they risked validating the racist conclusions of the white intellectuals who questioned their basic worth. But physiognomy and phrenology were ideologically flexible and potentially subversive disciplines—something that distinguished them from rigid theories like polygenesis, as well as from new sciences like craniometry. By the mid-nineteenth century, advocates of polygenesis were claiming that Black and white Americans belonged to different species. Meanwhile, skull collectors like Samuel Morton were measuring the cranial capacity of various races and producing "evidence" that the brains of Black people were smaller than those of their white counterparts. On its surface, Morton's method was quantitative, empirical, and systematic. He argued that bigger skulls held bigger brains, and that bigger brains signified superior minds. According to his "data," people of European descent simply possessed the largest craniums. Historians have since exposed the biases and faulty assumptions of Morton's experiments. At the time, though, Black Americans who sought to challenge this new permutation of scientific racism found themselves in a bind. Unlike white ethnologists, they did not possess extensive skull collections or well-funded laboratories. In fact, most of the nation's colleges and universities refused to admit Black scientists. Within this context, physiognomy and phrenology emerged as alternative, more accessible, and more ethical ways of thinking about human character and capacity.[4]

Craniometry rested on a simple presumption: the size of one's skull indicated the power of one's brain. Physiognomy and phrenology were less rigid. Physiognomists came to broad conclusions about the relative merits of individual faces, but the rules were squishy and modifiable. Phrenologists likewise advocated for a flexible and nuanced interpretation of the human mind.

They argued that every person's brain was unique, but also insisted that all brains were more similar than different. To be sure, white physiognomists and phrenologists invoked science to rationalize racism—sometimes subconsciously, and often deliberately. At the very least, though, popular scientists in the United States generally opposed slavery and believed in the common humanity of all people. They also maintained that all human beings could improve their minds and bodies, and they analyzed character in creative (if notoriously problematic) ways. Such realities made physiognomy and phrenology more appealing to Black Americans, who confronted an increasingly powerful cadre of white ethnologists who questioned their humanity. Black activists must have recognized that physiognomy and phrenology were imperfect. How could they not? Still, they held out hope that science might verify the reality of human equality. Some enthusiastically embraced physiognomy and phrenology. Others surely approached these sciences with more skepticism. Ultimately, though, regardless of whether they believed in the legitimacy of physiognomy and phrenology, Black intellectuals were willing to brandish these sciences as weapons in the fight for racial justice.[5]

It was free men of color in the antebellum North who became the most outspoken challengers of white ethnologists. Because of their social and economic status, these men sometimes advocated for the principles of "respectability" and "elevation," which alienated many of the laboring Black Americans they sought to include under their political umbrella. Moreover, when they embraced physiognomy and phrenology, Black intellectuals essentially conceded that the body could be scientifically scrutinized for signs of internal capacity. In the process, they sometimes reinforced class distinctions among Black people, buttressed harmful European beauty standards, or indirectly solidified a belief system that was tainted by the racist conjectures of white thinkers. Physiognomy and phrenology nevertheless gave Black thinkers a unique opportunity to address and challenge scientific racism, because these disciplines had provided an intellectual foundation for its development in the first place.[6]

By scientifically analyzing heads and faces, Black Americans both crafted and responded to the broader intellectual universe they inhabited. They were not passive imbibers of white bourgeois ideology, nor were they heroic radicals who were insulated from the hegemonic culture in which they existed. They were active co-fabricators and refashioners of predominant discourses—people who constructed unique ideas about science and race and then employed those ideas for their own political purposes. Black thinkers may have marshaled the master's tools in their efforts to dismantle the master's house, but they also crafted their own discursive strategies and cleverly refashioned the rhetoric that was available to them.[7] By emphasizing Black individuals who had already attained both mental and facial eminence, they challenged the claim that African Americans were an irremediably degraded class. And by speaking the same language as the white scientists and popular writers who challenged their humanity, people of color subverted theories of biological determinism from within. In doing so, they took an old—yet resilient—scientific tradition, wrested it away from white practitioners, and instead used it to advocate for the social, political, and economic advancement of Black Americans.

RACE, SCIENCE, AND VISUAL CULTURE

White physiognomists and phrenologists had long suggested that African Americans were inferior beings, couching their racist pronouncements in the rhetoric of scientific objectivity. They also relied on visual caricature to denigrate the moral and mental capacities of people of color. Between 1828 and 1830, the Philadelphia artist Edwards Williams Clay drew numerous engravings of Black Americans in his "Life in Philadelphia" series. Most of these images mocked free people of color by portraying them as uppity and self-centered dandies, attempting to rise above their station. These caricatures envisaged African Americans with distorted facial features, disproportioned bodies, and outlandish outfits. Clay's images were also sometimes accompanied by imagined conversations between mythical archetypes of Black Americans, littered with misspelled words and written in dialects that purportedly matched the speech patterns of Northern Blacks. As Jasmine Nichole Cobb has argued, "Clay's caricatures taught White viewers that free African Americans maintained unlearned and insurmountable racial deficiencies that would permanently bar them from national belonging." For white viewers, the joke was supposed to be clear: people of color were unaware of their inability to attain the "true" refinement that theoretically came naturally to whites.[8]

In most of these images, Clay marked his subjects with facial traits intended to signal their alleged inferiority. Physiognomists and phrenologists largely agreed that broad and high foreheads reflected intellectual greatness. If the lower regions in a person's face were more prominent than the upper regions, this meant that a person's animal traits had triumphed over their intellectual characteristics. Clay's drawings reified these racist assumptions, using physiognomic markers to make odious implications about Black beauty. In the cartoons, the white figures mostly have high, straight foreheads and small mouths. The Black figures, by contrast, have retreating foreheads and exaggerated mouths and lips. These were deliberate choices. By manipulating scientific ideas to visually caricature people of color, he conveyed a sinister message: Black Americans—no matter how refined they believed themselves to be—were individuals whose animalistic propensities predominated over their moral and mental faculties. These images, of course, were not accurate depictions of real Black faces. Nor was physiognomy ever an objective system for measuring moral character or intellectual cultivation. Instead, Clay's images trafficked in popular stereotypes, strategically using visual culture to reinforce a racist understanding of human nature.

Black intellectuals readily recognized such images as visual mockery, but they also viewed them as insidious statements about the inner capacities of an entire race. In his lecture on ethnology, Frederick Douglass claimed he had "never seen a single picture in an American work, designed to give an idea of the mental endowments of the negro." When artists and naturalists portrayed the "European face," he argued, they drew it "in harmony with the highest ideas of beauty, dignity, and intellect." By contrast, they drew Black Americans "with the features distorted, lips exaggerated, forehead depressed—and the whole expression of the countenance made to harmonize with the popular idea of negro imbecility." For this reason, Douglass sought to remedy the political reality of racial injustice through the power of pictures. He demanded that artists start depicting the faces of eminent Black intellectuals such as Henry Highland Garnet, William J. Wilson, and Martin Delany. Their heads, he contended, "indicate the presence of intellect more than any pictures I have seen" in American ethnological works.[9] As part of his political strategy, Douglass sat for dozens of daguerreotypes himself, ultimately becoming the most photographed American of the nineteenth century (for one example, see fig. 6.1). By carefully curating his own public image, he created a visual archive to refute racist imagery and serve as physiognomic proof of his own mental eminence.

Scholars of race and visual culture have often argued that Black Americans embraced visual culture to present a vision of human nature that subverted the racism of white artists. But they have less clearly answered a more fundamental question: why did Black activists believe that photographs visually conveyed human interiority to begin with? Answering that question requires historians to grapple with the cultural salience of popular sciences in antebellum America. During this period, images were meaningful because all Americans—regardless of race—inhabited a common cultural universe where countenances and craniums counted in the debate over Black capacity. When viewed in this context, Frederick Douglass's obsession with

FIGURE 6.1. Unknown maker, American. *Profile Portrait of Frederick Douglass*, ca. 1858. Daguerreotype, sixth-plate, image: 2 ½ × 2 inches (6.4 × 5.1 cm). The Nelson Atkins Museum of Art, Kansas City, Missouri. Gift of Hallmark Cards, Inc., 2005.27.42. © Nelson Gallery Foundation. Photo: Joshua Ferdinand. By sitting for portraits, Douglass consciously conveyed a different vision of Black physiognomy than prints that might be found in series like Edward Williams Clay's *Life in Philadelphia*, or in American and European ethnological works.

photography is far more than a fascination with a new form of visual representation. At a time when Americans increasingly saw intellectual capacity as one of the primary prerequisites for citizenship, Black thinkers saw visual depictions of facial features as scientific evidence with broader import in the fight for racial justice.[10]

BLACK THINKERS AS SCIENTIFIC PRACTITIONERS

....

Particularly between the late 1830s and the 1850s, African Americans confronted a veritable onslaught of scientific and political justifications for white supremacy. In the preceding decades, gradual emancipation laws had initiated the slow demise of slavery, and in doing so, had begun to dissolve the most significant barrier between enslaved Blacks and their free counterparts. This allowed white Americans to group all African Americans together as members of a uniformly degraded class, regardless of social or economic status. Black people also confronted a "rabid colonizationist mobilization" during this period that questioned their status as true Americans and sought to "return" them to their supposed African homeland. As the growth of the market economy accelerated and as a massive wave of poor European immigrants poured into northern cities, native-born white laborers grew increasingly resentful of Black workers. By the 1830s and 1840s, racially motivated mob violence had intensified in northern cities, just as the abolitionist movement was increasing in power and visibility. Particularly after a revitalized Fugitive Slave Act passed in 1850, sectional tensions reached a fever pitch, and the nation careened ever closer to civil war. Free people of color experienced these ominous developments with alarm, and they responded with a reenergized commitment to racial justice.[11]

Black writers often felt compelled to repudiate white ethnological theories, even though they bitterly resented this duty. As they did so, they turned to physiognomy and phrenology. In her article on the connections between phrenology and the abolitionist movement, Cynthia Hamilton argues that phrenology proved attractive for anti-slavery activists because of its flexible methodology. By analyzing the science's doctrinal inconsistencies, she explains how it could simultaneously appeal to both abolitionists and white supremacists. Yet Hamilton focuses primarily on white intellectuals. How were Black thinkers using these disciplines? And why did physiognomy and phrenology—sciences that white people regularly used to denigrate Black minds and bodies—develop into discourses that people of color used for their own purposes? As Hamilton points out, phrenology appealed to abolitionists because of its "mixed messages." Allowing people to believe that character was both "malleable and fixed," popular sciences gave Americans hope that their minds and bodies were improvable.[12]

Facial analysis was particularly appealing for Black writers because it proved more accessible than the "hard" ethnological sciences that solidified between the 1830s and 1850s. Samuel Morton's method of craniometrical analysis required an expensive and colossal collection of human skulls. To practice physiognomy or phrenology, all one needed was a discerning eye and a steady supply of heads. These disciplines did not require university training (which Black Americans did not have access to in the United States). They also allowed practitioners to agree on a general premise—that heads and faces revealed internal character—without having to agree on a set of specific, universal, or unchanging rules. Facial and cranial analysis ultimately depended on the individualized perceptions of the person doing the observing. For this reason, Black physiognomic observers had considerable latitude in their efforts to interpret visages.[13]

These disciplines also appealed to Black thinkers because they were ubiquitous within the visual, scientific, and literary cultures of antebellum America. If craniometry happened in the laboratory, popular sciences happened "in print, on stage, in the garden, church, parlor, and in other cultural spaces and productions."[14] Discussion about these disciplines was also common in the anti-slavery press. While *The Colored American* tried to expose the logical "fallacies" of phrenology, *The Liberator* regularly defended phrenology and advertised phrenological lectures.[15] Physiognomic sketches likewise appeared in abolitionist fiction by both white and Black authors. In *Uncle Tom's Cabin* (1852), Harriet Beecher Stowe provided physiognomic character sketches of both major and minor characters. When describing Uncle Tom, she pointed to his "truly African features," which were "characterized by an expression of grave and steady good sense." On another occasion, she focused on Lucy's face, which supposedly conveyed a "wild, painful and romantic history." Stowe argued that Lucy's "forehead was high, and her eyebrows marked with beautiful clearness." She pointed to her "straight, well-formed nose, her finely-cut mouth, and the graceful contour of her head and neck," writing that Lucy had a "fierce pride and defiance in every line of her face, in every curve of the flexible lip, in every motion of her body." By describing Lucy's forehead as "high," Stowe emphasized her impressive mental capacities. She also argued for her physical attractiveness, challenging those Americans who argued that Black women were incapable of achieving true beauty.[16]

Black novelists, too, used physiognomy to describe characters of color positively. In *The Garies and Their Friends* (1857), one of the earliest novels written by an African American, author Frank Webb describes Mr. Walters as a man who was "exceedingly well-proportioned; of jet-black complexion, and smooth glossy skin." In his description, Webb focused on the "attractiveness of his appearance," as well as his "broad but not very high forehead." For most physiognomists, broad foreheads indicated breadth of judgment, as well as the capacity to grapple with complicated material or engage in sustained intellectual deliberation. In emphasizing this trait, Webb defended Mr. Walters's mind. He also described Mr. Walters as having an "aquiline nose, thin lips, and broad chin." These were traits that white physiognomists typically associated only with white countenances. Physiognomists usually emphasized aquiline noses when they wanted to invoke the masculine bravery and boldness of the ancient Romans. Finally, Webb depicted Mr. Walters with small eyes, "black, and piercing, and set deep in his head." In doing so, he subtly challenged white scientific thinkers who dismissed African Americans as individuals with dull eyes and projecting eye sockets. Trying to answer a question that "slavery has raised in many thoughtful minds," the book's preface asked its readers, "Are the race at present held as slaves capable of freedom, self-government, and progress?" Through its descriptions of the minds, heads, and faces of Black people, Webb's novel answered this query in the affirmative.[17]

Black writers also employed physiognomic ideas when describing people of color who struggled to maintain their dignity in a white supremacist culture. The abolitionist William Still, for instance, used physiognomy to argue that enslaved people were capable of freedom and self-government. As he recorded encounters with fugitives on the Underground Railroad, he explained how refreshing it was "to observe in every countenance, determination, rare manly and womanly bearing, with remarkable intelligence." When recounting the stories of former slaves, Still highlighted individuals whose faces revealed their "ardent thirst for liberty." If someone had a particularly "intelligent countenance," he was sure to note it, similarly spotlighting those who had "marked intellectual features," a "large and high forehead, indicative of intellect," or "a countenance indicative of intelligence and spirit." In one instance, Still turned his focus to a formerly enslaved young woman named Hannah, who wore "a countenance that indicated that liberty was what she wanted and was contending for, and that she could not willingly submit to the yoke." Through such descriptions, Still used facial analysis as part of a larger assault on a slave system that tried—and failed—to turn autonomous beings into property.[18]

William Wells Brown similarly relied on popular sciences when arguing for the genius, history, and achievements of Black people in America. When describing the lawyer John Mercer Langston, for example, he wrote that Langston was an eloquent public speaker with a "high and well-formed forehead, eyes full, but not prominent, mild and amiable countenance, modest deportment, strong, musical voice, and . . . the air of a gentleman." Similarly, when recounting the mental and physical character of Charlotte Forten,

Brown described her as a skilled intellectual who "possesses genius of a high order," a trait visible in her "finely-chiselled features, well-developed forehead, countenance beaming with intelligence, and her dark complexion." Brown repeatedly argued that an intelligent mind could shine through any complexion, so long as a person exhibited the skeletal features that indicated refinement. If he did not believe that one's skin color reflected capacity, he nevertheless contended that people's faces were reliable indicators of their inner nature. As he recognized, white ethnologists, political thinkers, and popular writers were using physiognomic descriptions—alongside visual culture—to denigrate the mental and moral capacities of people of color. He responded by co-opting this language and using it to positively describe the bodies of Black women and men.[19]

Scholars are used to thinking of physiognomy and phrenology as the forerunners to scientific racism and biological essentialism. And it's true. These disciplines were, indeed, dangerous in the hands of white supremacists. But Black thinkers saw value in them, too. Recognizing the inherent instability and lack of systematic methodology of these sciences, Black intellectuals deployed them for radical purposes. Dr. J. J. Gould Bias, who had been born into slavery, combined his medical training at the Eclectic Medical College of Philadelphia with his knowledge of popular sciences to draft a treatise titled *Synopsis of Phrenology, and the Phrenological Developments* (1859). Although the book itself has not survived, Martin Delany lauded the work in *The Condition, Elevation, Emigration, and Destiny of the Colored People of the United States* (1852), describing Bias as a "practical phrenologist" and praising his "originality" as well as his "self-will, and determination of character." Delany also expressed admiration for the Black physician Lewis G. Wells, "a most successful practical phrenologist" who "lectured to large and fashionable houses of the first class ladies and gentlemen of Baltimore, and other cities."[20]

Black newspapers also advertised the work of Black phrenologists. In 1848, Frederick Douglass's *North Star* recounted the story of a "little genius" named Simon Foreman Laundrey, the son of a "poor man" who had trained himself in phrenology, anatomy, and dentistry. Laundrey was a "natural Phrenologist," the newspaper stated, "who examines heads, reads out the organs, and delivers lectures on the science." The author knew the fourteen-year-old boy had not yet fully mastered the discipline, but said he was just as promising as "competent professors of the science of phrenology," and that "if care be taken to give him an education, some day he will be at the head of his profession." Perhaps he could even embark on a national lecture tour and use his earnings to attend Oberlin, the author declared. In a similar way, the *North Star* advertised the "scientific lectures" of Dr. Henry Lewis, a Black man who enlightened audiences with "mental feasts" consisting of "phrenology, mesmerism, and other interesting branches of science." Then, in 1855, *Frederick Douglass' Paper* advertised the lectures of Mr. W. F. Johnson, a blind African American phrenologist and abolitionist. As these examples demonstrate, elite thinkers and ordinary Black people alike embraced popular sciences, using them to demonstrate their intellectual refinement.[21]

Some white authors were clearly threatened by Black phrenological knowledge. Attempting simultaneously to undermine phrenology and mock Black scientists, they printed cartoons portraying African Americans as bumbling quacks and as sexual threats to white womanhood. One almanac, for instance, parodied "Black Bumpology." This was both a critique of popular science and a racist indictment of Black practitioners. As Britt Rusert has contended, such cultural productions betrayed "a cultural anxiety that phrenology might be, or become, a black science." In a similar way, another political cartoon showed "Professor Pompey," a Black man, "Magnetizing an Abolition Lady." In the image, a Black advocate of "Mesmerism" practically mounts a white woman as other Black men conspire to seduce the female activists nearby. These cartoons were racist mockeries of Black Americans, but they betrayed a deeper anxiety that popular sciences, by their very nature, could not be limited to white practitioners.[22]

Despite the cultural criticism they engendered, Black thinkers continued to use popular sciences in their struggle for racial equality. Even as white artists portrayed them as intellectually deficient caricatures and bumbling practical scientists, Black intellectuals publicly discussed popular sciences and sat for portraits in an attempt to convey the power of Black minds. They were especially intent on highlighting African Americans' impressive foreheads, knowing that physiognomists, phrenologists, and craniologists mostly agreed that foreheads revealed one's mental capacity. For some Black thinkers, though, it was not enough to highlight Black people with "high brows" and "intelligent countenances." These individuals devised more systematic physiognomic paradigms to advocate for racial justice.

THE MUTABILITY OF MINDS AND FACES

In 1837, the abolitionist clergyman Hosea Easton became the first Black thinker to craft a comprehensive theory explaining the historical and scientific significance of race within the United States. After he and his parishioners faced the violence of white mobs on at least three occasions between

1834 and 1836, he published *A Treatise on the Intellectual Character, and Civil and Political Condition of the Colored People of the U. States* (1837).[23] For years, Easton had extolled the value of racial uplift, but in his 1837 *Treatise*, he divulged a sense of frustration and rage. No matter how "respectable" Black people had proven themselves to be, it seemed as if prejudice was ineradicable. Seeing white supremacism as "an almost insurmountable barrier," he decided to challenge it with "a direct intellectual assault." Although he had previously called on African Americans to "uplift" themselves, Easton penned his 1837 *Treatise* as a rebuke to white people for how thoroughly they had managed to degrade the minds and bodies of Black Americans.[24]

Like the white scientist Samuel Stanhope Smith, Easton viewed the body and mind as mutable. Both thinkers argued that slavery altered the brains and physiognomies of those in bondage. But while Smith had blamed African nations for their supposed mental and physical "degradation," Easton modified this interpretation, placing the blame entirely upon whites. To show "the lineal effects of slavery on its victims," Easton highlighted the "Contracted and sloped foreheads" of enslaved individuals, as well as their "prominent eyeballs; projecting under jaw; certain distended muscles about the mouth, or lower parts of the face; thick lips and flat nose." For Easton, these were physical traits that resulted from environmental forces unleashed by the greed and moral deprivation of white people, not from innate African inferiority.[25]

Of course, science is not and has never been a neutral technology for interpreting human intelligence and appearance. There is no evidence to support the notion that characteristics such as a "contracting and sloped forehead," "projecting under jaw," "thick lips," or "flat nose" reflect mental inferiority. Nor are these traits accurate descriptors of all Black people's appearances. These ideas, after all, were the products of a racist physiognomic discourse, which white scientists began propagating with gusto in the late eighteenth century. At the same time, it is important to remember that Hosea Easton was living and operating within the intellectual milieu of the 1830s. During this period, many thinkers—both Black and white—believed physiognomic features revealed important clues about the human mind. While Easton brazenly challenged certain aspects of this worldview, he also internalized other parts of it.

When engaging in public activism during the 1830s and 1840s, Easton and other Black intellectuals were often torn between two objectives. On the one hand, they wanted to demonstrate their own refinement and capacity for republican citizenship. This meant distinguishing themselves from the enslaved and from the Black working classes. On the other hand, they wanted to advocate for racial justice more broadly. This meant fighting against slavery and arguing for the social, political, and economic advancement of working-class African Americans. To reconcile these competing objectives, middle-class Blacks often argued for the "moral uplift" of their race, practicing a form of respectability politics that sometimes aligned more closely with the goals of white abolitionists than with those of working-class African Americans. As Leslie Harris has argued, Black reformers understood working-class Blacks better than white abolitionists did. "But they, too, viewed the mass of Blacks as inferior to whites, and perhaps to themselves, and believed that Blacks needed preparation and education for citizenship." Although they sought to emphasize unity among people of color, they also attempted to "reform," "uplift," and "educate" recently freed and working-class Blacks.[26]

This class- and status-based understanding of Black capacity translated into scientific descriptions. When Frederick Douglass complained that white artists did not focus on the refined physiognomies of the Black community, he focused on other middle-class intellectuals—men such as Alexander Crummel, Henry Highland Garnet, Charles Remond, James Pennington, and Martin Delany.[27] Similarly, when William Wells Brown compiled a volume on "the Black man, his genius, and his achievements," he focused much of his attention on individuals like Benjamin Banneker (a skilled astronomical observer who corresponded with Thomas Jefferson), Ira Aldridge (an internationally renowned actor), and James McCune Smith (a licensed medical doctor).[28] In a periodical series that provided sketches of both real and fictional people of color, the author William J. Wilson focused on the "finely formed head and ample brow" of the Reverend Peter Williams (cofounder of *Freedom's Journal*) and on the superior countenance of the Haitian leader Toussaint L'Ouverture. By contrast, he described a fictional group of enslaved people during a white minister's sermon, writing that they listened with "eyes dilated, mouths agape, nostrils distended and ears alert." Yet Wilson described other enslaved people—at the same fictional church service—as having more impressive visages: "These faces, in contrast with the others of the congregation, give a most striking effect to the picture," he wrote. Identifying signs of active intelligence and defiance in the faces of some enslaved people led Wilson to further conclusions about their inner selves. Admiring what he presented as their irreverence to the white minister's teachings and their resistance to the degradations of enslavement, Wilson suggested that militant enslaved people were more honorable and intelligent than their more passive counterparts. Their defiant spirit, Wilson claimed, was visible in their faces.[29]

Black intellectuals were invested in the project of racial justice, but they were also intent on proving their capacity for republic citizenship. As a result, they sometimes tried to distance themselves from the Black Americans who they saw as insufficiently refined. Easton's discussions of enslaved people's physiognomies should be viewed in this intellectual context. When grappling with racial difference, he contended that some Black countenances might, in fact, be unattractive. But he also insisted this disparity could be eradicated with the elimination of slavery and the moral and mental cultivation of all African Americans. For Easton, appearances were significant primarily because they reflected the workings of the mind. Throughout the *Treatise*, he repeated a single refrain: "Mind acts on matter." Building on the physiognomic premise that internal dispositions acted upon external features, he described the human body as a malleable entity, capable of change. He also believed that faces and bodies could change as the brain developed. He tied this claim directly to abolitionism by arguing that slavery imprisoned the minds of Black Americans and disfigured their appearances. If enslaved individuals attained liberty, he contended, their physiognomies would transform for the better:

> The countenance which has been cast down, hitherto, would brighten up with joy. Their narrow foreheads, which have hitherto been contracted for the want of mental exercise, would begin to broaden. Their eye balls, hitherto strained out to prominence by a frenzy excited by the flourish of the whip, would fall back under a thick foliage of curly eyebrows, indicative of deep penetrating thought.

Easton claimed, in other words, that when those in bondage became free, their foreheads would broaden, their eyes would recede, and their brows would begin to reflect "deep penetrating thought."[30] Of course, Easton's claims themselves reified white physiognomic theories, which suggested that broad and high foreheads were the symbols of intellectual excellence. If not purposefully, he also lent credence to white beauty standards, which maintained that Black people's eyes projected further than those of white people.

At the same time, by arguing that liberty transformed the mind as well as the body, Hosea Easton argued for emancipation while challenging the doctrines of pro-slavery physicians. During the antebellum era, slaveholders and physicians alike argued that emancipation was impossible because freedom would vitiate the minds and bodies of African Americans, causing both mental and physical disabilities. Both the white supremacist politician John C. Calhoun and the *New York Journal of Medicine*, for instance, claimed that deafness, blindness, and insanity afflicted free people of color more often than enslaved people. Easton, by contrast, argued that freedom would lead to mental and physiological improvement. Like many white scientists, Easton employed a form of positive physiognomic environmentalism that was premised on two notions: 1) that the body reflected the mind, and 2) that bodies could change as the mind improved. He wedded these ideas with political activism by working to secure voting rights for Northern Blacks and attending the early conventions of the American Society of Free Persons of Colour and the National Colored Convention. Though he died at the early age of thirty-five, he left an important legacy as the first African American to methodically theorize racial difference in a book-length work.[31]

As one of the only Black Americans to train as a professional medical doctor in the nineteenth century, James McCune Smith likewise challenged white scientists by claiming that climate, culture, and geographical position could transform both the mental and the physical aspects of mankind. In a critique of Jefferson's *Notes on the State of Virginia*, he argued that improvements in African Americans' mental faculties exerted osteological changes in their skulls and faces. Because African Americans were exposed to a more temperate climate than their African counterparts, Smith believed their appearances had already started changing. The slope of their foreheads had supposedly decreased, just as their jaws protruded less, their skin became lighter, and they became more attractive and intelligent. Like Easton, Smith highlighted the mutability of the Black body and demonstrated the importance of both environment and education on the human form.[32]

Smith ultimately had a complicated relationship with physiognomy, phrenology, and the "American School" of ethnology. Unlike the Black lecturers who embraced phrenology, he gave a public lecture on "the fallacy of Phrenology" in 1837, using a collection of skulls and "extemporaneous drawings" to challenge the legitimacy of this "so called" science. Intriguingly, though, the *Colored American* wrote that Smith's anti-phrenology position was "unpopular" with his audience. This did not stop Smith from mocking phrenologists through wry and satirical pieces on the "Heads of the Colored People" several years later. Despite his antipathy for phrenology, Smith did embrace comparative anatomy, which was rooted in many of the same physiognomic principles. He also seemed to accept the notion that higher facial angles signaled intellectual eminence.[33] Writing for the *Anglo-African Magazine* in 1859, he felt it necessary to prove that the "flat, retreating forehead" was not a physical trait defining every Black American. Within the "colored churches" of America, he argued, a careful observer "will find the low, retreating forehead to be the exception, and not the rule." In other words, Smith contended that observers would find a collection of high, intellectual foreheads in a congregation of "refined" Blacks, just like they might find in white churches. Even though Smith was conflicted about the scientific legitimacy of cranial analysis, it played an important role in his works. His ambivalence demonstrates just how central physiognomic ideas had been in shaping the debate over racial difference in America. Though he resented the need to engage with white ethnologists, Smith also knew he had to battle them using the language of the moment. Even so, his works demonstrate how the rhetoric of popular science functioned differently in pro-slavery and abolitionist thought.[34]

Even when African American writers agreed with white scientists in saying the face and skull reflected inner capacity, they rarely described heads and faces as permanent and heritable features, incapable of physical change. This allowed Black writers to analyze faces in creative ways. On the one hand, they regularly pointed to African Americans with impressive foreheads, chiseled jaws, and refined features. Yet they also sometimes argued that *certain* people of color were mentally and physically inferior beings. This was part of a larger process by which Black intellectuals sought to challenge white racism while simultaneously distinguishing themselves from the working poor and the enslaved. At the same time, they never wavered in their argument that minds could improve with cultivation, making the face more beautiful as the mental powers developed. Their strategy undermined the arguments of slavery's apologists, who often claimed that Black bodies were not suited for freedom and warned that the end of slavery would lead to the proliferation of mental insanity and physical degeneration among African Americans. Instead, they described Black bodies and minds as dynamic entities that were constantly transforming for the better.

THE LIBERATORY POTENTIAL OF FACIAL ANALYSIS

If faces or minds could change over time, then it stood to reason that there were African Americans who had already attained mental refinement, and thus exhibited the physical features that signified inner excellence. As another physiognomic strategy, Black writers highlighted the heads and faces of prominent members of their community to illustrate what was possible when the Black mind was nurtured and encouraged to develop. William J. Wilson, for instance, used physiognomy to assure Black readers that many African Americans had already achieved intellectual eminence, and that others were capable of further advancement. Wilson served as the headmaster of the African Free School in Brooklyn and regularly drafted pieces for *Frederick Douglass' Paper*, as well as the *Anglo-African Magazine*, usually writing under the pseudonyms "Ethiop" or the "Brooklyn Correspondent." Like Douglass, Wilson believed white artists often presented prejudicial images of Black faces. Yet Wilson was equally concerned with how the Black community interpreted the appearances of its own members. To remedy this problem, he drafted a submission to *Frederick Douglass' Paper* that reflected upon Philadelphia's Colored National Convention of 1855. Specifically, he focused on the appearances of two figures who spoke at the meeting: Isaiah C. Weare and Mary Ann Shadd.[35]

As a member of the Pennsylvania delegation at the Colored National Convention, Isaiah Weare was an intellectual giant, despite being "a young man, and very small." He was so slight in stature that those who observed him might have asked themselves: "how can a large mind be contained within so small a mould?" To answer this question, Wilson argued that Weare's intellectual eminence could be seen in his head, form, and features. He contended that Weare had fabulous "phrenological proportions" and that his "lower face, too, especially the under jaw, is wonderfully indicative of intellectual power." After spotlighting Weare's appearance, Wilson noted Mary Ann Shadd's "small and penetrating" eyes, "well moulded head," and "feminine" features. Melding the disciplines of physiognomy and phrenology, he suggested that the human exterior revealed the inner man or woman.[36]

Wilson's comments reveal a subversive form of physiognomic observation. By locating Isaiah Weare's "intellectual power" in his lower jaw, Wilson revised traditional physiognomic doctrines, articulated first by Lavater and adopted by physiognomists in both Europe and the United States. For physiognomists, it was the forehead and eyes that portrayed intellectual capacity. The jaw, by contrast, portrayed the baser instincts of humankind. By arguing that Wilson's intellect could be found in his jaw, Wilson not only undermined racist depictions of Black people's supposedly "prognathus" jaws; he also created his own physiognomic system, in which it was not merely the forehead and eyes that revealed intelligence, but also the "lower face." He also described Mary Ann Shadd as a beautiful feminine specimen with an expression of self-assurance that "white folks would [call] a very saucy look." As he saw it, physiognomic analysis was not merely a neutral and scientific process, but also a mode of aesthetic interpretation shaped by the biases and beliefs of the observer.[37]

Wilson knew that African Americans were intelligent and attractive, but he feared that many people of color preferred white facial features— mistakenly believing that only whites exhibited the physical signs of inner greatness. Because of this unconscious and internalized prejudice, he argued, some African Americans were incapable of recognizing specimens of genius within their own ranks. Fearing that people of color too often turned to white bodies for examples of physical and intellectual distinction, Wilson implored Black communities to turn inward when searching for faces of genius. If intellect could not be confined to white brains—and he believed it could not—then physical manifestations of intellect should be visible on Black features as well as white ones. By talking about the "beautiful" and "intellectual" countenances of Isaiah Weare and Mary Ann Shadd, he argued for the power of their minds.[38]

Wilson was so deeply invested in the liberatory potential of facial analysis that he dedicated an entire periodical series to describing fictional images of Black figures. Like Douglass, Wilson believed that white artists and writers could never fully represent the faces and minds of Black Americans. As a result, he argued that African Americans needed a gallery of artistic works where Black artists portrayed Black subjects. Such a gallery did not exist, so Wilson generated a fictional one in the pages of the *Anglo-African Magazine*, a periodical written by Black authors and specifically intended for Black readers. By touring through corridors that never materialized and analyzing artworks that did not exist, Wilson visualized a virtual art exhibit for his readers. He published these essays in serial form, over the course of nine months. In each installment, he guided readers through the gallery, encouraging them to meander and examine the pieces. Through the narrative voice of "Ethiop," the museum's fictitious curator, Wilson coached readers on how to interpret these mental images.[39]

This "picture gallery" never existed as a physical space, nor did Wilson draw the pictures he described. Instead, he provided his readers with "word paintings," in which he encouraged them to imagine a collection of artworks, even if they could not experience the gallery in person. As Ivy Wilson has argued, Wilson used the gallery to imagine "that which does not quite exist . . . as a way to present alternative ways of seeing for his black readers." Because the gallery was fictional, he could have included anything within its walls. The choices he made reveal just how deeply he believed in the power of facial analysis for African Americans' own perceptions of racial difference. Through a literary dramatization of an illusory art gallery, Wilson articulated artistic, scientific, and political messages that rejected the theories of Anglo-American ethnologists.[40]

Though Wilson used physiognomic descriptions throughout his entire public career, his meditation on Phillis Wheatley's "portrait" in the Afric-American Picture Gallery was one of his most detailed uses of this science. Wilson likely based his description on the only existing portrait of Wheatley: a profile engraving from the late eighteenth century that pictured her

FIGURE 6.2. "Phillis Wheatley, Negro servant to Mr. John Wheatley, of Boston," frontispiece for Phillis Wheatley, *Poems on Various Subjects, Religious and Moral* (London: Printed for Archibald Bell, Bookseller, 1773). The Library Company of Philadelphia.

in the act of poetic composition (fig. 6.2). Wilson had almost certainly seen Wheatley's portrait, and he likely imagined this image as he crafted a description of her countenance for the *Anglo-African Magazine*. Before Ethiop mentions anything about Wheatley's history, poetry, or achievements, he first gives readers a detailed description of her facial angle, forehead, and brain:

The facial angle contains a full ninety degrees; the forehead is finely formed, and the brain large; the nose is long, and the nostrils thin, while the eyes, though not large, are well set. To this may be added a small mouth, with lips prettily turned, and a chin—that perfection of beauty in the female face—delicately tapered from a throat and neck that are themselves perfection. The whole make-up of this face is an index of healthy intellectual powers, combined with an active temperament, over which has fallen a slight tinge of religious pensiveness. Thus hangs Phillis Wheatley before you in the Afric-American Picture Gallery.

To begin, Wilson's narrator focused on Wheatley's 90-degree facial angle. Here he referred to Petrus Camper's theory, which suggested that the Grecian face demonstrated facial angles between 95 and 100 degrees. Europeans were supposedly closest to this standard, purportedly with facial angles between 80 and 90 degrees, followed by other groups, such as "Moors," "Calmucks," and "Negroes," all with facial angles between 70 and 80 degrees.[41]

In the mid-nineteenth century, proponents of scientific racism would exaggerate Camper's theories, arguing that lower facial angles indicated animality, while facial angles between 90 and 100 degrees were the markers of human perfection. When placed in this context, Ethiop's insistence that Phillis Wheatley's portrait conveyed a 90-degree facial angle, an "intellectual" countenance, and a "finely formed" forehead becomes more significant. Using a fictional image of a Black woman, he argued that African Americans could display the physical features necessary for great mental accomplishment. In the process, Wilson invoked the same physiognomic theories that undergirded the rise of scientific racism, but he used them to argue for the eminence of Wheatley's brain.

In many ways, Wilson's use of the facial angle may have approached Camper's original intentions more closely than did the scientific racists who misinterpreted Camper's work. As Nell Painter has argued, Camper "insisted on the unity of mankind, even going so far as to suggest that Adam and Eve might well have been black, because no one skin color was superior to the others." By the middle decades of the nineteenth century, however, "scientific racists in Britain and the United States . . . went on reproducing his images as irrefutable proof of a white supremacy that Camper himself had never embraced." To be sure, Camper *did* arrange the skulls of white and Black individuals on a hierarchical continuum, with Black skulls positioned next to the skulls of apes. At the same time, he argued in favor of human variation and natural equality. When Wilson used the facial angle to describe Phillis Wheatley, he strategically invoked Camper's eighteenth-century vision of universal humanity—a theory that white ethnologists were distorting for their own purposes in the antebellum period.[42]

By the mid-nineteenth century, Wheatley's countenance became a battleground: one arena of a physiognomic conflict where ethnologists, abolitionists, and popular writers clashed in their interpretations of racial difference. For example, Robert Chambers, the famous British writer and anonymous author of the influential *Vestiges of the Natural History of Creation* (1844), provided extensive commentary on Wheatley's countenance and capacities. Citing the British ethnologists James Cowles Prichard and William Lawrence, he argued that intellectual "cultivation" was "capable of modifying even the shape of the skull." At the same time, he started from a belief in Black inferiority and argued for "a decided inferiority of cerebral organization in the Negro, attended with a corresponding inferiority of faculties." Chambers conceded that *some* Black individuals might exhibit great mental prowess and physical attractiveness, and cited Phillis Wheatley as his example. But he also suggested that the visages of the most intelligent African Americans had purportedly "white" features. Phillis Wheatley's portrait, he argued, demonstrated "not only a Caucasian brow and head, but those of the finest order." In Chambers's view, she was not a "typical" African at all. She was merely an exception that proved his racist rule.[43]

White authors often argued that Wheatley's face marked her as a unique individual, fundamentally unlike other African Americans. By contrast, both white and Black abolitionists claimed Wheatley's face displayed her intellectual eminence and hinted at the possibility for the mental refinement of the entire Black race. In *Narratives of Colored Americans* (1826), for instance, the white female abolitionist Abigail Mott suggested Wheatley's "countenance appears to have been pleasing, and her head highly intellectual." The British Quaker and abolitionist Wilson Armistead likewise used Wheatley to prove the moral and mental equality of the "coloured portion of mankind." This meant that when William J. Wilson "sketched" her portrait in 1859, he was building on a much larger tradition of Black and white activists who interpreted her portrait for political purposes. William Wells Brown then continued this tradition, copying the text of Wilson's description in *The Black Man, His Antecedents, His Genius, and His Achievements* (1863). Using Wheatley as an example, these authors argued that Blacks, too, could have high, "finely-formed" foreheads, "chiseled" features, and "beautiful" Roman or Grecian countenances. It was important for abolitionists to highlight African Americans who displayed the physiognomic features that supposedly signaled mental merit, because these were the criteria by which white scientists were judging the entire race.[44]

As anthropologists recognize, "race" is more a social construct than a biological reality. There is no such thing as a definitively "white" or "Black" countenance. Human variation is vast, and individuals of all backgrounds display a wide array of facial features. The messy reality of human diversity belies the existence of a racially standardized facial type. This is something that people of color recognized in the 1840s and 1850s, long before white intellectuals embraced the idea. Insisting people of color were distinct individuals, Black writers pointed out that all Black faces were unique. James McCune Smith, for instance, described race not as a biological reality, but rather as a way of seeing—a lens that shaped white viewers' perceptions of African Americans. White people did not see Black people for who they were, he complained. When they conceptualized a Black person, they did not imagine an "actual physical being of flesh and bones and blood." Instead, they relied on a stereotypical vision of Blackness, in which a person of color was not an individual but rather an exemplar of a type: "a hideous monster of the mind." Frederick Douglass similarly argued that white artists started with "a theory respecting the distinctive features of the negro physiognomy," rather than an inspection of the individualized faces of Black individuals. Rather than closely examining the physical features of the being in front of them, whites saw what their preconceptions had conditioned them to see.[45]

These insights are particularly important, because they reveal both the perils and the possibilities of physiognomy for advocates of racial equality. By encouraging people to use facial analysis, Black writers demanded that white Americans examine the "flesh and blood and bones" of African Americans, rather than relying on harmful stereotypes. Yet by themselves engaging in facial and cranial analysis, they gave credence to the idea that the physical form truly did reveal character. Even as they adapted physiognomic precepts for their own purposes, Black writers fashioned artificial distinctions between people of color who had "good" or "intelligent" physiognomies and those who, allegedly, did not. This meant that they sometimes reified white beauty standards, inadvertently legitimizing a system of skeletal analysis that had devalued them from its very inception.

One Black author, Martin H. Freeman, explicitly highlighted this problem. He argued that the "great want of the free colored race in this country" was its failure to recognize "its own intrinsic worthiness" and beauty. Freeman pointed to the "deplorable" practice in which every Black child "is taught directly or indirectly by its parents that he or she is pretty, just in proportion as the features approximate to the Anglo Saxon standard." When Black parents commented on the "good hair" or "good features" of their children, everyone knew this meant straight hair and "white" features. No wonder Black children were not proud of their racial heritage, he scoffed. How could they be proud when they spent so much time trying to "fix" their hair and features to approximate European standards of beauty? In a cultural climate like this—where "white" faces were "good" faces—how could young African Americans ever develop the self-respect necessary for racial advancement? Freeman argued this process had to start with Black parents, who should teach children not only to value and improve the Black mind, but also to love their Black bodies. Crafting a prescient critique of antebellum aesthetic standards, he highlighted the many problems that resulted when Black families reified a physiognomic value system created by white intellectuals.[46]

Freeman's focus on learning to love Black bodies predated the "Black is Beautiful" movement of the 1960s and 1970s, as well as twentieth-century Black feminist critiques of white beauty standards. But, as bell hooks once argued, there "has been little change" since the nineteenth century in how Black people are represented in popular culture. Most images of Black people, she argues, are either "constructed by white people who have not divested of racism, or by people of color/black people who may see the world through the lens of white supremacy." Freeman anticipated hooks's argument by over a century. If most African Americans were willing to criticize whites for how they portrayed Black appearances, Freeman was quite unique in his unapologetic and absolute refusal to adhere to a set of physiognomic standards created by white people and often reinscribed by Black writers.[47]

Engaging with the science of beauty could be an especially hazardous endeavor for Black women. As Mia Bay has argued, Black men regularly refuted and reformulated white ethnological discourses, but even the most educated Black women "remained conspicuously silent" about racial sciences. This could not have been accidental. Black women clearly resented white supremacy, and it would have been virtually impossible for them to be unfamiliar with physiognomy and phrenology. After all, their husbands, brothers, fathers, and friends were engaging with these sciences on a regular basis—and so were their white abolitionist allies. This nonetheless leaves several questions unanswered. Did Black women draw explicit connections between physiognomy, phrenology, and racist ethnological discourses? If so, why didn't they record those observations? And, as historians, how do we grapple with the fact that Black women were simultaneously experiencing racial discrimination and scientifically analyzing bodies, but mostly choosing not to publicly record their thoughts on the connections between these two subjects? In the words of Mia Bay, "how do you write the history of a silence?"[48]

Black feminist scholars have long recognized that Black women experienced a unique form of "double jeopardy," facing marginalization on account of both their race and their gender. The politics of female beauty were especially fraught for Black women. Since at least the eighteenth century, white transatlantic thinkers had crafted a physiognomic vision of intelligence that was both racialized and gendered. They not only argued that people of color were less attractive (and, by extension, less intelligent) than white individuals; they also suggested that Black individuals did not have the mental capacity to adequately recognize "perfect" female beauty. At a time when white women were battling to get people to value their intelligence *instead of* their beauty, Black women faced a white supremacist culture that refused to see them as beautiful at all.[49]

Popular science nevertheless shaped how Black women saw both themselves and others. Long after most Americans had abandoned their commitment to physiognomy and phrenology, Ida B. Wells continued using the language of popular science in her personal writings. In 1886, she recorded a flirtatious epistolary exchange with the Kentucky newspaperman Charles Morris, distinguishing between the man's intellectual abilities and his countenance. Wells gushed that Morris was "what I have long wished for as a correspondent, an interested, intellectual being who could lead & direct my wavering footsteps in intellectual paths." Yet she was disappointed by his countenance. "I told him I liked the face," she wrote, "but it is the face of a mere boy; whereas I had been led, from his writings, to suppose him a man." Wells sent Morris her picture in return, but the fact that he was younger than she was made her both hesitant and self-conscious. Still, she remained respectful of his mind: "If a boy, he certainly has a man's head and a man's thoughts in that head," she declared. A week later, Morris responded. He encouraged Wells to think of herself as a distinguished author, exhorting her to write a "work of fiction." He also evaluated her portrait. "He denominates my nose as my weakest feature," wrote Wells. "He phrenologizes my features minutely and unerringly, as well as amusingly."[50]

These exchanges were lighthearted and flirtatious, but they also illustrated how Black Americans struggled to address racial and gender stereotypes. Wells seemed struck by the apparent contradictions between Morris's mature prose, his young face, and his cranium. She also recognized that Morris had been justifiably "nettled" by her reference to him as "boy." Morris might have been two years her junior, but he was a twenty-two-year-old man. He then retaliated with his own critique. Though he praised the power of Wells's mind, he criticized her nose (the part of her countenance that most obviously deviated from racist beauty standards that privileged aquiline noses). Throughout the nineteenth century, white scientists and popular writers mocked the supposedly large noses and "distended nostrils" of Black Americans. In this context, Morris's critique came perilously close to validating the assumptions of white physiognomists. For her part, Wells did not seem particularly bothered by the comment, instead choosing to emphasize Morris's positive evaluation of her literary capacities. Still, the exchange demonstrates how some problematic legacies of popular sciences managed to infiltrate the intimate exchanges of even the most dedicated advocates of racial justice.

Popular sciences presented Black thinkers with a series of thorny ideological conundrums. Frederick Douglass, for instance, regularly engaged in scientific facial analysis and occasionally extolled the merits of phrenology. While he rejected the craniological theories of the "American School" of ethnology, he saw physiognomy and phrenology as alternative and more ethical scientific practices: disciplines allowing for the possibility of human improvement. Intellectuals like William Wells Brown and William J. Wilson were likewise committed to the project of corporeal analysis, even using it to suggest that Black people were superior to their white counterparts. But as people like M. H. Freeman rightly realized, whenever Black thinkers engaged with the physiognomic project, they ran the risk of legitimizing the very discourses that white Americans were using to rationalize white supremacy.

By the mid-1860s, Black writers had mostly stopped invoking physiognomy and phrenology when defending the mental capacities of African Americans. This was, in part, because these disciplines were being challenged by an increasingly organized group of professional scientists who rejected their legitimacy. As more and more Americans denounced practitioners of these sciences as "quacks" and "humbugs," physiognomic and phrenological defenses of the Black mind began to lose their intellectual purchase. Yet nineteenth-century Americans—both Black and white—continued to analyze faces for signs of human interiority, and we should not forget the cultural and scientific worldview that indelibly shaped their conceptions of race, beauty, and intellect.[51]

In particular, Frederick Douglass's ideas about photography were forged within this intellectual universe. By the 1860s, he had developed and delivered four lectures on "Pictures" that repudiated the arguments of white ethnologists, rather than reformulating them. Still, he never abandoned his commitment to visual culture, seeing photography as a liberatory technology that fostered democracy, equality, and a speculative imaginary of freedom. Douglass knew pictures served an important function: they could undermine the discourses of the "so called learned naturalists, archaeologists, and ethnologists" without giving credence to their racist diatribes. Images, he claimed, could "speak for themselves." Douglass eventually recognized that science and reasoned discourse alone would never cure racism. Although egalitarian logic was rational, it was cold, hard, and unconvincing. "The mighty fortress of the human heart silently withstands the assaults by the rifled cannons of reason," he declared. Pictures were different. When words and arguments failed, images captured people's emotions and sparked their imaginative faculties. They spoke to the soul.[52]

Above all else, Douglass had an unfailing faith in the universal equality of all mankind. He believed this elemental truth would eventually reveal itself to the world, and he thought new technologies like photography would facilitate the process. But he also suggested that people could not always trust their eyes. Douglass never saw images as fully objective reflections of reality, as some scholars have suggested. Instead, he hinted that the imaginative politics of vision might ultimately be more important than the images themselves. For Douglass, pictures—and bodies—were less meaningful than the way people read them. Even portraits of unattractive figures, he contended, would become beautiful as soon as the public decided those people were worthy of admiration. The physical realities of the portrait itself did not have to change. It was the viewer's mindset—their particular way of seeing—that mattered. As Douglass recognized, each person was equipped with a unique interpretive arsenal, allowing them to see the world in distinctive ways. During the first half of the nineteenth century, he was well aware that physiognomy and phrenology were important ideological weapons for Black and white Americans alike.[53]

Even today, art historians sometimes perpetuate the physiognomic rhetoric of the early nineteenth century. Particularly when analyzing portraits of Douglass, scholars identify traits like dignity, intelligence, defiance, and fortitude. For them, Douglass becomes the face of freedom. He is fearless, confident, and self-assured—an idealized vision of Black manhood. Douglass certainly knew that nineteenth-century Americans would have used popular sciences to interpret his portraits in this manner. But he could not have anticipated the extent to which future intellectuals would embrace this rhetoric, too. Silently imbibing some of the physiognomic language of the nineteenth century, scholars now look at portraits of Douglass and see precisely what he wanted people to see all along. In some ways, then, Douglass succeeded by teaching Americans to view race in a new way. Rejecting the white supremacist caricatures that Douglass despised, modern scholars have now embraced the alternative model of visual politics that he played a role in crafting.[54]

In the end, African American thinkers engaged in a radical act simply by reading heads and faces. By asserting their right to discern character scientifically, they not only used physiognomy and phrenology to challenge white supremacy; they also undermined many of the tenets supported by white thinkers. White physiognomists often hinted that only the most beautiful and intelligent individuals could be successful face decoders. If the racist caricatures of white artists and naturalists were to be believed, people of color not only lacked physical attractiveness; they also lacked the mental faculties to recognize it in others. The implication was unmistakable: wealthy, educated, white men were the best physiognomists. Beauty might be in the eye of the beholder, but only the truly beautiful could be skillful beholders. Black writers thus challenged a central element of existing thought, simply by asserting their claim to physiognomic knowledge. By insisting they, too, were scientific observers, they exhibited their influence over scientific and popular culture. To quote bell hooks: "There is power in looking."[55]

CONCLUSION

In 1862, the radical white abolitionist Gilbert Haven engaged in a form of anti-racist art criticism. After visiting the Great London Exposition of 1862, Haven zeroed in on two of the sculptures he had seen there: William Wetmore Story's *Cleopatra* and *Libyan Sybil*. Most of the works in the exhibition adhered to the reigning standard of beauty in the mid-nineteenth century. Big eyes. Aquiline noses. Small chins. Thin lips. Story's sculptures were different. Even though they sat "among a gallery full of Grecian faces," *Cleopatra* and *Sybil* were "evidently African, not Athenian." Haven was particularly taken with *Cleopatra*, whose full lips, high cheekbones, low forehead, and large features resembled "those on Egyptian monuments." Her countenance was "decidedly African," he declared, and she was stunning.[1]

Although abolitionists—both Black and white—occasionally described women of African descent as beautiful, they usually did so by holding up examples of women who embodied conventional physiognomic ideals. In this way, anti-slavery activists sometimes unintentionally reinforced the beauty standards of white ethnologists. They highlighted Black women's attractiveness, but they also described women of color as if they were Grecian beauties who incidentally happened to have dark skin. Black women thus became beautiful *in spite of* their Blackness, not *because* of it. Portrayed as dark-skinned figures with high foreheads, small mouths, and aquiline noses, they emerged as idealized counterparts to the Black women who did not have the "right" sorts of features. Haven rejected this approach. He argued that *Cleopatra* was beautiful precisely because she did not exemplify white beauty standards. In fact, it was her unmistakably African features that made her superior to her "Grecian rivals." Haven made his point by ventriloquizing the sculpture:

I belong to a despised race, but you shall feel that I am your superior. What if my forehead is low, and cheek-bones high, and lips thick, and nostrils not undistended—what if all the characteristics of the negro are stamped upon my countenance? You shall acknowledge that they are no drawback to my power, that they may even be the weapons of my conquest.

In Haven's mind, *Cleopatra* was not simply a sculpture. She was a marbleized repudiation of white supremacy (fig. C.1). This prompted him to describe the artist as "the bravest American except John Brown." At first, the comparison seems odd. Brown tried to spark a massive insurrection against enslavers. Story did no such thing. He simply carved a sculpture. But Haven argued that Story had "dared to do what the most courageous of Abolitionists have never yet dared to say. He has made a negress the model of beauty." To insist upon the reality of Black female beauty, Haven proclaimed, was not ancillary to the fight for racial justice. It was a profound and audacious act—something almost as courageous as an armed revolt.[2]

This example is important because it shows just how profoundly difficult it was for reform-minded Americans to escape the confines of their intellectual universe. On the one hand, Haven did something revolutionary. He rejected existing physiognomic standards and challenged the assumptions of white supremacists and scientific racists. On the other hand, he remained confined by the very discourses he sought to repudiate. To start, Haven described Story as phenomenally brave, merely for daring to depict a Black woman as beautiful. Was it really that audacious to envisage a Black woman as a model of beauty? Haven also never rejected the idea that heads, faces, and features conveyed internal worth. In fact, his own analysis of *Cleopatra*'s beauty resulted from an attempt to interpret the sculpture physiognomically. Haven might have discerned anti-racist messages in *Cleopatra*'s countenance, but he did not challenge the underlying assumption that appearances revealed character. Instead, he paired a vehement rejection of prevailing beauty standards with a continued commitment to popular sciences. Even as Haven made the case for racial equality, his evidence came from a sculpture's marble lineaments. By insisting on the importance of the head and face, Haven solidified the physiognomic project even as he rejected its most racist findings.

Historians have long recognized how scientists and political thinkers of the past have used science to develop and defend racist, sexist, and classist ideas. When scholars talk about physiognomy and phrenology, it is usually in an attempt to show how these disciplines laid the intellectual scaffolding for biological determinism and other forms of scientifically sanctioned bigotry. In these narratives, physiognomy and phrenology emerge as dangerous

FIGURE C.1. William Wetmore Story, *Cleopatra* (modeled 1858, carved 1860). Los Angeles County Museum of Art. Photo © Museum Associates/LACMA.

disciplines that treated bodies as infallible indicators of people's moral and intellectual value. For those who were invested in maintaining established power relationships, these sciences conveniently rooted inequities in the unchangeable features of the human body. Physiognomy and phrenology helped people believe that there were distinct and anatomically identifiable classes of superior and subordinate beings, that biology determined destiny, and that human character was a stable, measurable, and corporeal trait.

Physiognomy and phrenology were, indeed, problematic—but not solely because they were stepping-stones on the road to biological determinism. It's true that by the latter half of the nineteenth century, many elite and middle-class white Americans believed that hierarchies of race, class, and gender were physiological in nature, and largely unchangeable. It's also true that physiognomy and phrenology provided an ostensibly neutral way of excusing those hierarchies by furnishing scientific justifications for their existence. Yet popular sciences were not troubling merely because they were deterministic. These disciplines were dangerous in large part because they appealed to the very people who were damaged by them.

Abolitionists, women's rights activists, moral reformers, and ordinary people often rejected the conclusions of bigoted physiognomists and phrenologists, but they ended up embracing many of the same scientific ideas. Buying into the promise that physiognomy and phrenology could provide a more rational and objective way of evaluating human capacity, they put their faith in science, believing it would eventually validate their commitment to universal equality. When they did so, reformers lent credence to the physiognomic project and conceded that it might, indeed, be possible to measure human worth by analyzing physical beauty. They simply assumed that they would be able to read countenances and craniums more effectively—and more ethically—than their white supremacist and anti-feminist counterparts.

Physiognomy and phrenology were so broadly appealing to such a wide variety of people for a few reasons. First, they were accessible disciplines that traversed the boundaries between elite science and popular culture. Popular sciences were not limited to the highly educated or economically privileged. All sorts of people could envision themselves as practical scientists, invested with the authority to decipher the human body.[3] Second, these disciplines were doctrinally flexible. There were no reproducible methodologies and very few standardized or unbreakable rules. This created space for people to advance idiosyncratic interpretations and craft their own truths about human nature. Finally, physiognomy and phrenology validated people's belief in human perfectibility, encouraging them to believe that beautiful minds and bodies could be *earned* through hard work. Popular scientists suggested that everyone might take charge of their own destiny, so long as they embarked upon a project of personal betterment. While these sciences enshrined the notion that internal capacity telegraphed itself through the external form, they also allowed for the possibility that moral and intellectual cultivation could transform the physical features, making individuals more beautiful as their minds improved. Within this context, physiognomy and phrenology emerged as widely accessible tools that both progressive thinkers and defenders of the status quo could use against one another.

Physiognomy and phrenology became especially influential during the middle decades of the nineteenth century, an "Age of Reform" that seemed—to many people—like an exciting moment of progress and possibility. Evangelical ministers traveled from town to town, convincing ordinary people they were capable of spiritual regeneration. Abolitionists demanded the immediate eradication of slavery. Women's rights activists fought to expunge gendered inequities from the law. Temperance activists sought to create a less inebriated nation, and the country's most adventurous thinkers eschewed society entirely and instead cocooned themselves within utopian communities. Meanwhile, prison reformers devised an imaginatively exploitative carceral state and convinced themselves that they had developed a more rational system for punishing allegedly problematic citizens. Like many of the philosophies they adopted, these reformers were complex, contradictory, and creative. They disagreed about much, but they were mostly unified in their belief that Americans would be able to eradicate the irrationalities of the past and fashion a better, more vibrant, and more equitable nation—one that lived up to the lofty promises of its founding. Physiognomy and phrenology provided scientific support for the idea that change was both possible and desirable.

It was anathema to republican principles to believe that birth dictated destiny. But Americans were largely willing to assume that a person's inherited physical features might, indeed, hint at their internal capacities and their future trajectory. People simply conceptualized the connection between beauty and character differently depending on their racial, class, or gender identity. Even many abolitionists and women's rights activists embraced physiognomy and phrenology, co-opting these disciplines for their own purposes rather than rejecting them entirely or trying to refute their scientific legitimacy. White women used these sciences to argue that intellectual cultivation would magnify their beauty. African American thinkers used them to claim that emancipation, education, and political equality would improve the facial features of people of color. Moral reformers used them to insist that prisoners were capable of personal betterment. By stressing the mutability of the body and the mind, these groups espoused a scientific understanding of human nature that emphasized every individual's capacity for mental, moral, and physical improvement. In the process, they disputed emerging theories of biological determinism, which stressed the ineradicable nature of people's bodies and brains.

Still, when marginalized Americans engaged with popular sciences, it came at a price. As Audre Lorde so perceptively recognized, it is impossible to dismantle the master's house when relying on the master's tools.[4] Despite their eagerness to reformulate physiognomic and phrenological doctrines for their own aims, both white women and African Americans tacitly acceded to the notion that perhaps it *was* possible to discern internal character through an examination of the external form. Although they interpreted the human body in unique—and sometimes defiant—ways, they also helped solidify a broader commitment to the legitimacy of the human sciences and the invasive imperative that undergirded them: the desire to scrutinize people's bodies and scientifically justify existing social distinctions.

Even reformers' emphasis on mental and corporeal mutability provided a framework for rationalizing and reifying social distinctions. If people could beautify their bodies and their minds through hard work and an assiduous commitment to self-betterment, then they had no one to blame for their current inadequacies but themselves. One's faith in the body's changeability did not always translate into a demand for universal equality. Despite the pervasive cultural belief that both humans and society were improvable, many reformers, politicians, and medical professionals continued to believe that white women, people of color, poor people, incarcerated individuals, immigrants, and the mentally ill were simply second-rate beings with inferior bodies that justified their social, political, and economic subordination.

By the 1870s, most elite intellectuals had largely stopped taking physiognomy and phrenology seriously as sciences. Despite this fact, American and European scientists developed alternative disciplines for studying human difference, which included craniology, evolutionary biology, and physical anthropology (to name just a few). Thinkers such as Francis Galton published medical treatises proclaiming that traits such as genius and criminality were congenital realities with physical manifestations. Criminologists such as Cesare Lombroso insisted that prostitution, criminality, and mental illness were heritable characteristics, visible in the human face.[5] Anatomists and anthropologists feverishly collected skeletons and measured skulls, ranking human remains by race in hierarchical taxonomies. Meanwhile, evolutionary biologists such as Herbert Spencer attempted to compare the world's various populations by providing aesthetic evaluations of their bodies.[6] By the late nineteenth century, a new crop of sexual scientists—including thinkers such as George J. Romanes and Havelock Ellis—began fixating on the observable differences between male and female bodies. These men discovered tangible evidence of "truths" they already knew to be true. Women were suited for motherhood and not intellectual distinction. They were sure of it. And they would prove it through physiological and anatomical examinations of female bodies and brains.[7]

Few of these thinkers still fancied themselves physiognomists and phrenologists, yet all of them built on physiognomic theory, silently perpetuating some of its most basic doctrines even as they diligently distanced themselves from the "quackery" of the past. These men imagined that they were fashioning more quantitative—and thus more reliable—methodologies for studying human nature. In truth, they merely validated the physiognomical and phrenological imperative to root human difference in the body, stripping these sciences of their previous malleability while enshrining their most deterministic impulses. Eventually, scientists would come up with new metrics for measuring human worth, a project that would ultimately culminate in the eugenics movement of the late nineteenth and early twentieth centuries.

Even today, some anthropologists, psychologists, and art historians continue to recycle physiognomic ideas. Some researchers, for instance, have contended that people's criminal propensities, reproductive fitness, sexual orientation, or mental health status might be visible in their physical features. Unlike their forebears, contemporary scientists rely on new technologies like facial recognition software and functional neuroimaging to dispassionately prove their findings through empirical methodologies.[8] But we should remember that many eighteenth- and nineteenth-century scientists likewise believed that they were engaged in a pragmatic and objective endeavor. At the very least, the impulse that sustains modern research is largely the same as it was in the past: the effort to visualize intangible differences and make the world more legible in the process.

Nowadays, it might be tempting to look back at physiognomy and phrenology and dismiss them as both absurd and unreliable—the antiquated relics of an unenlightened past. Yet both disciplines were rooted in an assumption that remains with us today: the idea that science can make the invisible visible and reveal hidden truths about human nature. As the psychologist Alexander Todorov has argued, "We may poke fun at physiognomists, but we are all naïve physiognomists: we form instantaneous impressions and act on these impressions ... physiognomy has not disappeared and will not disappear from our lives." Social science research has consistently revealed that people are almost freakishly comfortable making snap judgments about character after analyzing photographs of individuals they have never met. In fact, they often come to firm conclusions after observing a face for just one-tenth of a second. Nor does more time staring change their opinion. It merely entrenches their gut reaction. Of course, observers are rarely correct when drawing conclusions at face value.

These ideas nonetheless have real ramifications in the world. When the public perceives certain political candidates as more physiognomically competent, those candidates are more likely to get elected. Facial features might not correlate with honesty, but legal defendants who *look* more trustworthy are more likely to get acquitted. "Baby-faced" people are often perceived as "weak, naïve, submissive, honest, kind, and warm," regardless of how they behave in their personal lives. When businesses interview CEO candidates who appear to have more competent countenances, those people end up getting job offers from the most successful firms, and they often bring in higher salaries—regardless of how they perform. To be clear, there's not much evidence that the people who are *perceived* as being competent are, in fact, better at their jobs. But there does seem to be a consensus about what competence looks like—and people make snap judgments based on those stereotypical ideas.[9]

It's easy to lambast the physiognomists and phrenologists as pseudoscientific quacks, but modern observers have also developed a broad social consensus about what a "trustworthy," "aggressive," or "competent" face looks like. Troublingly, this social consensus often matches up with the physiognomical and phrenological assumptions of the past. As the historian Courtney Thompson has argued, nineteenth-century popular sciences "helped average Americans to construct a visual language of good and bad heads," and that language is "still with us" today. Phrenologists, for instance, linked wide heads with a propensity for violence and criminality. Many modern researchers have done the same, using facial recognition software to compare people's facial width-to-height ratios (fWHR). Many of these studies replicate a central tenet of nineteenth-century phrenology: wide heads are associated with aggressive tendencies. Can this really be a coincidence? Are we to believe that there is, indeed, a link between wide heads and "badness"? Or might we consider that society's perceptions are still being shaped by the phrenological assumptions of the past? As Thompson puts it, "perhaps such studies in fact demonstrate not the correlation between head width and 'badness,' but the influence of phrenological thinking itself." In other words, perhaps phrenologists crafted such a successful vision of the "destructive" head that it continues to shape how we see people today.[10]

Physiognomic and phrenological ideas also occasionally rear their heads in popular culture—mostly in subtle ways, but sometimes more explicitly. The beauty magazine *Allure*, for instance, published an article that interprets the character traits of several A-list celebrities from their photos. The author describes Rihanna as having "a very high forehead," which apparently signifies that "she's intelligent and a forward thinker but a bit detached from real life and real problems." By contrast, she describes Jennifer Lopez as having a "rounded forehead," a sign that "she likes attention and is high-spirited" and that she has a "strong imagination." Evidently the physiognomical assumptions of the past are still with us.[11]

Researchers have also shown that the lighting or angle of a photograph can change how people perceive the person depicted in an image, regardless of features. So, when social media users go out of their way to photoshop pictures, use filters, or take photos from certain angles, it might seem vain. But it's practical. People do indeed judge others' personality based on first appearances—even when they think they're above that sort of thing. This has led some cultural critics to suggest that the old discipline of physiognomy is "gaining credibility" in the modern world. It's also captivated many psychologists and data scientists, who continue to search for signs of aggression, sexual orientation, and trustworthiness in the countenance.[12]

Science can be seductive. Human beings want to believe that if we commit ourselves to the dispassionate process of observation and experimentation, we will be able to discern objective truths about the natural world and our place within it. Empirical research holds out the promise of impartiality and rationality. It gives us hope that we might someday transcend the bounds of human fallibility and objectively understand the realities we inhabit. But science is—and always has been—an unreliable arbiter of human worth. When rationalized through its explanatory power, social hierarchies often seem like the natural byproducts of anatomical and physiological realities. This can obscure the fact that science is not neutral and never has been. Intellectuals have not always agreed on what counts as "good" science, and methodological standards have shifted dramatically over time, responding to a wide variety of social, economic, and political pressures. Science is a powerful tool, but it is also a capacious construct with a complicated and problematic history. If we want to use it responsibly, we should first reckon with its most unsavory legacies. Only then will we be able to craft more ethical methodologies for analyzing the world and the diverse populations who inhabit it.

ACKNOWLEDGMENTS

If you've enjoyed reading this book, it likely has less to do with my own highbrowed brilliance than with the generosity of the individuals and institutions who have supported my academic career over the past decade. First, I want to thank the editorial staff at the University of Chicago Press, particularly Tim Mennel, who invested in this project more than five years ago, when it was still a collection of half-finished chapters. My graduate advisor, Clare Lyons, also deserves special thanks. Thank you, Clare, for always pushing me to ask big questions and embrace the messy contradictions of the past. The advice and support of Richard Bell, Holly Brewer, and the late Ira Berlin likewise propelled this project from its early stages to its current form. I am grateful for their feedback, and for their constant advocacy on my behalf.

Many archives and institutions provided critical support for this project. Travel awards, research grants, and writing fellowships from the University of Hartford and the University of Maryland, College Park, made this book possible. I also benefited from short-term fellowships at the Virginia Historical Society, Massachusetts Historical Society, American Philosophical Society, Boston Athenaeum, Huntington Library, College of Physicians of Philadelphia, Center for the History of Science Technology and Medicine (CHSTM), Winterthur Library, and American Antiquarian Society.

This project was also funded by long-term fellowships from the Smithsonian Museum's National Portrait Gallery, Fred W. Smith National Library for the Study of George Washington at Mount Vernon, and the National Endowment for the Humanities (NEH). Despite the upheavals of the COVID-19 pandemic, this latter award allowed me to spend an invigorating and productive semester at the American Antiquarian Society. Any views, findings, conclusions, or recommendations expressed here do not necessarily reflect those of the NEH.

Through these fellowships, I uncovered the foundational sources for this book while meeting a collection of brilliant archivists and librarians. Special thanks to Paul Erickson, Nan Wolverton, and the spectacular staff at AAS. At the National Portrait Gallery, Brandon Brame Fortune, Ellen Miles, Asma Naeem, and Robyn Asleson shared their knowledge of physiognomy, silhouettes, and early American art. I'm especially grateful to Susan Garton for letting me reproduce the beautiful portrait of Amoret Gillett Austin, which appears in chapter 4. Finally, I'd like to single out Samantha Snyder at Mount Vernon. As a librarian, you've pointed me toward helpful manuscripts and enlightened me about John Dickinson's significance to the Founding Era (and his importance to Elizabeth Powel). As a friend, you've been a real catch. Thanks for your help—and your unfailing humor.

I also feel particularly blessed to have held long-term fellowships from the Library Company of Philadelphia and McNeil Center for Early American Studies. These institutions brought me into Philadelphia's vibrant and collegial interdisciplinary community for two years. My experiences in the city shaped the trajectory of this project in foundational ways. I am thankful for the support and advice of Jim Green, Connie King, Linda August, and Krystal Appiah, who helped me scour the collections of the Library Company. I was also lucky to work alongside Dan Richter at the McNeil Center, benefiting from his thoughtful guidance, cheerful encouragement, and witty repartee.

More personally, I am immensely grateful for the relationships that I developed while in Philadelphia. At the Library Company, Randy Browne became both a valuable editor and a trusted confidant, while Jessica Linker became my go-to reference for obscure questions about material texts, gender history, and turtles. Don James McLaughlin became one of my dearest friends, as well as my most incisive editor (and, I might add, the world's most spectacular dance partner). Thanks, also, to Tommy Richards for the year we spent sharing both an office and our hottest political takes. I also feel particularly fortunate to have met Liz Ellis, Daniel Couch, and Max Dagenais at the McNeil Center for Early American Studies. These three have continued to inject a steady stream of humor into my life. More importantly, they've never once judged my love of Kerrygold butter. I'm also indebted to Nora Slonimsky for generously agreeing to share her bungalow during our Huntington Library research trip. I hope you never scram, you old groat. Finally, thanks to Alan Noonan. Entering my life as I ensconced myself in the archives, you were one of the best discoveries of my research.

I have also amassed a collection of treasured colleagues and valued friends at the University of Hartford. Warren Goldstein and Steven Rosenthal not only read portions of this manuscript; they also provided unflinching support and encouragement in their roles as Department Chairs. Thanks to both of you—and to Shelly Duford and Amy Weiss—for crafting a departmental culture of wit, snark, and genuine friendship. I'd also like to extend my thanks to Kat Owens and Bryan Sinche, who helped me revise parts of this book; to T. Stores, who has worked hard to create a supportive environment for Junior Faculty; and to Nicholas Ealy, who has provided me with a model of what rigorous scholarship and transformative teaching in the humanities can look like. Special thanks go to Michael Robinson, my fellow historian of science, who not only read this entire manuscript in an early form but also provided consistent feedback as I tore it apart and created something new. Thank you for always navigating analytical logjams with me. I am also appreciative of my Dean, Katherine Black, who supported me taking a year of research leave to finish this book. Finally, I want to thank all the friends, colleagues, and happy hour compatriots who brought joy to my life during my time here in Hartford (particularly Katie Lance, who not only gave me feedback on my introduction, but also took care of my cat—and my plants—when I traveled for research).

It would be difficult to adequately convey my gratitude to all the phenomenal friends and colleagues who have filled my life with moments of levity and laughter as I worked on this book. I'm grateful for all the people who read early drafts of this work, including Sarah Gronningsater, Amy Sopcak-Joseph, Carla Bittel, Lindsay Keiter, Alisa Wade, Lauren Duval, and Christopher Lukasik (who also generously shared his enormous physiognomy database with me). Thank you to Alicia, Kelly, and Abby (for sharing a home with me during my chaotic graduate school years), to Chesley (for being *mon petit chou*), and to Amanda (for being one of my oldest and dearest friends). Thanks, also, to Stephanie Hinnershitz (for bullying me into productivity during our semi-annual writing months), and to Lindsey and Caitlin Humphreys (who have helped me think, laugh, and learn since our days in the IC History Department). I am also enormously grateful for Jacqueline Beatty, who sustained me through the horrors of the academic job market (even though we had most of the same interviews!). Our friendship could so easily have been competitive and stressful. Instead, it's been full of humor, generosity, and grace. Thanks for being a decent human. To Ashley Towle and Caitlin Haynes: I'm not sure if I can adequately express just how much your friendship means to me, so I'll just say this: thank you for humoring my "deep thoughts" on topics both trivial and monumental, for making me laugh so hard that I'm silently squealing, and for never judging me—even when I deserve it. You are my confidants when I'm in crisis, and my companions in degeneracy. I am absurdly grateful for you both.

I also want to thank a group of female mentors who have nourished and sustained me over the course of my life. In New Hampshire, Anne Englert allowed me to be a teen while helping me to become an adult. Thanks, Anne, for entrusting me with responsibility, teaching me how to be a professional, and encouraging me to present myself with poise and confidence. Vicki Etchings also took me under her wing, stimulating my mind while encouraging me to use my voice. Perhaps you don't know this, Vicki, but it was at your kitchen table that I first became enraptured by intellectual debates about music, politics, scholarship, and the world. Thank you for being one of my most treasured mentors and friends. I also want to thank my mentors from Ithaca College: the place where I first learned to think like a historian. Karin Breuer had a larger impact on this project than she probably realizes. As a sophomore, I was desperate to get out of the music school and needed a new major. By chance, I took one of her courses and liked it so much that I haphazardly decided on history. The rest is, well, history. I also want to extend my deepest gratitude to Vivian Conger, who started as my academic advisor but has since become an invaluable mentor, colleague, and friend. Thank you for sparking my love of early American women's history, and for being the first person who believed I could succeed in this field.

Finally, I owe an enormous debt to my family, whose love and support propelled me throughout my professional career. My cousin Jecoliah Ellis deserves special praise. Thank you for letting me swoop into your apartment, colonize your living space, and become "part of your world" for months at a time. At a time when I was living on a grad student's salary, your generosity made my research possible. I also want to thank my siblings, Gabrielle and David, and especially my parents, David and Dulcenia Walker. With an almost evangelical fervor, my parents have always insisted that their children were the smartest, most talented people on planet Earth. As I was growing up, they told anyone who would listen that I was destined to be a star. I don't think an academic career is precisely what they had in mind, but they never wavered in their pride or in their support.

Writing a book can feel like a solitary endeavor sometimes, but when I think about all the people and places that have supported me throughout this process, I am overcome with gratitude. Thank you to all the people who shaped this project, in ways both large and small. I appreciate you all.

An earlier version of chapter 6 was previously published as "Facing Race: Popular Science and Black Intellectual Thought in Antebellum America," *Early American Studies* 19, no. 3 (Summer 2021): 601–40. I would like to thank the editors for allowing me to republish it here, and for awarding this essay the Murrin Prize for the Best Article published in *EAS* in 2021.

NOTES

INTRODUCTION

1. Winthrop Sargeant, "In Defense of the High-Brow," *LIFE* (New York), 11 April 1949. The classic scholarly work on these distinctions is Lawrence Levine's *Highbrow/Lowbrow: The Emergence of Cultural Hierarchy in America* (Cambridge: Harvard University Press, 1988), which traces the transition from "a rich shared public culture" in the nineteenth century to a more "rigid" and hierarchal culture by the twentieth century (9). Besides one brief paragraph on phrenology, Levine does not examine the connection between concepts like "highbrow" and "lowbrow" and the sciences that popularized them (221–22).

2. Physiognomists and phrenologists were not the first people to associate high brows with intelligence and goodness. As art historians have shown, Renaissance portraitists valued "the high, domed forehead which was seen as a mark of elegance and female beauty during this period." See Paola Tinagli, *Women in Italian Renaissance Art: Gender, Representation and Identity* (Manchester, UK: Manchester University Press, 1997), 52. For works emphasizing the prevalence of high foreheads in Renaissance art, see Wolfgang Bruhn and Max Tilke, *A Pictorial History of Costume from Ancient Times to the Nineteenth Century* (1955; repr. London: Dover Publications, 2004); Norbert Schneider, *The Art of the Portrait: Masterpieces of European Portrait Painting, 1420–1670* (Köln: Taschen, 2002); Victoria Sherrow, *Encyclopedia of Hair: A Cultural History* (London: Greenwood Press, 2006), 329; and Margaret A. Sullivan, "Alter Apelles: Dürer's 1500 Self-Portrait," *Renaissance Quarterly* 68, no. 4 (Winter 2015), 1174. William Shakespeare also used this trait in his plays to signal refinement. In the late eighteenth and early nineteenth centuries, though, physiognomists and phrenologists popularized this long-standing belief and invested it with new scientific meaning. For them, high brows were not just signs of beauty and refinement, but anatomical signifiers of mental excellence that could be empirically evaluated.

3. Throughout this book, I have chosen to capitalize "Black" and not "white." I made this decision for two major reasons. First, Black scholars, journalists, and activists have rightfully pointed out that using a lowercase "b" reduces Blackness to a descriptive term, tied to skin color. In reality, Black Americans are a complex group of people united by a collective history and unique cultural heritage—and also by the shared experience of discrimination in a world shaped by white supremacy. Capitalizing the "B" takes Blackness seriously as a vibrant cultural identity. Second,

I have chosen *not* to capitalize "white" primarily because white nationalists and white supremacists have demanded that we do so as a way to signal—and enshrine—the alleged superiority of "whiteness" as an identity. Rather than reify this racist assumption, I have continued to use the lowercase "w" when referring to white people and white supremacy. With that said, many scholars and journalists of color have argued that capitalizing "Black" and not "white" is misguided, because it standardizes whiteness as the norm while portraying Blackness as a sort of cultural deviation. I take that criticism seriously, and I also recognize that language is constantly evolving. My opinions may well change in the future. For now, though, I have decided to capitalize "Black" as a way of affirming that "Blackness" signifies a complex cultural identity—one that deserves the linguistic respect that a capital letter confers. By contrast, I have chosen *not* to capitalize "white," in part because I do not want to imply that "whiteness" is a special or superior cultural identity, and in part because I recoil at the prospect of giving white nationalists the linguistic recognition for which they have been clamoring.

4. "To John Adams from Thomas Jefferson, 28 October 1813," *Founders Online*, National Archives, https://founders.archives.gov/documents/Adams/99-02-02-6189. Numerous scholars have debated the meaning of the American Revolution, emphasizing both its lofty promises and its failure to achieve them. For those who emphasize the revolutionary nature of republican ideals, see Bernard Bailyn, *The Ideological Origins of the American Revolution* (1967; repr. Cambridge: Harvard University Press, 2017); Pauline Maier, *From Resistance to Revolution: Colonial Radicals and the Development of American Opposition to Britain, 1765–1776* (New York: Alfred A. Knopf, 1972); and Gordon S. Wood, *The Radicalism of the American Revolution* (New York: Vintage Books, 1991). For work that emphasizes the impact of republican ideology on women, African Americans, and other marginalized groups, see Alfred F. Young and Gregory H. Nobles, *Whose American Revolution Was It? Historians Interpret the Founding* (New York: New York University Press, 2011); Woody Holton, *Unruly Americans and the Origins of the Constitution* (New York: Hill and Wang, 2007); Alfred F. Young, *Liberty Tree: Ordinary People and the American Revolution* (New York: New York University Press, 2006); Gary Nash, *The Unknown American Revolution: The Unruly Birth of Democracy and the Struggle to Create America* (New York: Viking, 2005); and Holton, *Forced Founders: Indians, Debtors, Slaves, & the Making of the American Revolution in Virginia* (Chapel Hill: University of North Carolina Press, 1999). For an excellent book focused on how the French and American republics tried to reconcile their egalitarian beliefs with their commitment to meritocracy, see John Carson, *The Measure of Merit: Talents, Intelligence, and Inequality in the French and American Republics, 1750–1940* (Princeton, NJ: Princeton University Press, 2007).

5. Courtney Thompson has shown that "phrenological enthusiasm" was initially "elite and medical in nature." See Courtney E. Thompson, *An Organ of Murder: Crime, Violence, and Phrenology in Nineteenth-Century America* (New Brunswick, NJ: Rutgers University Press, 2021), 43.

6. The world's most famous physiognomist argued that "All men who have eyes and ears, have talents to become physiognomists." Yet he also suggested that only those with good visages could truly master the science. In his words, "No one whose person is not well formed, can become a good physiognomist." See Lavater, *Essays on Physiognomy* (London: G. G. J. and J. Robinson, 1789), 1:116. For the argument that even babies could be successful physiognomists, see John Neal, "Phrenology and Physiognomy," *American Phrenological Journal* [hereafter *APJ*] 44, no. 3 (New York), September 1866.

7. On the history of physiognomy in Europe, see John Graham, "Lavater's Physiognomy in England," *Journal of the History of Ideas* 22 (1961): 561–72; Ellis Shookman, ed., *The Faces of Physiognomy: Interdisciplinary Approaches to Johann Caspar Lavater* (Columbia, SC: Camden House, 1993); Dror Wahrman, *The Making of the Modern Self: Identity and Culture in Eighteenth-Century*

England (New Haven, CT: Yale University Press, 2004), 294–305; and Sharrona Pearl, *About Faces: Physiognomy in Nineteenth-Century Britain* (Cambridge, MA: Harvard University Press, 2010). For foundational scholarship on phrenology, see Steven Shapin, "Phrenological Knowledge and the Social Structure of Early Nineteenth-Century Edinburgh," *Annals of Science* 32, no. 3 (1975): 219–43; Roger Cooter, *The Cultural Meaning of Popular Science: Phrenology and the Organization of Consent in Nineteenth-Century Britain* (Cambridge: Cambridge University Press, 1984); T. M. Parssinen, "Popular Science and Society: The Phrenology Movement in Early Victorian Britain," *Journal of Social History* 8, no. 1 (Autumn 1974): 1–20; John van Wyhe, *Phrenology and the Origins of Victorian Scientific Naturalism* (Burlington, VT: Ashgate Publishing Company, 2004); and Sherrie Lynne Lyons, *Species, Serpents, Spirits, and Skulls: Science at the Margins in the Victorian Age* (Albany: SUNY Press, 2009), chapter 3. For a recent *tour de force* placing phrenology in a global context, see James Poskett, *Materials of the Mind: Phrenology, Race, and the Global History of Science, 1815–1920* (Chicago: University of Chicago Press, 2019).

8. On popular sciences as "legitimate knowledge systems," see Britt Rusert, "Delaney's Comet: Fugitive Science and the Speculative Imaginary of Emancipation," *American Quarterly* 65, no. 4 (December 2013): 801. For scholarship that cautions against using the term "pseudoscience," see Daniel Patrick Thurs, *Science Talk: Changing Notions of Science in American Culture* (New Brunswick, NJ: Rutgers University Press, 2007); Lyons, *Species, Serpents, Spirits, and Skulls*; and Michael D. Gordin, *The Pseudoscience Wars: Immanuel Velikovsky and the Birth of the Modern Fringe* (Chicago: University of Chicago Press, 2012).

9. Richard Yeo, *Defining Science: William Whewell, Natural Knowledge and Public Debate in Early Victorian Britain* (Cambridge: Cambridge University Press, 1993), 111. Katherine Pandora sees the late nineteenth century as the moment where "the rupture occurred" between "popular" and "proper" science. See Katherine Pandora, "Popular Science in National and Transnational Perspective: Suggestions from the American Context," *Isis* 100, no. 2 (June 2009): 358.

10. Ralph O'Connor, "Reflections on Popular Science in Britain: Genres, Categories, and Historians," *Isis* 100, no. 2 (2009): 339. On the intersections of art, science, and literature in Europe, see Graeme Tytler, *Physiognomy in the European Novel: Faces and Fortunes* (Princeton, NJ: Princeton University Press, 1982); Mary Cowling, *The Artist as Anthropologist: The Representation of Type and Character in Victorian Art* (Cambridge: Cambridge University Press, 1989); Juliet McMaster, *The Index of the Mind: Physiognomy in the Novel* (Lethbridge, Alberta: University of Lethbridge Press, 1990); Joan K. Stemmler, "The Physiognomical Portraits of Johann Caspar Lavater," *The Art Bulletin* 75, no. 1 (March 1993): 151–68; Christopher Rivers, *Face Value: Physiognomical Thought and the Legible Body in Marivaux, Lavater, Balzac, Gautier, and Zola* (Madison: University of Wisconsin Press, 1994); Barbara M. Benedict, "Reading Faces: Physiognomy and Epistemology in Late Eighteenth-Century Sentimental Novels," *Studies in Philology* 92, no. 3 (Summer 1995): 311–28; Melissa Percival, *The Appearance of Character: Physiognomy and Facial Expression in Eighteenth-Century France* (Leeds: W. S. Maney & Son Ltd., 1999); Lucy Hartley, *Physiognomy and the Meaning of Expression in Nineteenth-Century Culture* (Cambridge: Cambridge University Press, 2001); David Bindman, *Ape to Apollo: Aesthetics and the Idea of Race* (Ithaca, NY: Cornell University Press, 2002); and Rhonda Boshears and Harry Whitaker, "Phrenology and Physiognomy in Victorian Literature," *Progress in Brain Research* 205 (2013): 87-112. For the early American context, see Brandon Brame Fortune, "Portraits of Virtue and Genius: Pantheons of Worthies and Public Portraiture in the Early American Republic, 1780–1820," (PhD diss., University of North Carolina, Chapel Hill, 1986); Charles Colbert, *A Measure of Perfection: Phrenology and the Fine Arts in America* (Chapel Hill: University of North Carolina Press, 1997); Shawn Michelle Smith, *American Archives: Gender, Race, and Class in Visual Culture* (Princeton, NJ: Princeton University

Press, 1999); Wendy Bellion, *Citizen Spectator: Art, Illusion, and Visual Perception in Early National America* (Chapel Hill: University of North Carolina Press, 2011); Marcy Dinius, *The Camera and the Press: American Visual and Print Culture in the Age of the Daguerreotype* (Philadelphia: University of Pennsylvania Press, 2012); Jasmine Nichole Cobb, *Picture Freedom: Remaking Black Visuality in the Early Nineteenth Century* (New York: New York University Press, 2015); and Catherine E. Kelly, *The Republic of Taste: Art, Politics, and Everyday Life in Early America* (Philadelphia: University of Pennsylvania Press, 2016).

11. Carla Bittel, "Testing the Truth of Phrenology: Knowledge Experiments in Antebellum American Cultures of Science and Health," *Medical History* 63, no. 3 (2019): 352.

12. Both James Secord and Katherine Pandora have argued that the historiography on popular science in Victorian Britain is well developed, while the scholarship on popular science in antebellum America is largely nonexistent. See James A. Secord, "Knowledge in Transit," *Isis* 95, no. 4 (December 2004), 669; and Pandora, "Popular Science in National and Transnational Perspective," 349. On popular science in Britain, see O'Connor, "Reflections on Popular Science in Britain"; Bernard Lightman, *Victorian Popularizers of Science: Designing Nature for New Audiences* (Chicago: University of Chicago Press, 2007); James A. Secord, *Victorian Sensation: The Extraordinary Publication, Reception, and Secret Authorship of* Vestiges of the Natural History of Creation (Chicago: University of Chicago Press, 2000); and Anne Secord, "Science in the Pub: Artisan Botanists in Early Nineteenth-Century Lancashire," *History of Science* 32 (1994): 269–315.

13. In 1980, even the pathbreaking historian of science Roger Cooter initially suggested that "Whig historians of phrenology were perfectly justified in calling phrenology pseudoscientific." See Marsha P. Hanen, Margaret J. Osler, and Robert G. Weyant, eds. *Science, Pseudo-Science, and Society* (Waterloo, Ont.: Wilfrid Laurier University Press, 1980), 242. For scholarship that refers to physiognomy and/or phrenology as pseudosciences, see Arthur Wrobel, ed., *Pseudoscience and Society in 19th-Century America* (Lexington: University of Kentucky Press, 1987); Jon Butler, "Magic, Astrology, and the Early American Religious Heritage, 1600-1760," *American Historical Review* 84, no. 2 (April 1979): 317–46; Christopher Grasso, "Skepticism and American Faith: Infidels, Converts, and Religious Doubt in the Early Nineteenth Century," *Journal of the Early Republic* 22, no. 3 (Autumn 2002): 465–508; Christopher J. Beshara, "Moral Hospitals, Addled Brains, and Cranial Conundrums: Phrenological Rationalisations of the Criminal Mind in Antebellum America," *Australasian Journal of American Studies* 29, no. 1 (July 2010): 36–60; Matthew Dennis, "Natives and Pioneers: Death and the Settling and Unsettling of Oregon," *Oregon Historical Quarterly* 115, no. 3 (Fall 2014): 286; and Allison B. Kaufman and James C. Kaufman, eds. *Pseudoscience: The Conspiracy against Science* (Cambridge: MIT Press, 2018), 9–10.

14. Physiognomy and phrenology are largely absent from the two flagship journals in the field, the *American Historical Review* and *Journal of American History*. The *AHR* published just two articles focused on phrenology between 1895 and 2015 (one in 1933, the other in 1952). Between 1964 and 2015, the *JAH* published just one: Cameron B. Strang, "Violence, Ethnicity, and Human Remains during the Second Seminole War," *Journal of American History* 100, no. 4 (March 2014): 973–94. Historians occasionally make passing references to physiognomy and phrenology in articles about other subjects, but they often dismiss them as "pseudosciences" or "fads." Influential US History textbooks rarely discuss these sciences. Eric Foner's *Give Me Liberty!* Textbooks do not mention them, nor does *The American Yawp*; Harry L. Watson's *Building the American Republic*; or John M. Murrin et al. *Liberty, Equality, Power: A History of the American People*. Only one brief note about the links between physiognomy, phrenology, and scientific racism appears in Mary Beth Norton et al., *A People and a Nation: A History of the United States*, 10th edition (Stamford, CT: Cengage Learning, 2015), 569.

15. Christopher Lukasik's *Discerning Characters: The Culture of Appearance in Early America* (Philadelphia: University of Pennsylvania Press, 2011) is the only book-length work on physiognomy in America. This book brilliantly demonstrates how physiognomy shaped ideas about social distinction in the early republic. Rather than providing a social history of science in action, it focuses primarily on the relationship between science and the nation's literary culture. The classic works on phrenology remain John D. Davies, *Phrenology, Fad and Science: A 19th-Century American Crusade* (New Haven, CT: Yale University Press, 1955); and Madeleine Stern, *Heads and Headlines: The Phrenological Fowlers* (Norman: University of Oklahoma Press, 1971). Both books primarily focus on the Fowlers, a family of entrepreneurial phrenologists who turned phrenology into a big business and widespread social practice in the United States. In more recent years, scholars have written specialized monographs on specific aspects of phrenology's cultural and political influence. On the relationship between art and science, see Colbert, *A Measure of Perfection*. For a sweeping discussion of phrenology and the common school movement, see Stephen Tomlinson, *Head Masters: Phrenology, Secular Education, and Nineteenth-Century Social Thought* (Tuscaloosa: University of Alabama Press, 2005). For a fascinating examination of how phrenology shaped American ideas about crime and punishment, see Thompson, *An Organ of Murder*. For the global history of phrenology, see Poskett, *Materials of the Mind*. This book argues that any history of phrenology must be a global history, since phrenological knowledge and materials regularly traversed national borders. This is certainly true. In this book, though, I acknowledge that phrenology was a global science while keeping my focus on how popular sciences shaped debates about society and politics in the United States.

16. Karen Halttunen's *Confidence Men and Painted Women: A Study of Middle-Class Culture in America, 1830–70* (New Haven, CT: Yale University Press, 1982) effectively shows how nineteenth-century Americans became preoccupied with distinguishing between hypocrites and liars on the one hand, and virtuous and sensible Americans on the other. These concerns became especially pressing with the growth of cities and the emergence of the market economy. Halttunen's book briefly mentions—but does not focus on—physiognomy and phrenology. By studying these disciplines, it is possible to recover the scientific logic that people used to assuage all the cultural anxieties that Halttunen so lucidly depicted.

17. On the connections between physiognomy, phrenology, and other more essentialist disciplines, see Elizabeth Fee, "Nineteenth-Century Craniology: The Study of the Female Skull," *Bulletin of the History of Medicine* 53, no. 3 (December 1978): 415–33; Nancy Leys Stepan, *The Idea of Race in Science: Great Britain 1800–1960* (London: Palgrave Macmillan, 1982), 20–46; Cynthia Eagle Russett, *Sexual Science: The Victorian Construction of Womanhood* (Cambridge, MA: Harvard University Press, 1989), 16–48; Cathy Boeckmann, *A Question of Character: Scientific Racism and the Genres of American Fiction, 1892–1912* (Tuscaloosa: University of Alabama Press, 2000), 44–49; Bruce Dain, *A Hideous Monster of the Mind: American Race Theory in the Early Republic* (Cambridge, MA: Harvard University Press, 2002), 197–226; John Wood Sweet, *Bodies Politic: Negotiating Race in the American North, 1730–1830* (Philadelphia: University of Pennsylvania Press, 2003), 295–301; Martin S. Staum, *Labeling People: French Scholars on Society, Race, and Empire, 1815–1848* (Montreal: McGill-Queen's University Press, 2003); Richard T. Gray, *About Face: German Physiognomic Thought from Lavater to Auschwitz* (Detroit, MI: Wayne State University Press, 2004); Elizabeth Ewen and Stuart Ewen, *Typecasting: On the Arts and Sciences of Human Inequality* (New York: Seven Stories Press, 2006), especially chapters 7, 11, 12, and 13; Nell Irvin Painter, *The History of White People* (New York: W. W. Norton & Company, 2010), 65–71, 190–200; Kris Belden-Adams, *Eugenics, "Aristogenics," Photography: Picturing Privilege* (New York: Routledge, 2020), 2–4.

18. Ever since the 1950s, historians of popular science have pointed out that phrenology was

intimately entangled with the reform causes of the nineteenth century. US historians who study those reforms, though, often mention phrenology in passing or ignore it entirely. Only in recent years have scholars begun to study how early American abolitionists and women's rights activists embraced this science. See Cynthia S. Hamilton, "'Am I Not a Man and a Brother?' Phrenology and Anti-Slavery," *Slavery & Abolition* 29, no. 2 (June 2008): 173–87; Susan Branson, "Phrenology and the Science of Race in Antebellum America," *Early American Studies* 15, no. 1 (Winter 2017): 164–93; and Carla Bittel, "Woman, Know Thyself: Producing and Using Phrenological Knowledge in 19th-Century America," *Centaurus* 55, no. 2 (2013): 104–30. For a global perspective on phrenology and reform, see Poskett, *Materials of the Mind*. For an interpretation that rejects the idea that phrenology was a reform science, see John van Wyhe, "Was Phrenology a Reform Science?" *History of Science* 42, no. 3 (September 2004): 313–31.

19. Francis Hutcheson, *A Short Introduction to Moral Philosophy: In Three Books; Containing the Elements of Ethics and the Law of Nature* (Glasgow: Robert Foulis, 1747), 2.

20. Stephanie M. H. Camp, "Black Is Beautiful: An American History," *Journal of Southern History* 81, no. 3 (August 2015): 679; Londa Schiebinger, "The Anatomy of Difference: Race and Sex in Eighteenth-Century Science," *Eighteenth-Century Studies* 23, no. 4 (Summer 1990): 387–405; and Painter, *The History of White People*, especially chapters 5 and 6.

21. Lavater, *Essays on Physiognomy*, 1:68.

22. Phrenology's intellectual debt to physiognomy was reflected in the titles of early phrenological works. In the early nineteenth century, Gall's research partner, Johann Gaspar Spurzheim, published *The Physiognomical System of Drs. Gall and Spurzheim* (London: Baldwin, Cradock, and Joy, 1815) and *Phrenology, in Connexion with the Study of Physiognomy* (London: Treuttel, Würtz, and Richter, 1826). These books argued that the face and head revealed character, but they tried to "do more" than Lavater had done by systematizing the physiognomical method. For the American edition, see *Phrenology, in Connexion with the Study of Physiognomy* (Boston: Marsh, Capen, and Lyon, 1833), 8–9, 22. For an American article that draws a direct line between Lavater and Gall, see Lucerna [pseud.], "Phrenology," *New Monthly Magazine* 7 (Boston), 1824. For work that shows "just how important physiognomy was to 19th-century phrenology," see Richard Twine, "Physiognomy, Phrenology and the Temporality of the Body," *Body & Society* 8, no. 1 (2002): 68.

23. On physiognomy's prevalence in American literature and culture, see Lukasik, *Discerning Characters*; Dana Brand, *The Spectator and the City in Nineteenth-Century American Literature* (Cambridge: Cambridge University Press, 1991), esp. 34–49 and 118–20; James V. Werner, "The Detective Gaze: Edgar A. Poe, the Flaneur, and the Physiognomy of Crime," *American Transcendental Quarterly* 15, no. 1 (March 2001): 5–21; Kevin J. Hayes, "Visual Culture and the Word in Edgar Allan Poe's 'The Man of the Crowd,'" *Nineteenth-Century Literature* 56, no. 4 (March 2002): 445–65 (esp. 457 and 465); and Kevin J. Hayes, "Poe, The Daguerreotype, and The Autobiographical Act," *Biography* 25, no. 3 (Summer 2002): 477–92.

24. Thompson, *An Organ of Murder*.

25. "The Doctrine of the Temperaments," *Ladies' Magazine* 2, no. 8 (Boston), August 1829; "What Good Will Phrenology Do the Ladies?" *Ladies' Magazine and Literary Gazette* 5, no. 10 (Boston), October 1832; "The Gatherer," *Lady's Book* (New York), 1 September 1830; Emma C. Embury, "Our Jessie, or, The Exclusives," *Godey's Lady's Book* 20, no. 5 (New York), January 1840; and "A Series of Papers on the Hair: Chapter V. Modes of Wearing the Hair," *Godey's Lady's Book and Magazine* 50 (New York), May 1855.

26. Carla Bittel, "Cranial Compatibility: Phrenology, Measurement, and Marriage Assessment," *Isis* 112, no. 4 (December 2021): 795–803.

27. For just a few examples, see A Reporter, "Glances at Congress," *United States Magazine*

and Democratic Review 1, no. 1 (New York), October 1837; "Political Portraits: Theodore Sedgwick," *United States Magazine and Democratic Review* 7, no. 26 (New York), February 1840; and C. Montgomery, "The Presidents of Texas," *United States Magazine and Democratic Review* 26, no. 81 (New York), March 1845.

28. See *Proceedings and Debates of the Virginia State Convention of 1829–30* (Richmond, VA: Samuel Shepherd & Co., 1830), 27. For secondary work on these transformations, see Sean Wilentz, *The Rise of American Democracy: Jefferson to Lincoln* (New York: W. W. Norton & Company, 2005); Alexander Keyssar, *The Right to Vote: The Contested History of Democracy in the United States* (New York: Basic Books, 2000); Rogers M. Smith, *Civic Ideals: Conflicting Visions of Citizenship in US History* (New Haven, CT: Yale University Press, 1997), especially chapter 8; Jacob Katz Cogan, "The Look Within: Property, Capacity, and Suffrage in Nineteenth-Century America," *Yale Law Journal* 107, no. 2 (November 1997): 473–98; and Wood, *The Radicalism of the American Revolution*.

29. On women and politics in the early republic, see Linda K. Kerber, *Women of the Republic: Intellect and Ideology in Revolutionary America* (Chapel Hill: University of North Carolina Press, 1980); Jan Lewis, "The Republican Wife: Virtue and Seduction in the Early Republic," *William and Mary Quarterly* 44, no. 4 (October 1987): 689–721; Elizabeth Varon, "Tippecanoe and the Ladies, Too: White Women and Party Politics in Antebellum Virginia," *Journal of American History* 82 (September 1995): 494–521; and Susan Branson, *These Fiery Frenchified Dames: Women and Political Culture in Early National Philadelphia* (Philadelphia: University of Pennsylvania Press, 2001). On women voting in New Jersey, see Judith Apter Klinghoffer and Lois Elkis, "'The Petticoat Electors': Women's Suffrage in New Jersey, 1776–1807," *Journal of the Early Republic* 12, no. 2 (Summer 1992): 159–93; and Rosemarie Zagarri, *Revolutionary Backlash: Women and Politics in the Early American Republic* (Philadelphia: University of Pennsylvania Press, 2007), 30–37.

30. Physiognomical descriptions of American politicians often suggested that republican virtue was visible in the visage. See Christopher Lukasik, "The Face of the Public," *Early American Literature* 39, no. 3 (2004): 413–64; Wendy Bellion, "Heads of State: Profile and Politics in Jeffersonian America," in *New Media, 1740–1915*, ed. Lisa Gitelman and Geoffrey B. Pingree (Cambridge, MA: MIT Press, 2003), 31–59; and Fortune, "Portraits of Virtue and Genius."

31. Henry McNeal Turner, "Speech on the Eligibility of Colored Members to Seats in the Georgia Legislature," 3 September 1868, The Henry McNeal Turner Project, http://www.thehenrymcnealturnerproject.org/2019/03/speech-on-eligibility-of-colored.html. For a good analysis of this speech, see Michele Mitchell, "'Lower Orders,' Racial Hierarchies, and Rights Rhetoric: Evolutionary Echoes in Elizabeth Cady Stanton's Thought during the Late 1860s," in Ellen Carol DuBois, ed. *Elizabeth Cady Stanton, Feminist as Thinker: A Reader in Documents and Essays* (New York: New York University Press, 2007), 141.

CHAPTER ONE

1. Abigail Smith Adams to John Adams, 16 July 1775; and Abigail Smith Adams to John Adams, 5 November 1775, in the *Adams Family Papers: An Electronic Archive*, MHS, http://www.masshist.org/digitaladams/. George Washington's friend, Elizabeth Powel, similarly analyzed the "benignity" of his countenance. See Elizabeth Powel to Ann Bolling Randolph Fitzhugh, 18 May 1784, Box 3, Folder 4, Series III, Powel Family Papers (Collection 1582), HSP. Thank you to Samantha Snyder, Reference Librarian at Mount Vernon, for making me aware of the Powel family's engagement with physiognomy.

2. John Adams to Abigail Smith Adams, 18 November 1775, MHS. Interestingly, John Adams analyzed his own face and found it lacking. "There is a Feebleness and a Languor in my Nature," he wrote. "My Mind and Body both partake of this Weakness. By my Physical Constitution, I am but an ordinary Man. The Times alone have destined me to Fame—and even these have not been able to give me, much. When I look in the Glass, my Eye, my Forehead, my Brow, my Cheeks, my Lips, all betray this Relaxation." See John Adams Diary 47, 26 April 1779, MHS.

3. On physiognomy and political virtue, see Brandon Brame Fortune, "Portraits of Virtue and Genius: Pantheons of Worthies and Public Portraiture in the Early American Republic, 1780–1820," (PhD diss., University of North Carolina, Chapel Hill, 1986); Christopher Lukasik, "The Face of the Public," *Early American Literature* 39, no. 3 (2004): 413–64; and Wendy Bellion, "Heads of State: Profile and Politics in Jeffersonian America," in *New Media, 1740–1915*, ed. Lisa Gitelman and Geoffrey B. Pingree (Cambridge, MA: MIT Press, 2003), 31–59.

4. Lavater began publishing on physiognomy in 1772, but he released his first major volume in 1775. According to John Graham, "By 1810, there had been published sixteen German, fifteen French, two American, one Dutch, one Italian, and no less than twenty English versions—a total of fifty-five editions in less than forty years." Christopher Lukasik also argues that "By 1826, no fewer than 677 articles referring to physiognomy had been published in newspapers and periodicals from Maine to Florida, Massachusetts to Illinois." This count understates physiognomy's impact, because Lukasik only counts articles where the words "Lavater" or "physiognomy" appear explicitly, and not articles in which authors simply used Lavaterian techniques to describe real individuals and fictional characters. See Graham, "Lavater's Physiognomy in England," *Journal of the History of Ideas* 22, no. 4 (1961): 562; and Lukasik, *Discerning Characters: The Culture of Appearance in Early America* (Philadelphia: University of Pennsylvania Press, 2011), 33.

5. See "Lavater," *The Scots Magazine, or, General Repository of Literature, History, and Politics* 63, no. 2 (Edinburgh), February 1801, quoted in John Graham, "Lavater's 'Physiognomy': A Checklist," *Papers of the Bibliographical Society of America* 55, no. 4 (1961): 297. For the phrase "celebrated Lavater," see "Account of a Visit to the Celebrated Lavater, and the Present Religion in France," *Vergennes Gazette and Vermont and New-York Advertiser* (Vergennes, VT), 27 December 1798; "A Letter Received at Baltimore," *Newport Mercury* (Newport, RI), 16 July 1799; "Philadelphia, July 10," *New Hampshire Gazette* (Portsmouth, NH), 23 July 1799; "New York, July 29. By the Brig Trio, Capt. Hill," *The Pennsylvania Gazette*, 31 July 1799; "Boston, May 1. From France-Direct," *American Citizen and General Advertiser* (New York), 6 May 1800; "Bern, Aug. 16," *Political Repository* (Brookfield, MA), 4 November 1800; "Bern, August 16," *The Oracle of Dauphin and Harrisburgh Advertiser* (Harrisburg, PA), 17 November 1800; "Observations on the Philosophy of Kant," *The New England Quarterly Magazine* 2, no. 3 (Boston), October-December 1802; "Extracts from the Letters of an American Traveller in Europe, written in 1800 and 1801," *The Port-Folio* 4, no. 8 (Philadelphia), 25 February 1804; "Extract from One of the Letters of an American Traveller in Europe," *Northern Post* (Salem, NY), 18 October 1804; "Communications. For the Balance," *The Balance, and Columbian Repository* (Hudson, NY), 30 September 1806; Thomas Branagan, *The Excellency of the Female Character Vindicated* (New York: Samuel Wood, 1807), 106; "The Ladies' Toilette; or, Encyclopaedia of Beauty," *Lady's Weekly Miscellany* 6, no. 15 (New York), 6 February 1808; Benjamin Rush, *Medical Inquiries and Observations, Upon the Diseases of the Mind* (Philadelphia: Kimber and Richardson, 1812), 260; "Anecdotes," *Religious Remembrancer* 1, no. 9 (Philadelphia), 30 October 1813; "Buffon, Klopstock, and Gessner," *The Port-Folio* 2, no. 5 (Philadelphia), November 1813; and "Dress," *The Parlour Companion* 2, no. 14 (Philadelphia), 4 April 1818. One newspaper claimed that Lavater "enjoys not an hour of leisure" because travelers from all over the

world kept coming "to see a man of such celebrity." See "Anecdote of Lavater," *Concord Observer* (New Hampshire), 11 October 1819.

6. For early examples of American periodicals writing about physiognomy, see "Extract from a Treatise on Physiognomy," *New Haven Gazette, and the Connecticut Magazine* 1, no. 41, 23 November 1786; "London," *The Massachusetts Centinel and the Republican Journal* 6, no. 28 (Boston), 23 December 1786; "Physiognomy," *The Pennsylvania Herald and General Advertiser* 5, no. 62 (Philadelphia), 25 August 1787; "Jack Flash," *The Columbian Magazine* 1, no. 15 (Philadelphia), November 1787; and "To the Editor," *The Columbian Magazine* 2 (Philadelphia), March 1788. For a novel that ruminates on the legitimacy and usefulness of Lavaterian theories, see Louis-Sébastien Mercier, *The Night Cap* (Philadelphia: W. Spotswood, 1788).

7. The most popular edition in the United States was Johann Caspar Lavater, *Essays on Physiognomy; for the Promotion of the Knowledge and the Love of Mankind*, trans. Thomas Holcroft (London: G. G. J. and J. Robinson, 1789). Unless otherwise noted, I will cite from this version. There were also translations by Henry Hunter and an abridged version by Reverend C. Moore. For the first American edition, see *Essays on Physiognomy; for the Promotion of the Knowledge and the Love of Mankind* (Boston: William Spotswood & David West, 1794). The advertisement for that edition noted that Lavater's works were already "so universally known and celebrated" by 1794 that it was "unnecessary to attempt their eulogium." The editors complained about "literary pirates" who shamelessly plagiarized the books and "mangled" Lavater's original arguments.

8. Editions of Lavater's "Aphorisms on Man" were published in Boston, New York, and Philadelphia in 1790, Newburyport in 1793, and Catskill, New York in 1795. They were also reprinted alongside other essays in *The New Complete Letter Writer* (Philadelphia, 1792); and *The Gentleman's Pocket Library* (Boston, 1794). Three editions of the "Pocket Lavater" were published in New York City between 1817 and 1818, as well as in New Haven, Connecticut in 1829, and New York City in 1839. Editions of the "Juvenile Lavater" were published in New York City in 1815.

9. Dror Wahrman, *The Making of the Modern Self: Identity and Culture in Eighteenth-Century England* (New Haven: Yale University Press, 2004), 294.

10. "Lavater," *Lady's Monitor* 1, no. 13 (New York), 14 November 1801. This quotation was reprinted widely after Lavater's death. Original versions did not include "America" in the list of places enraptured by physiognomy. It seems as if the *Lady's Monitor* made this addition themselves. For European versions of the same article, see "Additions and Corrections.—Obituary," *Gentleman's Magazine and Historical Chronicle* (London), 1801; and "Deaths Abroad," *The Monthly Magazine, or, British Register* (London), 1801. For an American example, see *Eccentric Biography; or, Sketches of Remarkable Characters, Ancient and Modern* (Boston: B. & J. Homans, 1804), 175.

11. Americans did not universally accept physiognomy as an exact science, but "few denied physiognomy's pervasiveness as a social practice or its general truth as an imperfect and still evolving science of moral character." Lukasik, *Discerning Characters*, 35.

12. John Adams to Abigail Smith, 14 April 1764, MHS.

13. Americans recognized that Lavater, despite his popularity, was not the first to come up with a physiognomical system. See "The Gleaner: No. II," *The Literary Magazine, and American Register* 5, no. 32 (Philadelphia), May 1806.

14. See entry for July 1794, in Elaine Forman Crane, ed., *The Diary of Elizabeth Drinker* (Boston: Northeastern University Press, 1991), 1:573–74. The obituary appeared in *Poulson's American Daily Advertiser* on 2 December 1807, and in *Bronson's United States Gazette*. Quoted in Crane, ed., *The Diary of Elizabeth Drinker*, 304n11. See also "Mortuary," *The Port Folio* 4, no. 23 (Philadelphia), 5 December 1807; and Charles Brockden Brown and Robert Walsh, eds., *The American Register, or General Repository of History, Politics, and Science* (Philadelphia: A. Conrad & Co., 1808), 98–99.

15. Joseph Bartlett, *Physiognomy: A Poem Delivered at the Request of The Society of ΦBK in the Chapel of Harvard University on the Day of Their Anniversary, July 18th, 1799* (Boston: John Russell, 1799); and Richard Brown, *An Essay on the Truth of Physiognomy and its Application to Medicine* (Philadelphia: Thomas T. Stiles, 1807), CPP. For an advertisement of Brown's graduation, see *The United States' Gazette* (Philadelphia), 11 April 1807. For a positive review of his "interesting dissertation," see the *Philadelphia Medical Museum* 4, no. 4, 1807. One scholar has suggested that late-eighteenth-century lawyers in Edinburgh relied on physiognomy to stop a man named John Wright from practicing law. See John W. Cairns, "The Face that Did Not Fit: Race, Appearance, and Exclusion from the Bar in Eighteenth-Century Scotland," *Fundamina* 9, no. 11 (2003): 11–43.

16. Benjamin Rush initially delivered this lecture before the American Philosophical Society in 1786. For a reprint, see Rush, *An Inquiry into the Influence of Physical Causes Upon the Moral Faculty* (Philadelphia: Haswell, Barrington, and Haswell, 1839), 3.

17. See "Chapman's Notes, between circa 1810 and 1830," and "Chapman's Lecturs [sic], after 1810," in the Nathaniel Chapman Papers, ca. 1810–1853, MSS 10a Chapman, CPP.

18. "Likenesses of the Insane, taken for Dr. Jn. K. Mitchell during his Service at the Pennsylvania Hospital," MSS 10a/133, CPP.

19. Isaac Ray ("A Physician"), *Conversations on the Animal Economy* (London: Longman, Rees, Orme, Brown, and Green, 1827), 56–65, 164–70. For another didactic physiognomic lesson for young girls, see *The Children's Companion* (New York: S. King, 1825), 29–40.

20. Catharine Maria Sedgwick to Katharine Sedgwick Minot, 2 May 1852, in Mary E. Dewey, ed., *Life and Letters of Catharine M. Sedgwick* (New York: Harper & Brothers, 1871), 339. Sedgwick was referring to James Redfield, author of *Outlines of a New System of Physiognomy* (New York: J. S. Redfield, 1849).

21. Ronald J. Zboray, "Reading Patterns in Antebellum America: Evidence in the Charge Records of the New York Society Library," *Libraries & Culture* 26, no. 2 (Spring 1991): 311–12.

22. "H," "The Eloquence of Eyes," *New Monthly Magazine and Literary Journal* 4, no. 19 (Boston), 1 July 1822.

23. Louisa Catherine Johnson Adams to John Adams, 21 July 1822, *Founders Online*, National Archives, https://founders.archives.gov/documents/Adams/99-03-02-4091.

24. On physiognomical pantheons, see Fortune, "Portraits of Virtue and Genius," 34–86; David C. Ward, *Charles Willson Peale: Art and Selfhood in the Early Republic* (Berkeley: University of California Press, 2004), 82–89; Lukasik, *Discerning Characters*, 127–40; and Catherine E. Kelly, *The Republic of Taste: Art, Politics, and Everyday Life in Early America* (Philadelphia: University of Pennsylvania Press, 2016), 159–94.

25. For printed galleries, see James Hardie, *The New Universal Biographical Dictionary* (New York: Johnson & Stryker, 1801–1805); John Sanderson, *Biographies of the Signers of the Declaration of Independence* (Philadelphia: R. W. Pomeroy, 1823–27); James Thacher, *American Medical Biography* (Boston: Richardson & Lord, 1828); James Herring and James Barton Longacre, *The National Portrait Gallery of Distinguished Americans* (Philadelphia: Henry Perkins, 1834–1839); Stephen W. Williams, *American Medical Biography* (Greenfield, MA: L. Merriam and Co., 1845); and Rufus Wilmot Griswold, *The Republican Court; or American Society in the Days of Washington* (New York: D. Appleton and Company, 1856).

26. Charles Willson Peale to the Representatives of the State of Massachusetts in Congress, 14 December 1795, in Lillian B. Miller, ed., *Selected Papers of Charles Willson Peale and His Family: The Artist as Museum Keeper, 1791–1810*, vol. 2, pts. 1 and 2 (New Haven, CT: Yale University Press, 1983), 136. Peale owned his own copy of Lavater's *Aphorisms on Man*. In the marginalia, he wrote

that Lavater was a skilled and careful observer of character. On Peale's efforts to convey character through portraiture, see Fortune, "Portraits of Virtue and Genius," 166–82.

27. Typescript version of C. W. Peale Diaries, 1817–1818, 2 December 1817, Series 7, Volume 14, Peale-Sellers Family Collection, 1686–1963, Mss.B.P31, APS.

28. Charles Willson Peale to Rembrandt Peale, 16 June 1817, quoted in Lillian B. Miller, *In Pursuit of Fame: Rembrandt Peale, 1778–1860* (Seattle: University of Washington Press, 1992), 123.

29. According to David R. Brigham, "One of Peale's central ideas was that nature is simultaneously hierarchical and harmonious." He sought to portray this order in his museum. See Brigham, "'Ask the Beasts, and They Shall Teach Thee': The Human Lessons of Charles Willson Peale's Natural History Displays," *Huntington Library Quarterly* 59, no. 2/3 (1996): 183. David C. Ward also points out that Peale's "collection included no women, African Americans, working people, or 'ordinary' Americans, let alone 'outcasts' like the poor or criminals." See Ward, *Art and Selfhood in the Early Republic*, 83. One exception appears to be the portrait of Joseph Brandt, the Mohawk leader, who appeared in Peale's gallery after 1797. See Elizabeth Hutchinson, "From Pantheon to Indian Gallery: Art and Sovereignty on the Early Nineteenth-Century Frontier," *Journal of American Studies* 47, no. 2 (2013): 333–34.

30. Baron de Beaujour, *Sketch of the United States of North America*, trans. William Walton (London: J. Booth, 1814), 151.

31. For reprintings, see "The English and the Americans Compared," *The Mirror of Literature, Amusement, and Instruction* 4, no. 102 (London), 4 September 1824; "Speculations of a Traveller concerning the People of the United States," *The Museum of Foreign Literature, Science, and Art* 5, no. 28 (Philadelphia), 1 October 1824; "England and the United States," *Daily National Intelligencer* (Washington, DC), 22 October 1824; "Speculations of a Traveller concerning the People of the United States; with Parallels," *The Athenaeum; or, Spirit of the English Magazines* 2, no. 3 (Boston), 1 November 1824; and "The English and the Americans Compared," *United States Catholic Miscellany* 13, no. 32 (Charleston, SC), 8 February 1834.

32. Joseph Delaplaine, *Repository of the Lives and Portraits of Distinguished American Characters* (Philadelphia, 1815–1816), iv–v (for "transcendent greatness," 81).

33. *Prospectus of Delaplaine's National Panzographia: For the Reception of the Portraits of Distinguished Americans* (Philadelphia: William Brown, 1818), 12.

34. See "Declaration of Independence," *The Port-Folio* 18, no. 270 (Philadelphia), October 1824; "Review of New Books," *The Literary Gazette; or, Journal of Criticism, Science, and the Arts* 1, no. 1 (Philadelphia), 6 January 1821; "ART. XII.—Biography of the Signers to the Declaration of Independence," *The North American Review* 13, no. 38 (Boston), January 1823.

35. Robert T. Conrad, ed., *Sanderson's Biography of the Signers to the Declaration of Independence* (Philadelphia: Thomas, Cowperthwait, & Co., 1847), 377, 389, 714, 789.

36. Lavater, *Essays on Physiognomy*, 2:28.

37. "Portrait of General Washington," *The Columbian Magazine* 1, no. 5 (Philadelphia), January 1787.

38. John Bell, "Sketches of the President of the United States," *The Massachusetts Magazine; or, Monthly Museum* 3, no. 3 (Boston), March 1791.

39. Samuel Powel, Accounts and Memorandum (1786–1796), box 2, vol. 32, Powel Family Papers, HSP. Thank you to Samantha Snyder at the Fred W. Smith National Library for the Study of George Washington for making me aware of this source. The silhouette, along with a small note explaining its creation, is currently displayed at the Powel House in Philadelphia. For more on the silhouette, see Kelly, *Republic of Taste*, 2–3. Washington's own letters suggest that he was familiar with physiognomy. Yet when Francis Bailey sent him an "english translation of Lavater's treatise,"

his secretary, Tobias Lear, responded, "the President's time is so much occupied by business" that he would not have time to read it. See Tobias Lear to Francis Bailey, 26 January 1791, *Founders Online*, National Archives, https://founders.archives.gov/documents/Washington/05-07-02-0159.

40. "Memoirs of George Washington, Esq.," *The Philadelphia Monthly Magazine; or, Universal Repository of Knowledge and Entertainment* 1, no. 6, June 1798.

41. "Washington," *The Rural Magazine* 1, no. 26 (Newark), 11 August 1798.

42. *Delaplaine's Repository*, 1:81, 105.

43. Rembrandt Peale, *Portrait of Washington: An Account of a Portrait by Rembrandt Peale, with Letters Addressed to the Artist, Testifying to the Accuracy of the Likeness* (Philadelphia, 1824), 3.

44. Lavater, *Essays on Physiognomy*, trans. Henry Hunter (London: Murray and Highly, 1798), 3:135–37.

45. Lavater, *Essays on Physiognomy*, trans. Hunter, 3:135–37.

46. Lavater, *Essays on Physiognomy*, 1:16. Physiognomists, phrenologists, ethnologists, and anthropologists deployed this tripartite division of the countenance throughout the late eighteenth and nineteenth centuries. See, for instance, Herbert Spencer, *Essays: Scientific, Political and Speculative* (London: Longman, Brown, Green, Longmans, and Roberts, 1858), 1:422. For a scholarly discussion of the tripartite division of the face, see Lucy Hartley, "A Science of Beauty? Femininity and the Nineteenth-Century Physiognomic Tradition," *Women: A Cultural Review* 12, no. 1 (2001): 23–24.

47. Lukasik refers to this as the "physiognomic fallacy." Physiognomy rested on the idea "that a person has one essential character over time and that a face can express it." In the end, though, Americans placed their hope in a "false opposition between a model of character read from performance and one read from the face." See Lukasik, *Discerning Characters*, 21, 30.

48. Christopher Rivers, *Face Value: Physiognomical Thought and the Legible Body in Marivaux, Lavater, Balzac, Gautier, and Zola* (Madison: University of Wisconsin Press, 1994), 74. Brandon Brame Fortune has likewise argued that "Lavater's theory was not terribly systematic in nature." See Fortune, "Portraits of Virtue and Genius," 192.

49. "Physiognomy," *New England Galaxy* 1, no. 22 (Boston), 13 March 1818.

50. Lavater, *Essays on Physiognomy*, first American edition (1794), 48–49.

51. See, for example, Prince de Broglie's 1782 description of Washington, in William Spohn Baker, *Character Portraits of Washington as Delineated by Historians, Orators and Divines* (Philadelphia: Robert M. Lindsay, 1887), 18; Robert Hunter Jr. Travel Diary, 16 November 1785, in John P. Kaminski, ed., *George Washington: A Man of Action* (Madison: Wisconsin Historical Society Press, 2017), 41; Edward Thornton to Sir James Bland, 2 April 1792, in "The United States through English Spectacles in 1792–1794," *Pennsylvania Magazine of History and Biography* 9, no. 2 (July 1885): 214; and George Washington Parke Custis, *Recollections and Private Memoirs of Washington* (Philadelphia: J.W. Bradley, 1861), 527.

52. Letter from Abigail Adams to John Adams, 28 January 1797, MHS.

53. Emma Buckley Howard Edwards, Scrapbooks [ca.1820–ca.1880], fol. 255, Winterthur Library. Edwards also cut out physiognomical engravings from the *Columbian Magazine* and inserted them into her scrapbook.

54. Lukasik, *Discerning Characters*; and Karen Halttunen, *Confidence Men and Painted Women: A Study of Middle-Class Culture in America, 1830–1870* (New Haven, CT: Yale University Press, 1982).

55. For examples of writers who argued for a connection between physiognomy and social status, see "Sketch of Lavater," *Lady's Magazine and Musical Repository* (New York), May 1801; Jonathan Truepenny, "To Peter Sketch," *The Providence Gazette*, April 1803; "The Gleaner: No. II,"

The Literary Magazine, and American Register 5, no. 32 (Philadelphia), May 1806; Thomas Branagan, *The Charms of Benevolence and Patriotic Mentor* (Philadelphia: Johnston and Patterson, 1813), 109–80; "American Biography: Life of Henry Laurens, Esq.," *The Port-Folio* 4, no. 3 (Philadelphia), September 1814; "Article VI. Tracts Published by the Christian Tract Society, London, and Republished by Wells and Lilly," *The Christian Disciple and Theological Review* 2, no. 8 (Boston), March/April 1820; "Old Times," *Carolina Observer* 353 (Fayetteville, NC), 4 March 1824. For satirical examples, see "Musaeus's Physiognomical Journal," *The Literary Mirror* 1, no. 8 (Portsmouth), 9 April 1808; and Musaeus, "Article 1—Answer," *The Literary Mirror* 1, no. 11 (Portsmouth), 23 April 1808.

56. Noah Webster, *The Prompter; or, A Commentary on Common Sayings and Subjects* (Boston: Thomas and Andrews, 1794), 17. For the countenance as a "letter of recommendation," see E. W. Robbins, "A Chapter on Physiognomy," *The American Literary Magazine* (Hartford), May 1849.

57. Mathew Carey, "A Fragment," *The New York Magazine, or Literary Repository* (New York), August 1796. For a similar story, see "Estelle Aubert: A Tale," *American Ladies' Magazine* 7, no. 7 (Boston), July 1834. For scholarship on how such stories were supposed to inspire "sympathy" and "sensibility" in middle-class readers, see Sarah Knott, *Sensibility and the American Revolution* (Chapel Hill: University of North Carolina Press, 2009), especially chapters 1 and 2; Nicole Eustace, *Passion Is the Gale: Emotion, Power, and the Coming of the American Revolution* (Chapel Hill: University of North Carolina Press, 2008); Julia A. Stern, *The Plight of Feeling: Sympathy and Dissent in the Early American Novel* (Chicago: University of Chicago Press, 1997); and G. J. Barker-Benfield, *The Culture of Sensibility: Sex and Society in Eighteenth-Century Britain* (Chicago: University of Chicago Press, 1992).

58. Lukasik, *Discerning Characters*, 38.

59. Africanus, "For the New-York Journal, &c. Negroes Interior to the Whites," *New-York Journal and Patriotic Register* (New York), 4 February 1792.

60. See Petrus Camper, *The Works on the Connexion between the Science of Anatomy and the Arts of Drawing, Painting, Statuary*, trans. T. Gogan (London, 1794), 42. Miriam Meijer points out that Camper argued for the universal humanity of mankind. Nineteenth-century scientific racists, however, purposefully misinterpreted Camper's theories for their own ends. See Meijer, *Race and Aesthetics in the Anthropology of Petrus Camper* (Amsterdam: Rodopi, 1999). On physiognomy's relationship to scientific racism, see John Wood Sweet, *Bodies Politic: Negotiating Race in the American North, 1730–1830* (Philadelphia: University of Pennsylvania Press, 2003), 296–98; Roxann Wheeler, *The Complexion of Race: Categories of Difference in Eighteenth-Century British Culture* (Philadelphia: University of Pennsylvania Press, 2000); Judith Wechsler, "Lavater, Stereotype, and Prejudice," in *The Faces of Physiognomy: Interdisciplinary Approaches to Johann Caspar Lavater*, ed. Ellis Shookman (Columbia, SC: Camden House, 1993), 104–26; Kay Flavell, "Mapping Faces: National Physiognomies as Cultural Prediction," *Eighteenth-Century Life* 18 (1994–95): 8–22; and Richard T. Gray, *About Face: Physiognomic Thought from Lavater to Auschwitz* (Detroit, MI: Wayne State University Press, 2004). On physiognomy's relationship to anthropology, see Mary Cowling, *The Artist as Anthropologist: The Representation of Type and Character in Victorian Art* (Cambridge: Cambridge University Press, 1989).

61. See Lavater, *Essays on Physiognomy*, 2:140–44, 2:221, 2:235–36, 3:199, 3:205–6.

62. These remarks were influential because they infiltrated the periodicals that Americans read regularly. As Susan Branson has argued, "It was magazines, more than any other medium, that helped to develop an American public discourse on gender roles and gender relations." See Branson, *These Fiery Frenchified Dames*, 23. In 1789, Lavater's remarks on women were published under the title "Male and Female" in the London periodical the *Analytical Review, or History of Literature, Domestic and Foreign*. See also "Characteristic Differences of the Male and Female of

the Human Species," *Edinburgh Magazine, or Literary Miscellany* (Edinburgh), January 1790. The *New York Magazine* lifted its comments directly from the *Edinburgh Magazine*. See "Characteristic Differences of the Male and Female of the Human Species: A Word on the Physiognomonical Relation of the Sexes," *New York Magazine, or Literary Repository* 1, no. 6, June 1790. See also "General Remarks on Women," *Massachusetts Magazine; or, Monthly Museum* 6, no. 1 (Boston), January 1794; "General Remarks on Women," *Rural Magazine; or, Vermont Repository* 2, no. 11 (Rutland, VT), November 1796; and "From Lavater's *Essays on Physiognomy*," *Weekly Visitor, or Ladies' Miscellany* 1, no. 28 (New York), 16 April 1803.

63. For the first edition, see *Encyclopaedia Britannica: or, A Dictionary of Arts and Sciences*, first edition, vol. 3 (Edinburgh: A. Bell and C. Macfarquhar, 1771), 581.

64. See *Encyclopaedia Britannica: or, A Dictionary of Arts, Sciences, and Miscellaneous Literature; Enlarged and Improved*, third edition, vol. 17 (Edinburgh: A. Bell and C. Macfarquhar, 1797), 327–29. The 1860 edition removed the entry for "Sex" entirely. The entry reappeared in the ninth edition, published in 1886, but Lavater's comments were excised and replaced with a longer and more technical discussion of reproductive organs in insects, fish, mammals, and humans. This new entry synthesized the work of evolutionary biologists like Charles Darwin and Herbert Spencer. See *Encyclopaedia Britannica: or, A Dictionary of Arts, Sciences, and General Literature*, ninth edition, vol. 21 (Edinburgh: Adam and Charles Black, 1886), 721–25.

65. See Thomas Dobson's *Encyclopaedia; or, A Dictionary of Arts, Sciences, and Miscellaneous Literature* (Philadelphia: Thomas Dobson, 1798), 17:327–29; *New Encyclopædia; or, Universal Dictionary of Arts and Sciences* (London: Vernor, Hood, and Sharpe, 1807), 20:520–22; and *Encyclopaedia Perthensis; or Universal Dictionary of the Arts, Sciences, Literature, &c.*, second edition (Edinburgh: John Brown, 1816), 20:520–21. Sydney Owenson repeated these alleged truisms about the female mind in a letter to Mrs. Lefanu in 1803, but misidentified "Lavater" as "Salvater." See *Lady Morgan's Memoirs: Autobiography, Diaries, and Correspondence* (Leipzig: Bernhard Tauchnitz, 1863), 1:211. Lavater's quotation—about men being more "profound" thinkers than women—was also quoted on numerous occasions: John Corry, *A Satirical View of London at the Commencement of the Nineteenth Century* (London: G. Kearsley, 1801), 173; *Encyclopaedia Britannica* (Edinburgh: Andrew Bell, 1810), 29:203–4; John Platts, *The Book of Curiosities; or, Wonders of the Great World* (London: Caxton Press, 1822), 34–46; *Memoirs of a Deist, Written First A.D. 1793–4* (London: J. Hatchard and Son, 1824), 85; William H. Porter, *Proverbs: Arranged in Alphabetical Order* (Boston: James Munroe and Company, 1845), 74–77; and "Woman Superior to Man," in *Everybody's Book: or, Gleanings Serious and Entertaining, in Prose and Verse, from the Scrap-Book of a Septuagenarian*, ed. John Henry Freese (London: Longman, Green, Longman, and Roberts, 1860), 298–301. Lavater's remarks continued to be reprinted in European and American periodicals until at least 1882, when they were directly copied into the entry for "Curiosities Respecting Man.—Difference Between the Sexes," in I. Platt, ed., *Cyclopedia of Wonders and Curiosities of Nature and Art, Science and Literature* (New York: Hurst & Company, 1882), 1:34–35.

66. For critiques of Lavater's ideas about women, see "Defence of the Female Claim to Mental Equality," *Lady's Monthly Museum* 6, no. 7 (London), February 1801; and "Defence of the Female Claim to Mental Equality," *Lady's Magazine and Musical Repository* (New York), December 1801.

67. For scholarship on gender, virtue, and politics in the early republic, see Kerber, *Women of the Republic*; Mary Beth Norton, *Liberty's Daughters: The Revolutionary Experience of American Women, 1750–1800* (Boston: Little, Brown, 1980); Ruth H. Bloch, "The Gendered Meanings of Virtue in Revolutionary America," *Signs* 13, no. 1 (Autumn 1987): 37–58; Jan Lewis, "The Republican Wife: Virtue and Seduction in the Early Republic," *William and Mary Quarterly* 44, no. 4 (October 1987): 689–721; and Rosemarie Zagarri, *Revolutionary Backlash: Women and Politics in*

the Early American Republic (Philadelphia: University of Pennsylvania Press, 2007). On women's increased access to education, see Mary Kelley, *Learning to Stand and Speak: Women, Education, and Public Life in America's Republic* (Chapel Hill: University of North Carolina Press, 2006); and Lucia McMahon, *Mere Equals: The Paradox of Educated Women in the Early American Republic* (Ithaca, NY: Cornell University Press, 2012).

68. Branagan, *The Charms of Benevolence*, 120, 128. John Quincy Adams once used physiognomy to distinguish between two female companions. Although he admired the countenance of a woman named Miss Robertson, he noted that she wore a patch on her face, which convinced him she might be "a coquet." See the Diary of John Quincy Adams, 26 June 1787, Adams Papers: Digital Edition, MHS.

69. Branagan, *The Charms of Benevolence*, 123 (for "she fiend," see 119). Lavater's description of the "noble spotless maiden" also appeared in *Medical Abstracts: On the Nature of Health, with Practical Observations; and the Laws of the Nervous and Fibrous Systems* (London: J. Johnson, 1797), 809; Robert John Thornton, *The Philosophy of Medicine: or, Medical Extracts on the Nature of Health and Disease*, 4th edition (London: C. Whittingham, 1799), 48; "Danger of Sporting with the Affections," *The New England Quarterly* 2, no. 1 (Boston), April-June 1802; and "Danger of Sporting with the Affections," *Philadelphia Repository and Weekly Register* 3, no. 34, 20 August 1803. For Lavater's description of the ideal woman, see *Essays on Physiognomy*, first American edition (1794), 174.

70. Mason Locke Weems, *The Lover's Almanac* (Fredericksburg, VA: T. Green, 1798). It appears that Weems plagiarized large sections of this text from Samuel Jackson Pratt, *The Pupil of Pleasure*, 2nd edition (London: G. Robinson and J. Bew, 1777), 219. Other authors also urged men to look beyond the beauties of the body while still describing women's virtue and intelligence in visual terms. See "An Essay on Woman," *The Maryland Journal* (Baltimore), 24 January 1786; "On Beauty and Flattery," in *The Pleasing Instructor, or Entertaining Moralist* (Boston: Joseph Bumstead, 1795), 65–68; and George Wright, *The Lady's Miscellany* (Boston: William T. Clap, 1797), 9–12.

71. Mason Locke Weems, *The Bad Wife's Looking Glass; or, God's Revenge Against Cruelty to Husbands*, 2nd edition (Charleston: Printed for the author, 1823), 4–5

72. Branagan, *The Charms of Benevolence*, 124–30.

73. Lukasik, *Discerning Characters*, 32.

CHAPTER TWO

1. Samuel Wells to John Brown Jr., 2 March 1854, Charles E. Frohman Collections, FR-5, Rutherford B. Hayes Presidential Library.

2. Nelson Sizer to John Brown Jr., 26 March 1860, Charles E. Frohman Collections, FR-5, Rutherford B. Hayes Presidential Library.

3. Christopher G. White, "Minds Intensely Unsettled: Phrenology, Experience, and the American Pursuit of Spiritual Assurance, 1830–1880," *Religion and American Culture: A Journal of Interpretation* 16, no. 2 (Summer 2006): 245.

4. Harriet Beecher Stowe, *Uncle Tom's Cabin; or, Life among the Lowly* (Boston: John P. Jewett & Company; Cleveland: Jewett, Proctor, and Worthington, 1852), vol. 1: 13–14, 16, 40–41, 98, 133, 184, 211–12, 229, and 269; and vol. 2: 12, 32, 70, 102, 112, 114, 164, 166, 181, and 189.

5. Stephen Tomlinson, *Head Masters: Phrenology, Secular Education, and Nineteenth-Century Social Thought* (Tuscaloosa: University of Alabama Press, 2005), chapter 11.

6. Garrison commissioned a phrenological reading from Lorenzo Fowler and published the

results in his newspaper. "Mr. Fowler was entirely ignorant of the person whose head he was examining," the paper explained, "and yet captured Garrison's personality with "striking accuracy." The newspaper cited this fact as evidence that phrenology was "founded in truth." See "Miscellaneous," *The Liberator* (Boston), 21 September 1838. See also "Prospectus of the American Phrenological Journal & Miscellany," *The Liberator* (Boston), 16 November 1838; and "Phrenology," *The Liberator* (Boston), 29 November 1839. For more on Garrison's interest in phrenology, see Goldwin Smith, ed., *The Moral Crusader: William Lloyd Garrison—A Biographical Essay* (New York: Funk and Wagnalls, 1892), 115.

7. Lucretia Coffin Mott to George Combe, 13 June 1839, in Beverly Wilson Palmer, ed., *Selected Letters of Lucretia Coffin Mott* (Urbana and Chicago: University of Illinois Press, 2002), 53; and Carla Bittel, "Woman, Know Thyself: Producing and Using Phrenological Knowledge in 19th-Century America," *Centaurus: An International Journal of the History of Science and its Cultural Aspects* 55, no. 2 (May 2013): 120.

8. Gall initially called his science *Schädellehre* (doctrine of the skull) or *Organologie*. It was not until around 1815 that people began using the word "phrenology." See John van Wyhe, "The Authority of Human Nature: The *Schädellehre* of Franz Joseph Gall," *British Journal of the History of Science* 35 (2002): 22–24.

9. J. G. Spurzheim, *Phrenology, in Connexion with the Study of Physiognomy* (London: Treuttel, Würtz and Richter, 1826), 22.

10. For descriptions of Spurzheim, see "Dr. Spurzheim," *Boston Medical and Surgical Journal* 8, no. 7, 27 March 1833; and "Obituary Notice of Dr. Gaspar Spurzheim," *American Journal of Science and Arts* 23, no. 2 (New Haven, CT), 2 January 1833. On his visit to Yale, see Nahum Capen, *Reminiscences of Dr. Spurzheim and George Combe* (New York: Fowler and Wells, 1881), 9.

11. "Dr. Spurzheim," *Ladies' Magazine and Literary Gazette* 5, no. 12 (Boston), December 1832.

12. John D. Davies, *Phrenology, Fad and Science: A 19th-Century American* Crusade (New Haven, CT: Yale University Press, 1955), 26. For a firsthand account of Combe's lectures, see Madeleine Stern, "A Letter from New Haven—1840," *Yale University Library Gazette* 49, no. 3 (January 1975): 288–92.

13. W. Emerson Wilson, ed., *Phoebe George Bradford Diaries* (Wilmington: Historical Society of Delaware, 1975), 88.

14. Madeleine Stern, *Heads and Headlines: The Phrenological Fowlers* (Norman: University of Oklahoma Press, 1971), 32.

15. For the publication history of the *American Phrenological Journal and Miscellany*, see Stern, *Heads and Headlines*, 26. Stern argues that by the 1850s, "Fowler and Wells were manipulators not only of heads but of big business. Their publishing department, under Samuel Wells, claimed to have the largest mail-order list in the city" (84).

16. Some scholars have suggested that phrenology "supplanted" physiognomy in the 1830s. See, for example, Cathy Boeckmann, *A Question of Character: Scientific Racism and the Genres of American Fiction, 1892–1912* (Tuscaloosa: University of Alabama Press, 2006), 45; and Jonathan Smith, *Charles Darwin and Victorian Visual Culture* (Cambridge: Cambridge University Press, 2006). As Smith recognizes, though, physiognomical works continued to be popular throughout the nineteenth century, and phrenologists remained preoccupied with the face. In 1863, the *American Phrenological Journal* added the word "physiognomy" to its title. The Fowler and Wells press also published numerous explicitly physiognomical works: Samuel R. Wells, *New Physiognomy: or, Signs of Character as Manifested through Temperament and External Forms, and Especially in "The Human Face Divine"* (New York: Fowler and Wells, 1866); Wells, *Outline of the Science of Man, According to Phrenology, Physiology, Physiognomy, and Psychology* (New York: S. R.

Wells, 1871); Nelson Sizer and Henry Shipton Drayton, *Heads and Faces, and How to Study Them: A Manual of Phrenology and Physiognomy for the People* (New York: Fowler and Wells, 1885); and Lorenzo N. Fowler, *Revelations of the Face: A Study in Physiognomy* (New York: Fowler and Wells, 1895). Scholars sometimes argue that physiognomy was a *visual* science while phrenology was more *tactile*. Yet as Richard Twine has argued, this overstates the differences between the disciplines and elides the importance of the face within phrenological discourses. See Twine, "Physiognomy, Phrenology and the Temporality of the Body," *Body & Society* 8, no. 1 (2002): 68.

17. Orson Squire Fowler and Lorenzo Niles Fowler, *The Phrenological Almanac, and Physiological Guide, for the Year of Our Lord 1845* (New York: O. S. Fowler, 1844), 30; and Samuel Wells, *New Physiognomy*, xxv.

18. Ann Fabian, *The Skull Collectors: Race, Science, and America's Unburied Dead* (Chicago: University of Chicago Press, 2010), 92.

19. T. J. Ellinwood, ed., *Autobiographical Reminiscences of Henry Ward Beecher* (New York: Frederick A. Stokes Company, 1898), 38.

20. For "cheap literature revolution," see Adam Gordon, "Review: The Rise of Print Culture Canon," *Early American Literature* 49, no. 2 (2014): 547. By 1850, the literacy rate for white Americans was 90 percent for the country writ large and 97 percent for the northeastern states. See Ronald Zboray, *A Fictive People: Antebellum Economic Development and the American Reading Public* (New York: Oxford University Press, 1993), 83; and Mary Heininger, *At Home with a Book: Reading in America, 1840–1940* (Rochester, NY: Strong Museum, 1986), 3. On public lectures and lyceum circuits, see Donald M. Scott, "The Popular Lecture and the Creation of a Public in Mid-Nineteenth-Century America," *Journal of American History* 66, no. 4 (March 1980): 791–809; Angela Ray, *The Lyceum and Public Culture in the Nineteenth-Century United States* (East Lansing: Michigan State University Press, 2005); and John van Wyhe, "The Diffusion of Phrenology through Public Lecturing," in *Science in the Marketplace: Nineteenth-Century Sites and Experiences*, ed. Aileen Fyfe and Bernard Lightman (Chicago: University of Chicago Press, 2007). For scholarship showing how popular writers doubled as propagators of science, see Shelley R. Block and Etta M. Madden, "Science in Catharine Maria Sedgwick's *Hope Leslie*," *Legacy* 20, no. 1/2 (2003): 22–37; Geoffrey Cantor and Sally Shuttleworth, eds., *Science Serialized: Representations of the Sciences in Nineteenth-Century Periodicals* (Cambridge: MIT Press, 2004); Jan Pilditch, "'Fashionable Female Studies': The Popular Dissemination of Science in 'Godey's Lady's Book,'" *Australasian Journal of American Studies* 24, no. 1 (July 2005): 20–37; and Britt Rusert, "The Science of Freedom: Counterarchives of Racial Science on the Antebellum Stage," *African American Review* 45, no. 3 (Fall 2012): 291–308.

21. See Georgiana Souther Barrows Diary, 1842, Mss. Octavo Vols. B, AAS; and Papers of Lucy Chase, Folder 3 Diary fragments, 1841–1847, in Chase Family, Papers, c. 1787–c. 1915, Mss. Boxes C, AAS. For secondary scholarship on public lectures and the appeal of popular science for lay practitioners, see van Wyhe, "The Diffusion of Phrenology through Public Lecturing," 60–96.

22. Mary Ferguson Diaries, #6405, Division of Rare and Manuscript Collections, Cornell University Library (see entries for 2 November 1858 and 4 November 1858).

23. Harriet Jane Hanson Robinson, *Loom and Spindle; or, Life among the Early Mill Girls* (New York: Thomas Y. Crowell & Company, 1898), 88.

24. Harriot F. Curtis papers, Sophia Smith Collection of Women's History, Smith College.

25. Lucy Larcom, *A New England Girlhood, Outlined from Memory* (Boston and New York: Houghton Mifflin Company, 1889), 248. Despite Larcom's complaints, phrenologists did not, in fact, give positive readings to everyone. Lorenzo Fowler admonished one client for being "too impulsive and impudent" and said he needed to be "more circumspect, cautious, and mindful

of consequences." The phrenologist also said the man was "deficient in financial talent," warning him to "perfect your character as much as possible." See "Phrenological Description of Mr. James Terry," James Terry Papers, 1848–1916, Folder 19, Joseph Downs Collection of Manuscripts and Printed Ephemera, Winterthur Library.

26. James C. Whorton, *Crusaders for Fitness: The History of American Health Reformers* (Princeton, NJ: Princeton University Press, 1982), 124.

27. John Davies has argued that "in contrast to Gall's cynical pessimism, Spurzheim looked forward to the perfection of the race by the aid of phrenology." Davies, *Phrenology: Fad and Science*, 8. For "quintessentially American," see Britt Rusert, *Fugitive Science Empiricism and Freedom in in Early African American Culture* (New York: New York University Press, 2017), 122.

28. Orson Squire Fowler, *Fowler's Practical Phrenology* (Philadelphia: O. S. Fowler, 1840), 22. This book went through numerous editions throughout the rest of the nineteenth century.

29. Fowler, *Fowler's Practical Phrenology*, 24–25.

30. "Individual vs. General Reforms," *American Phrenological Journal* [hereafter *APJ*] 23, no. 2 (New York), February 1856. Orson Fowler described phrenology as "reformatory work" that was "effectively shaping the public mind." In one letter, he praised a family member and fellow phrenological lecturer, writing, "Glad that your soul still burns with phrenological and reformatory fire." His business partner and brother-in-law, Samuel Wells, similarly viewed phrenologists as "mental Luminaries" and bold-minded reformers who were "glad to meet ignorance, and superstition, even face to face, and put them both to shame." See Orson S. Fowler to Paul C. Howe, 14 December 1849; Orson S. Fowler to Paul C. Howe, 20 April 1849; and Samuel R. Wells to Paul Howe, 25 January 1847, all in Box 4, Fowler and Wells Families Papers, Division of Rare and Manuscript Collections, Cornell University Library.

31. Lavater attempted to clarify his conflicting assertions by saying that visages might change *marginally*, while also suggesting that they could never *dramatically* transform in ways that would undercut their original nature. See Lavater, *Essays on Physiognomy*, trans. Thomas Holcroft (London: G. G. J. and J. Robinson, 1789), 1:197, 2:52, and 2:209.

32. C. K. to Anne Nelson, Kensington [PA?], 15 December 1823, Eliza K. Nelson papers, 1823–1867, David M. Rubenstein Rare Book and Manuscript Library, Duke University, quoted in Courtney E. Thompson, *An Organ of Murder: Crime, Violence, and Phrenology in Nineteenth-Century America* (New Brunswick, NJ: Rutgers University Press, 2021), 30.

33. Abby Kelley to Stephen S. Foster, 30 January 1843, Box 1, Folder 13, Abigail Kelley Foster Papers, 1836–1891, Mss. Boxes F, AAS.

34. "Hereditary Descent: Its Laws and Facts Applied to Human Improvement," *APJ* 10, no. 4 (New York), 1 April 1848; "Idiocy and Superior Talents, Hereditary," *APJ* 5, no. 11 (New York), November 1843; and "Intellect Hereditary—As to Both Kind and Amount," *APJ* 5, no. 12 (New York), December 1843. For secondary scholarship, see Bittel, "Cranial Compatibility."

35. "Phrenology and Physiology Applied to Education and Self-Improvement," *APJ* 4, no. 6 (New York), June 1842; "Self-Improvement," *APJ* 8, no. 5 (New York), May 1846; "Education, Phrenologically Considered," *APJ* 15, no. 1 (New York), January 1852; and "Culture of the Mental Faculties," *APJ* 27, no. 3 (New York), September 1857.

36. This story was first translated into English in 1788 and then republished in several different volumes, in both Great Britain and the United States. For the first English-language edition, see: *The Children's Friend* (London: J. Stockdale; J. Rivington and Sons; B. Law; J. Johnson; C. Dilly; J. Murray; J. Sewell, 1788). For nineteenth-century US editions, see Increase Cooke, *Sequel to the American Orator; or, Dialogues for Schools* (New Haven, CT: Increase Cooke & Co., 1813); *The Children's Companion* (New York: S. King, 1825); *The Beauties of the Children's Friend* (Boston:

Lincoln & Edmands, 1827); and *The Children's Friend* (Boston: Munrow and Francis, and New York: Charles S. Francis, 1833). On the politics of aging in early America, see Corinne T. Field, *The Struggle for Equal Adulthood* (Chapel Hill: University of North Carolina Press, 2014).

37. Daniel Harrison Jacques, *Hints toward Physical Perfection; or, The Philosophy of Human Beauty* (New York: Fowler and Wells, 1859), 90.

38. Sylvester Graham to "Friend Wells," Northampton, Massachusetts, 13 March 1857, Box 1, Folder 1-1, Fowler and Wells Families Papers.

39. On eighteenth-century beauty ideals and their connection to Grecian statuary, see Nell Irvin Painter, *The History of White People* (New York: W. W. Norton & Company, 2010), esp. chapter 5; Charles Colbert, *A Measure of Perfection: Phrenology and the Fine Arts in America* (Chapel Hill: University of North Carolina Press, 1997), chapter 3; and David Bindman, *Ape to Apollo: Aesthetics and the Idea of Race* (Ithaca, NY: Cornell University Press, 2002), chapter 2.

40. Lavater, *Essays on Physiognomy*, 1:16. Physiognomists, phrenologists, ethnologists, and anthropologists deployed this tripartite division of the countenance throughout the late eighteenth and nineteenth centuries. See Lucy Hartley, "A Science of Beauty? Femininity and the Nineteenth-Century Physiognomic Tradition," *Women: A Cultural Review* 12, no. 1 (2001): 23–24.

41. See William H. Aspinwall to Hiram Powers, 10 April 1865; Hiram Powers to William H. Aspinwall, 12 April 1865; and William H. Aspinwall to Hiram Powers, 7 June 1865, Box 1, Folder 17, Hiram Powers papers, AAA.

42. Lavater, *Essays on Physiognomy*, 1:64, 1:117.

43. For examples of these distinctions, see John Gibson Lockhart, *Peter's Letters to His Kinsfolk*, 2nd American edition (New York: James and John Harper, 1820), 469; "Washington City: New Members of the 25th Congress," *The Huntress* 2, no. 17 (Washington), 31 March 1838; Emily Chubbuck, *Alderbrook: A Collection of Fanny Forester's Village Sketches, Poems, Etc.* (Boston: Ticknor and Company, 1849), 206; E. W. Robbins, "A Chapter on Physiognomy," *Frank Leslie's Ladies' Magazine* 10, no. 5 (New York), April 1862; and "Timothy Tot: A Prose Story with Poetic Passages," *The Radical* 10, no. 2 (Boston), February 1872.

44. Clara Crowninshield, *Diary: A European Tour with Longfellow, 1835–1836* (Seattle: University of Washington Press, 1956), 59, 77, 288.

45. Brenda Stevenson, ed., *The Journals of Charlotte Forten Grimké* (New York: Oxford University Press, 1988), 84.

46. Ednah D. Cheney, ed., *Louisa May Alcott: Her Life, Letters, and Journals* (Boston: Little, Brown, and Company, 1898), 9.

47. Elizabeth Blackwell, *Pioneer Work in Opening the Medical Profession to Women: Autobiographical Sketches* (London: Longmans, Green, and Co., 1895), 38, 48, 189.

48. *Journals and Letters of Emma Cullum Cortazzo* (Meadville, PA: E. H. Shartle, 1919), 255.

49. Robinson, *Loom and Spindle*, 153.

50. Despite his comments, Dickens sat for a phrenological examination in Massachusetts. See "The Phrenological Developments of Charles Dickens" (1842), Charles Dickens Collection, Mss. Misc. Box D, AAS. For a few physiognomic character sketches of US politicians, see "Etchings of the Senate," *New-England Magazine* (Boston), November 1834; "The Heads of Our Great Men," *American Monthly Magazine* 11 (New York), April 1838; and Dr. N. Wheeler, *The Phrenological Characters and Talents of Henry Clay, Daniel Webster, John Quincy Adams, William Henry Harrison, and Andrew Jackson* (Boston: Dow & Jackson, 1845).

51. John Adams to John Quincy Adams, 31 March 1817, *Founders Online*, National Archives, https://founders.archives.gov/documents/Adams/99-03-02-3283. For Combe's evaluations of US politicians, see Charles Gibbon, *The Life of George Combe* (London: Macmillan and Co., 1878), 2:52.

52. James Madison to Charles Caldwell, 20 September 1826, *Founders Online*, National Archives, https://founders.archives.gov/documents/Madison/99-02-02-0736.

53. *The Register of Debates; Being a Report of the Speeches Delivered in the Two Houses of Congress* (Washington: Duff Green, 1834), 2:608.

54. See, for example, "Phrenological and Physiological Organization and Character of Andrew Jackson," *APJ* 7, no. 7 (New York), July 1845; and Mirabeau, "Political Portraits," *The New-England Galaxy and the United States Literary Advertiser* (Boston), 4 February 1825.

55. "Mr. Webster," *American Monthly Magazine* 1, no. 12 (Boston), March 1830; "Sketches of Eminent Americans: Mr. Webster," *Cincinnati Mirror* 1, no. 3, 29 October 1831; "A Glance at the United States Senate of 1834," *Southern Literary Journal and Magazine of Arts* (Richmond, VA), December 1834; Erastus Brooks, "Sketches of Our Public Men," *Literary Examiner and Western Monthly Review* 1, no. 2 (Pittsburgh), June 1839; "Portraits of Public Characters: Mr. Daniel Webster," *New World* 1, no. 24 (New York), 14 November 1840; and "The Phrenological Character of Daniel Webster, with a Likeness," *APJ* 5, no. 3 (New York), March 1843.

56. Brooks, "Sketches of Our Public Men"; "Daniel Webster," *Western Monthly Magazine* 2, no. 8 (Cincinnati), August 1833; and "The Phrenological Character of Daniel Webster, with a Likeness."

57. For broad overviews of the political, economic, and technological transformations of the early nineteenth century: Daniel Walker Howe, *What Hath God Wrought: The Transformation of America, 1815–1848* (New York: Oxford University Press, 2007); John Lauritz Larson, *The Market Revolution in America: Liberty, Ambition, and the Eclipse of the Common Good* (New York: Cambridge University Press, 2010); and Brian P. Luskey, *On the Make: Clerks and the Quest for Capital in Nineteenth-Century America* (New York: New York University Press, 2010).

58. For foundational scholarship on "true womanhood," "separate spheres," and nineteenth-century gender conventions, see Barbara Welter, "The Cult of True Womanhood: 1820–1860," *American Quarterly* 18, no. 2 (Summer 1966): 151–74; Nancy F. Cott, *The Bonds of Womanhood: "Woman's Sphere" in New England, 1780–1835* (New Haven, CT: Yale University Press, 1977); and Jeanne Boydston, *Home and Work: Housework, Wages, and the Ideology of Labor in the Early Republic* (New York: Oxford University Press, 1990). For scholars who have challenged the idea that nineteenth-century men and women were indeed confined to separate spheres, see Linda K. Kerber, "Separate Spheres, Female Worlds, Woman's Place: The Rhetoric of Women's History," *Journal of American History* 75, no. 1 (June 1988): 9–39; Carol Lasser, "Beyond Separate Spheres: The Power of Public Opinion," *Journal of the Early Republic* 21, no. 1 (Spring 2001): 115–23; and Cathy N. Davidson and Jessamyn Hatcher, eds., *No More Separate Spheres! A Next Wave American Studies Reader* (Durham, NC: Duke University Press, 2002).

59. Cameron B. Strang, *Frontiers of Science: Imperialism and Natural Knowledge in the Gulf South Borderlands, 1500–1850* (Chapel Hill: University of North Carolina Press, 2018), 313–14.

60. Richard H. Colfax, *Evidence against the View of the Abolitionists: Consisting of Physical and Moral Proofs of the Natural Inferiority of the Negroes* (New York: James T. M. Bleakley, 1833), 25. In a similar way, one European pamphlet—which was translated by an American physician and published in New York in 1853—discussed the "negro physiognomy," describing Africans and African Americans in derogatory ways and hinting that external characteristics correlated with internal nature. See Hermann Burmeister, *The Black Man: The Comparative Anatomy and Psychology of the African Negro* (New York: William C. Bryant, 1853), 11.

61. Douglass Baynton, "Disability and the Justification of Inequality in American History," in *The New Disability History: American Perspectives*, ed. Paul K. Longmore and Lauri Umansky (New York: New York University Press, 2001), 37. Mia Bay has also argued that "the Negro's capacity was becoming the central issue for spokesmen on both sides of the slavery debate." See Mia Bay,

The White Image in the Black Mind: African-American Ideas about White People, 1830–1925 (New York: Oxford University Press, 2000), 20. Frederick Douglass himself made a similar argument. See Douglass, "Selections. Woman's Rights Convention," *The North Star* (Rochester), 11 August 1848, quoted in Baynton, "Disability and the Justification of Inequality in American History," 44.

62. For phrenology's relationship to both scientific racism and anti-slavery politics, see Susan Branson, "Phrenology and the Science of Race in Antebellum America," *Early American Studies* 15, no. 1 (Winter 2017): 164–93; James Poskett, *Materials of the Mind: Phrenology, Race, and the Global History of Science, 1815–1920* (Chicago: University of Chicago Press, 2019); and Cynthia S. Hamilton, "'Am I Not a Man and a Brother?' Phrenology and Anti-Slavery," *Slavery & Abolition* 29, no. 2 (June 2008): 173–87.

63. On Morton, see William R. Stanton, *The Leopard's Spots: Scientific Attitudes toward Race in America, 1815–1859* (Chicago: University of Chicago Press, 1960); George M. Fredrickson, *The Black Image in the White Mind: The Debate on Afro-American Character and Destiny, 1817–1914* (New York: Harper and Row, 1971); Stephen J. Gould, *The Mismeasure of Man* (1981; repr. New York: W. W. Norton, 1996); and Fabian, *The Skull Collectors*. On the complex relationship between phrenology and ethnology, see James Poskett, "National Types: The Transatlantic Publication and Reception of *Crania Americana* (1839)," *History of Science* 53, no. 3 (2015): 265–95.

64. Painter, *The History of White People*, 68.

65. Josiah Clark Nott and George Robins Gliddon, *Types of Mankind; or, Ethnological Researches* (Philadelphia: Lippincott, Grambo & Co., 1854). This book quickly became a touchstone for white supremacists. Nott and Gliddon claimed that "Phrenology is no longer a science," but their declaration belied their own reliance on physiognomic and phrenological techniques. For examples of facial and cranial analysis, see pages 124–79, 184–207, 213–16, 219–27, and 246–71.

66. Orson Fowler, *The Practical Phrenologist* (Boston: O. S. Fowler, 1869), 13.

67. *The Illustrated Annuals of Phrenology and Physiognomy for 1865* (New York: S. R. Wells & Company, 1864), 36.

68. Anthony J. La Vopa, *The Labor of the Mind: Intellect and Gender in Enlightenment Cultures* (Philadelphia: University of Pennsylvania Press, 2017), 38. On Enlightenment era debates about the powers and capacities of the female mind, see Thomas Laqueur, *Making Sex: Body and Gender from the Greeks to Freud* (Cambridge, MA: Harvard University Press, 1992); Londa Schiebinger, *Nature's Body: Sexual Politics and the Making of Modern Science* (London: Pandora, 1994); Londa Schiebinger, *The Mind has No Sex?: Women in the Origins of Modern Science* (Cambridge, MA: Harvard University Press, 1996); and Geneviève Fraisse, *Reason's Muse: Sexual Difference and the Birth of Democracy* (Chicago: University of Chicago Press, 1994).

69. On gendered ideas about "sensibility" and how they connected to medical theories about the nervous system, see Sarah Knott, *Sensibility and the American Revolution* (Chapel Hill: University of North Carolina Press, 2009), chapter 2.

70. For juxtapositions of women's vivacity, perception, and sensibility with men's rationality, see Thomas Bell, *Kalogynomia, or The Laws of Female Beauty: Being the Elementary Principles of that Science* (London: Walpole Press, 1819), 61–62; Alexander Walker, *Physiognomy, Founded on Physiology, and Applied to Various Countries, Professions, and Individuals* (London, 1834), 64; "Essay on Woman," *The Souvenir* 1, no. 2 (Philadelphia), 11 July 1827; Joseph Turnley, *The Language of the Eye: As Indicative of Female Beauty, Manly Genius, and General Character* (London: Partridge and Co., 1856), 33; Jacques, *Hints toward Physical Perfection*, 39; Wells, *New Physiognomy*, 113; James McGrigor Allan, "On the Real Differences in the Minds of Men and Women," *Journal of the Anthropological Society of London* 7 (1869); and Orson Fowler, *Creative and Sexual Science; or, Manhood, Womanhood, and Their Mutual Interrelations* (New York: Fowler and Wells, 1870), 160.

71. For positive descriptions of bluestockings, see "Catharine Talbort," *Ladies' Magazine* 1, no. 6 (Boston), June 1828; and Oliver Oldboy, *George Bailey: A Tale of New York Mercantile Life* (New York: Harper and Brothers, 1880), 154.

72. Charles Caldwell, *Phrenology Vindicated, and Antiphrenology Unmasked* (New York: Samuel Colman, 1838), 73.

73. White male intellectuals almost always focused their attention on Black men and white women. "It was these two groups—and not African women—who were contenders for power," and thus most immediately threatening. See Londa Schiebinger, "The Anatomy of Difference: Race and Sex in Eighteenth-Century Science," *Eighteenth-Century Studies* 23, no. 4 (Summer 1990): 389.

74. J. R. Buchanan, M.D., "Article V. On the Faculty of Language and Its Cerebral Organs," *APJ* 3, no. 5 (New York), 1 February 1841. For a similar example, see "Natural History: The Natural History of Man," *APJ* 18, no. 1 (New York), July 1853.

75. Alexander Walker, *Beauty; Illustrated Chiefly by an Analysis and Classification of Beauty in Woman* (New York: J. & H. G. Langley, 1840), 21–29. For the first London edition, see Alexander Walker, *Beauty: Illustrated Chiefly by an Analysis and Classification of Beauty in Woman* (London: Effingham Wilson, 1836). Unless otherwise noted, I will cite from the New York edition.

76. Walker, *Beauty*, 185.

77. Bell, *Kalogynomia*, 67–68, 77, 96–98. Alexander Walker lifted many ideas and passages from Bell in his books on female beauty and physiognomy. The Fowler and Wells Press then plagiarized from Alexander Walker. In this way, racist ideals of female beauty circulated between popular and scientific texts.

78. Wells, *New Physiognomy*, 537.

79. Wells, *New Physiognomy*, 537–38, 653.

80. Scholars have called this period of religious revival the "Second Great Awakening." Jon Butler refers to nineteenth-century America as the "antebellum spiritual hothouse." See Butler, *Awash in a Sea of Faith: Christianizing the American People* (Cambridge, MA: Harvard University Press, 1990), chap. 8.

81. On the intersections of religion and popular science, see White, "Minds Intensely Unsettled."

82. "The Different Races of Men," *National Anti-Slavery Standard* (New York), 5 January 1842.

83. William Goodell, "The Brotherhood of the Human Race," *The Emancipator* (New York), 18 November 1834.

84. Catherine H. Birney, ed., *The Grimké Sisters: Sarah and Angelina Grimké, the First American Women Advocates of Abolition and Woman's Rights* (Boston: Lee and Shephard, 1885), 166.

85. "From the St. Louis Bulletin," *Farmers' Gazette, and Cheraw Advertiser* (Cheraw, SC), 23 September 1840. For other stories in which elite or middle-class individuals try to distinguish between the virtuous and vicious poor through physiognomy, see T. S. Arthur, "Giving and Witholding; or What Is Charity?" *Godey's Lady's Book* 23 (New York), September 1841; "Sketcher," "The Purse: An American Tale," *The Rural Repository* 18, no. 25 (Hudson, NY), 21 May 1842; "A New Contributor," "The Alms House," *The Knickerbocker; or New York Monthly Magazine* 23, no. 3 (New York), March 1844; Reverend Edward Price, "Sick Calls: From the Diary of a Missionary," *Brownson's Quarterly Review* 6, no. 1 (Boston), 1 January 1852; and George Combe, "Criminal Legislation and Prison Discipline: Chapter X," *APJ* 22, no. 6 (New York), December 1855. For secondary scholarship on reformers' attitudes toward the "virtuous" vs. "vicious" poor, see Paul Lewis, "'Lectures or a Little Charity': Poor Visits in Antebellum Literature and Culture," *The New England Quarterly* 73, no. 2 (June 2000): 246–73; Christine Stansell, *City of Women: Sex and Class in New York, 1789–1860* (Chicago: University of Illinois Press, 1987), especially chapters 1 and 2;

and Monique Bourque, "Poor Relief 'Without Violating the Rights of Humanity,'" in *Down and Out in Early America*, ed. Billy G. Smith (University Park: Pennsylvania State University Press, 2004), 189–212.

86. Letter from Harriet Wadsworth Winslow, 3 September 1817, in *Memoir of Mrs. Harriet L. Winslow, Thirteen Years a Member of the American Mission in Ceylon* (New York: American Tract Society, 1840), 111.

87. "A Doctor's Ana, No. 1," *The American Monthly Magazine* 10 (New York), October 1837.

88. "Tracts Published by the Christian Tract Society, London, and Republished by Wells and Lilly," *The Christian Disciple and Theological Review* 2, no. 8 (Boston), March/April 1820.

89. Jacques, *Hints toward Physical Perfection*, 114–17.

CHAPTER THREE

1. Papers of Lucy Chase, Folder 3 Diary fragments, 1841–1847, AAS. See entries for 1, 5, and 25 February 1842; 2 March 1842; and 24 August 1842. More than a year before Fowler arrived, Chase was already using phrenological language. On December 31, 1841, she described one of her contemporaries, Mrs. Bigelow, as "a *splendid* woman, she has a *glorious* head, & is very talented & intellectual."

2. This chapter is based on the letters, diaries, and writings of 189 women who wrote between the 1770s and 1880s. These records include both published and manuscript sources. Of these women, 61 engaged in one or more of these activities: explicitly discussed physiognomy or phrenology in their letters, diaries, or published works; got their heads examined; attended popular science lectures; or referenced the works of famous physiognomists and phrenologists (such as Johann Caspar Lavater, Johann Gaspar Spurzheim, George Combe, or the Fowler brothers); 112 of these women used physiognomical and phrenological language to discern character from heads and faces, even if they did not discuss popular sciences explicitly; 16 of the women discussed the relationship between female beauty, intellect, and character, but did not do so in physiognomical or phrenological ways. These records are centered on—but not limited to—women in the Northeast and Mid-Atlantic states. The women range in age from their teenage years to adulthood, although much of the most compelling evidence comes from women between adolescence and their early thirties. As women's historians have pointed out, women's diaries are more verbose before marriage and before child-rearing. See Martha Tomhave Blauvelt, *The Work of the Heart: Young Women and Emotion, 1780–1830* (Charlottesville: University of Virginia Press, 2007), 176–78; Lucia McMahon, *Mere Equals: The Paradox of Educated Women in the Early American Republic* (Ithaca, NY: Cornell University Press, 2012), 126–27; and Judy Nolte Lensink, *"A Secret to Be Buried": The Diary and Life of Emily Hawley Gillespie, 1858–1888* (Iowa City: University of Iowa Press, 1989), 180. For a list of the published and manuscript sources that served as the foundation for this chapter, consult the Selected Bibliography.

3. For an article that discusses the "beauty of expression" that arose on the physiognomies of cultivated individuals, see "For the Balance: A Cultivated Mind the Beauty of Expression," *Balance and Columbian Repository* 5, no. 9 (Albany, NY), 4 March 1806.

4. For foundational scholarship on women and science in the late eighteenth and nineteenth centuries, see Sally Gregory Kohlstedt, "Parlors, Primers, and Public Schooling: Education for Science in Nineteenth-Century America," *Isis* 81, no. 3 (September 1990): 424–45; Ann B. Shteir, *Cultivating Women, Cultivating Science: Flora's Daughters and Botany in England, 1760–1860* (Bal-

timore, MD: Johns Hopkins University Press, 1996); Nina Baym, *American Women of Letters and the Nineteenth-Century Sciences: Styles of Affiliation* (New Brunswick, NJ: Rutgers University Press, 2002); and Kim Tolley, *The Science Education of American Girls: A Historical Perspective* (New York: Routledge Falmer, 2003).

5. Letter from Rebecca Gratz to Maria Gist Gratz, 9 March 1834, in David Philipson, ed., *Letters of Rebecca Gratz* (Philadelphia: Jewish Publication Society, 1929), 316.

6. Georgiana Bruce Kirby journal, 1852–1860, mssFAC 2017, Huntington Library. See entries for 25 January 1854, August 1855, and undated entry (after December 1857). For instances where Kirby analyzes her own phrenological faculties, see entries for 3 February 1853 and August 1855. A Mormon schoolteacher in Nauvoo, Illinois similarly used phrenology to study "the characters and dispositions of [his] scholars." After reading works by Spurzheim and Fowler, he wrote: "I think I must make out a chart of their heads with a description of their organs, and then concoct a plan for their education in accordance with those principles." See entries for 24-25 April 1845, in James M. Monroe Diary, 1841–1845, mssFAC 1679, Huntington Library.

7. Frank Shuffelton, "Margaret Fuller at the Greene Street School: The Journal of Evelina Metcalf," *Studies in the American Renaissance* (1985): 35; Mary Ware Allen, School Journal, 11 May 1838, Mss. Octavo Vol. 25, in Allen-Johnson Family Papers, AAS; and Anna D. Gale, Journal A, 6 February 1838, Octavo Volume 1, Gale Family Papers (1828–1854), Mss. Octavo Vols. G, AAS. For a published version of the diary, see Paula Kopacz, "The School Journal of Hannah (Anna) Gale," *Studies in the American Renaissance* (1996): 67–113. For more on Fuller's exchanges with the phrenological Fowlers, see Laraine R. Fergenson, "Margaret Fuller as a Teacher in Providence: The School Journal of Ann Brown," *Studies in the American Renaissance* (1991): 73–74; and Madeleine B. Stern, "Margaret Fuller and the Phrenologist Publishers," *Studies in the American Renaissance* (1980): 229.

8. Lucy Larcom, *A New England Girlhood, Outlined from Memory* (Boston and New York: Houghton Mifflin Company, 1889), 248; and Harriet Jane Robinson, *Loom and Spindle; or, Life among the Early Mill Girls* (New York: Thomas Y. Crowell & Company, 1898), 88.

9. On Spurzheim's visit to Sarah Paul's school, see Kabria Baumgartner, *In Pursuit of Knowledge: Black Women and Educational Activism in Antebellum America* (New York: New York University Press, 2019), 177; and Nahum Capen, *Reminiscences of Dr. Spurzheim and George Combe* (New York: Fowler & Wells, 1881), 24.

10. Diary of Charlotte Sheldon, 20 May 1796, in *Chronicles of a Pioneer School from 1792 to 1833, Being the History of Miss Sarah Pierce and Her Litchfield School*, ed. Elizabeth C. Barney Buel, comp. Emily Noyes Vanderpoel (Cambridge, MA: Harvard University Press, 1903), 11; Letter from Margaret G. Cary to George Blankern Cary, 14 April 1843, in Caroline G. Curtis, ed., *The Cary Letters* (Cambridge: Riverside Press, 1891), 36; Ruth Henshaw Bascom Diaries, October 1833, Mss. Octavo Vol. 39, AAS; and Madeleine B. Stern, "A Letter from New Haven—1840," *The Yale University Library Gazette* 49, no. 3 (January 1975): 288–92.

11. Caroline H. Hance Long Diary, 1852, Mss. Am.0712, HSP.

12. Daniel Dulany Addison, ed., *Lucy Larcom: Life, Letters, and Diary* (Boston: Houghton, Mifflin, and Co., 1894), 119; and *Glimpses of Fifty Years: The Autobiography of an American Woman* (Chicago: Woman's Temperance Publication Association, 1889), 157.

13. Ella Sterling Mighels, *Life and Letters of a Forty-Niner's Daughter* (San Francisco: Harr Wagner Publishing Company, 1929), 35–36.

14. Hannah Crafts, *The Bondwoman's Narrative*, ed. Henry Louis Gates (New York: Warner Books, 2002), 26–27, 63, 91, 123. Gates argues that "Hannah Crafts wrote what she read, as is abundantly obvious from her uses of conventions from gothic and sentimental novels" (xxxiv). For a

catalog of the books in her enslaver's library, see Appendix C, compiled by Bryan Sinche (321–330). Hannah Crafts was the pen name for Hannah Bond, who was identified by Gregg Hecimovich.

15. Dorothy Sterling, ed., *We Are Your Sisters: Black Women in the Nineteenth Century* (New York: W. W. Norton & Company, 1984), 462–63.

16. Sterling, ed., *We Are Your Sisters*, 464–72.

17. "Sarah Kinson, or Margru," *APJ* (New York), 1 July 1850. This description was also printed in *The Illustrated Phrenological Almanac for 1851* (New York: Fowler and Wells, 1850), 30.

18. "Sarah Kinson, or Margru," *APJ* (New York), 1 July 1850. Montgomery also read the physiognomist Samuel Wells's book, *Self-Culture*. See Loren Schweninger, *Black Property Owners in the South, 1790–1915* (Urbana: University of Illinois Press, 1990), 229.

19. Lucretia Fiske Farrington, "Notebook for 1834," Mss. S143, BA. On one occasion, she transcribed a physiognomical description of John C. Calhoun, which was printed in "Etchings of the Senate," *New-England Magazine* (Boston), November 1834.

20. See Journal of Hannah Margaret Warton, 15[?] October 1812; 10[?] February 1813; and 15 February 1813, HSP. See also Wharton's "Recollections of 1817-18," 1 December 1817, HSP. Walter Scott was an especially prolific user of physiognomy in his fictional works. See Graeme Tytler, "Lavater's Influence on Sir Walter Scott: A Tacit Assumption?" in *Physiognomy in Profile: Lavater's Impact on European Culture*, ed. Melissa Percival and Graeme Tytler (Newark: University of Delaware Press, 2005), 109–20; and Tytler, "'Faith in the Hand of Nature': Physiognomy in Sir Walter Scott's Fiction," *Studies in Scottish Literature* 33, no. 1 (2004): 223–46.

21. See entry for 13 July 1854, in Brenda Stevenson, ed., *The Journals of Charlotte Forten Grimké* (New York: Oxford University Press, 1988), 86. Forten also read books such as Mrs. Crosland's *Memorable Women* (1854) and Mary Langdon's anti-slavery novel, *Ida May* (1854), which provided detailed physiognomical sketches of real and fictional characters. See entries for 24 November 1854, 26 November 1854, and 22 May 1857.

22. Robert Turnbull, *The Genius of Scotland; or, Sketches of Scottish Scenery, Literature and Religion*, 2nd ed. (New York: Robert Carter, 1847), 77–78.

23. Anna D. Gale, Journal A, 28 February 1838, AAS. Rachel van Dyke also used physiognomic language to critique the frontispiece of a book she was reading. See Lucia McMahon and Deborah Schriver, eds., *To Read My Heart: The Journal of Rachel Van Dyke* (Philadelphia: University of Pennsylvania Press, 2000), 87–88.

24. Proteus Echo, "Essays," *The Emerald, or, Miscellany of Literature* 1, no. 9 (Boston), 19 December 1807; and Samuel Wells, *The Illustrated Annual of Phrenology and Physiognomy* (New York: Samuel R. Wells, 1868), 20. For other works drawing connections between reading and physiognomic investigation, see E. W. Robbins, "A Chapter on Physiognomy," *The American Literary Magazine* 4, no. 5 (Hartford), May 1849; "Review of *The South West*," *The Knickerbocker; or New York Monthly Magazine* 6, no. 6 (Boston), December 1835; and "Old Books," *The New Monthly Magazine and Literary Journal* 2, no. 7 (Boston), 1821.

25. "Reminisces by Emerson," in *Love-letters of Margaret Fuller: 1845–1846* (New York: Appleton and Company, 1903), 195. For nineteenth-century authors who quoted or echoed Emerson's comments about Fuller's appearance, see Robert Chambers, ed., *The Book of Days: A Miscellany of Popular Antiquities* (London & Edinburgh: W. & R. Chambers, 1832), 2:68; William Channing, ed., *Memoirs of Margaret Fuller Ossoli* (Boston: Phillips, Sampson and Company, 1851), 1:337; George Ripley and Charles A. Dana, eds., *The New American Cyclopaedia: A Popular Dictionary of General Knowledge* (New York: D. Appleton and Company, 1863), 15:598. For modern writers who comment on Fuller's supposed plainness, see Charles Capper, *Margaret Fuller: An American Roman-*

tic Life (New York: Oxford University Press, 1992), 297–98; Joel Myerson, ed., *Fuller in Her Own Time: A Biographical Chronicle of Her Life* (Iowa City: University of Iowa Press, 2008), xxii–xxiii; and Meg McGavran Murray, *Margaret Fuller, Wandering Pilgrim* (Athens: University of Georgia Press, 2012), 91, 311, 363. For unflattering descriptions of Fuller's beauty, see "Pen Portraits of Some Noted Women," *Godey's Lady's Book* (Philadelphia), September 1876; "Pen Portraits of Some Noted Women," *Frank Leslie's Illustrated Newspaper* (New York), 2 September 1876; "Pen Portraits of Some Noted Women," *The Australian Journal* 12 (Melbourne), May 1877; and "Noted Women," *Manford's New Monthly Magazine* 34, no. 10 (Chicago), October 1890.

26. Thomas Wentworth Higginson recounted this story in *Margaret Fuller Ossoli* (Boston: Houghton, Mifflin, and Company, 1884), 299.

27. "Pen Portraits of Some Noted Women," *Godey's Lady's Book*, September 1876; Diary of Caroline Healey Dall, 8 March 1841, in Helen R. Deese, ed. *Daughter of Boston: The Extraordinary Diary of a Nineteenth-Century Woman* (Boston: Beacon Press, 2005), 24; Anna Brackett, "Margaret Fuller Ossoli," *The Radical* (Boston), December 1871; and *The Works of Edgar Allan Poe*, vol. 3 (New York: W. J. Widdleton, 1849), 79. Intriguingly, Caroline Healey Dall once described one of the Fowler brothers as "one of many quacks," after one of her friends suggested that she get her head examined. Even so, she proceeded to have a detailed discussion with her dentist about her own phrenological organs, demonstrating that she was intimately familiar with phrenological principles. She remained unsure if phrenology was a legitimate discipline or a "charlatan faith," but this did not stop her from using phrenological language or speculating about her own organs and propensities. See Helen R. Deese, ed., *Selected Journals of Caroline Healey Dall, Volume 1: 1838–1855* (Charlottesville: University of Virginia Press, 2006), 79.

28. Letter from Sallie Holley to Mrs. Porter, 2 August 1855, in John White Chadwick, ed., *A Life for Liberty: Anti-Slavery and Other Letters of Sallie Holley* (New York: G. P. Putnam's Sons, 1899), 167-168.

29. "Female Biography: A Sketch of Miss Neville's Character," *The Ladies' Literary Portfolio* 1, no. 20 (Philadelphia), 29 April 1829.

30. Emma Embury, "Characters . . . No. II," *American Ladies' Magazine* 7, no. 3 (Boston), March 1834.

31. Embury, "Characters . . . No. II." Other authors made similar arguments. See James Garnett, *Seven Lectures on Female Education* (Richmond, VA: T. W. White, 1824), 70; and John Burton, *Lectures on Female Education and Manners*, 2nd edition (London: J. Johnson, 1793), 141.

32. See entry for June 1839, in *Daniel Webster in England: Journal of Harriette Story Paige, 1839* (Boston: Houghton, Mifflin, & Co., 1917), 65. For a similar example, see *Diary of Sarah Connell Ayer* (Portland, ME: Lefavor-Tower Co., 1910), 35, 114.

33. Letter from Margaret Bayard Smith to Susan B. Smith, 26 December 1802, in Gaillard Hunt, ed., *The First Forty Years of Washington Society in the Family Letters of Margaret Bayard Smith* (New York: Frederick Ungar Publishing, 1906), 34.

34. Mrs. Newton Crosland, ed., *Memorable Women: The Story of Their Lives* (Boston: Ticknor and Fields, 1854), 341–42; Chambers, ed., *The Book of Days*, 2:68; and Julia Ward Howe, *Margaret Fuller* (Boston: Little, Brown, and Co., 1905), 19.

35. Howe, *Margaret Fuller*, 191–92.

36. Lowell's diaries show that she used physiognomic language to find appealing mental and moral traits in the faces of women she did not find pretty. Anna Cabot Lowell II Papers/Diaries, 1818–1894, Ms. N-1512, MHS. See entries for 26 December 1829 (Box 2, Volume 15); 30 June 1830 (Box 2, Volume 16); 26 August 1830 (Box 2, Volume 16); 24 October 1830 (Box 2, Volume 16); and 8 March 1832 (Box 3, Volume 20).

37. See the October 1849? entry in "Diary of Susan Fenimore Cooper," in *Journal of a Naturalist in the United States* (London: Richard Bentley & Son, 1855), 2:105. Catharine Maria Sedgwick also distinguished between stereotypically beautiful faces and countenances which conveyed more meaningful personality traits. Though she was skeptical about whether physiognomy and phrenology could provide a truly scientific measure of character, Sedgwick nonetheless interpreted the faces of friends and loved ones on a regular basis. See, for instance, Catharine Maria Sedgwick to Frances Watson, February 1822; and Catharine Maria Sedgwick to Katharine Maria Sedgwick Minot, 13 October 1849; in Mary E. Dewey, ed., *Life and Letters of Catherine M. Sedgwick* (New York: Harper & Brothers, 1872), 149, 317.

38. Rev. L. P. Hickok, *An Address Before the Canton Female Seminary at their Annual Examination, June 8, 1843* (Canton, OH: Daniel Gotshall, 1843), 15.

39. "Address Delivered on the Opening of the New-York High-School for Females," *American Journal of Education* 1, no. 5 (May 1826): 273.

40. Johann Caspar Lavater, *Essays on Physiognomy; for the Promotion of the Knowledge and the Love of Mankind*, trans. Thomas Holcroft (London: G. G. J. and J. Robinson, 1789), 1:186.

41. Cyrus A. Bartol Diary, 1 January 1834, Ms. N-1812, MHS.

42. *The Journal of Rachel Van Dyke*, 204–5.

43. Lavater, *Essays on Physiognomy*, 1:18; and Emma Willard, *Journal and Letters from France and Great-Britain* (Troy, NY: N. Tuttle, 1833), 374.

44. Willard, *Journal and Letters*, 380.

45. Willard, *Journal and Letters*, 383.

46. Willard, *Journal and Letters*, 375, 121.

47. Sarah Wentworth Morton, *My Mind and Its Thoughts*, in *Sketches, Fragments, and Essays* (Boston: Wells and Lilly, 1823), 186–87.

48. Morton, *My Mind and Its Thoughts*, 186–90.

49. Sarah Josepha Hale, *Northwood; or, Life North and South* (Boston: Bowles and Dearborn, 1827; repr., New York: H. Long & Brother, 1852), 403.

50. Almira Hart Lincoln Phelps, *Lectures to Young Ladies: Comprising Outlines and Applications of the Different Branches of Female Education* (Boston: Carter, Hendee, & Co., 1833), 118, 199.

51. "Influence of Moral Sentiment in Producing Personal Beauty," *The Universal Asylum and Columbian Magazine* (Philadelphia), May 1792. This article was reprinted numerous times under a series of different titles: "The Art of Being Pretty," *Weekly Museum* 7, no. 360 (New York), 4 April 1795; "The Art of Being Pretty," *The Philadelphia Minerva* 1, no. 10, 11 April 1795; "Personal Beauty Produced by Moral Sentiment," *The New York Weekly Magazine* 1, no. 43, 27 April 1796; "Personal Beauty Produced by Moral Sentiment," *The Philadelphia Minerva* 2, no. 70, 4 June 1796; "Personal Beauty Produced by Moral Sentiment," *The Baltimore Weekly Magazine*, 28 January 1801; M.S., "Personal Beauty Produced by Moral Sentiment," *The New England Quarterly Magazine* 2, no. 3 (Boston), October-December 1802; "The Ladies' Toilette: Feminine Beauty," *The Freemasons' Magazine and General Miscellany* 1, no. 1 (Philadelphia), 1 April 1811; and "Feminine Beauty," *The Monthly Magazine and Literary Journal* 1, no. 4 (Winchester), 1 August 1812.

52. Thomas Broadhurst, *Advice to Young Ladies on the Improvement of the Mind and the Conduct of Life*, 2nd ed. (London: Longman, Hurst, Rees, and Orme, 1810 [first ed., 1808]), 84. This book went through at least nine English editions between 1808 and 1822. Parts of it were reprinted in John Platts, ed., *The Female Mentor; or, Ladies' Class-Book* (Derby, England: Henry Mozley, 1823). Lucretia Farrington copied this passage into her commonplace book. See "Notebook for 1834," BA.

53. Samuel Galloway, "An Address," *Ladies Repository, and Gatherings of the West* (Cincin-

nati), March 1841. This address was reprinted in the April 1841 edition of the *Ladies Repository, and Gatherings of the West.*

54. Letter from Rebecca Gratz to Benjamin Gratz and Ann Boswell Gratz, 13 May 1844, Gratz Family Papers, 1750–1974, Mss.Ms.Coll.72, APS.

55. Letter from Mary Putnam Jacobi to Victorine Haven Putnam, 2 August 1868, in Ruth Putnam, ed., *Life and Letters of Mary Putnam Jacobi* (New York: G. P. Putnam's Sons, 1925), 188.

56. Bathsheba H. Morse Crane, *Life, Letters, and Wayside Gleanings, for the Folks at Home* (Boston: J. H. Earle, 1880), 264; and Samuel Roberts Wells, *New Physiognomy; or, Signs of Character as Manifested through Temperament* (New York: Fowler and Wells, 1866), 659.

57. Anna Cabot Lowell diaries, 17 September 1825, Box 2, Volume 10, Ms. N-1512, MHS. For a similar example, see her discussion of Elizabeth Jackson on 8 March 1832, Box 3, Volume 20.

58. "Biographical Sketches of Irish Authoresses and Heroines. No. IV.: Margaret Derenzy," *The Irish Shield: A Historical and Literary Weekly Paper* 3, no. 3 (Philadelphia), 28 January 1831; "For the Juvenile Port-Folio, by a Young Lady of the City," *The Juvenile Port-Folio, and Literary Miscellany* 1, no. 57 (Philadelphia), 13 November 1813; and James Northcote, *Memoirs of Sir Joshua Reynolds* (Philadelphia: M. Carey & Son, 1817), 6. For examples of the eye being described as the "window" of the soul, see Richard Brown, *An Essay on the Truth of Physiognomy and Its Application to Medicine* (Philadelphia: Thomas T. Stiles, 1807), 23; Sarah Anderson Hastings, *Poems on Different Subjects* (Lancaster, PA: William Dickson, 1808), 102; "The Blind Girl," *The Juvenile Miscellany* 2, no. 1 (Boston), March 1827; "The Mirror of the Graces," *Lady's Book* (New York), February 1832; "Hints to Students on the Use of the Eyes," *The Biblical Repository* 3, no. 11 (Andover, MA), 1 July 1833; "The Five Senses," *The Family Magazine* 3 (New York), May 1836; The Editor, "Hints to Youthful Readers," *Ladies Repository, and Gatherings of the West* 4 (Cincinnati), September 1844; and Joseph Turnley, *The Language of the Eye: As Indicative of Female Beauty, Manly Genius, and General Character* (London: Partridge and Co., 1856), 19.

59. Letter from Margaret Bayard Smith to Susan B. Smith, in Hunt, ed., *The First Forty Years of Washington Society*, 34; Mary Ware Allen, School Journal, 11 May 1838, Mss. Octavo Volume 25, in Allen-Johnson Family Papers, 1759–1992, AAS; Journal of Anna D. Gale, 28 February 1838, AAS; and Catharine Maria Sedgwick to Katharine Sedgwick Minot, 16 April 1837, in Mary Dewey, ed., *Life and Letters of Catharine M. Sedgwick* (New York: Harper & Brothers, 1871), 264. For women who provided descriptions of eyes and faces, "beaming" with traits of mind or soul, see the Papers of Lucy Chase, Folder 3, Diary Fragments, 14 May 1844 to June 1844, in Chase Family Papers, AAS; Elizabeth C. Clemson, Commonplace Book, 1824–1828, HSP; and the *Letters of Anna Seward: Written Between the Years 1784 and 1807* (Edinburgh: George Ramsay & Company, 1811), 3:262. For examples in women's magazines and novels, see "For the Juvenile Port-Folio, by a Young Lady of the City," *The Juvenile Port-Folio, and Literary Miscellany* 1 no. 57 (Philadelphia), 13 November 1813; "Resignation, an Original Tale: Chapter XV," *The Ladies' Literary Cabinet* 2, no. 4 (New York), 3 June 1820; "Flirtation," *Ladies' Magazine* 1, no. 8 (Boston), August 1828; "Miscellaneous Communications: Distinguished Persons in England," *Christian Spectator* 2, no. 11 (New Haven, CT), 1 November 1828; and Sarah Josepha Hale, *Northwood*, 65, 290.

60. As Graeme Tytler has argued, "It is not that the ideas were new but, rather, that they had been only perfunctorily treated in eighteenth-century fiction." See Tytler, "Lavater's Influence on Sir Walter Scott," 112.

61. Anna Cabot Lowell Extract Book, Ms.N-1513, MHS; and Anna Cabot Lowell II Papers/Diaries, 1818-1894, Ms. N-1512, MHS. See entries for 19 October 1832 and 30 October 1832.

62. For female authors playing with Walter Scott's phrase, "mantling blood in ready play," see Amelia Opie's novel, *Adeline Mowbray*, 2nd edition (London: Longman, Hurst, Rees, and Orme,

1805), 47; "Louisa Worthington," *Ladies' Magazine and Literary Gazette* 3, no. 1 (Boston), January 1830; and Almira Hart Lincoln Phelps, *Ida Norman* (Baltimore, MD: Cushing & Brother, 1848), 239. For women who do so in their manuscripts, see Sally Bridges, Autograph Book, 1849–1863, Mss. Am.8702, HSP; Journal of Hannah Margaret Wharton (after 15 February 1813), 43–44, HSP; and entry of February 1853, in "Diary of Sara Jane Lippincott," in *Haps and Mishaps of a Tour of Europe* (Boston: Ticknor & Co., 1854), 233.

63. Thomas Jefferson, *Notes on the State of Virginia* (London: John Stockdale, 1787), 230; and Charles White, *An Account of the Regular Gradation in Man, and in Different Animals and Vegetables; and from the Former to the Latter* (London: C. Dilly, 1799), 135, quoted in Nell Irvin Painter, *The History of White People* (New York: W. W. Norton & Company, 2010), 70–71.

64. *The Journals of Charlotte Forten Grimké*, 398, 33, 315, 156, 373, 300.

65. *The Journals of Charlotte Forten Grimké*, 64, 84, 357, 246. For descriptions of Black women as "pretty" or "intelligent," see 139, 236, 398, 410.

66. Mary Hayden Pike ["Mary Langdon"], *Ida May: A Story of Things Actual and Possible* (London: Sampson, Low, Son, & Co., 1854), 46, 129.

67. Pike, *Ida May*, 146. In addition to using physiognomic language, Pike explicitly invoked another popular science of the day: phreno-magnetism (63–64).

CHAPTER FOUR

1. For some early critiques of women's high foreheads, see Leigh Hunt's commentaries on female beauty, which were frequently reprinted in Europe and the United States throughout the 1820s and 1830s. See "Criticism on Female Beauty," *Mirror of Literature, Amusement, and Instruction* 10, no. 55 (London), 1 January 1825; "Criticism on Female Beauty," *The New Monthly Magazine and Literary Journal* 10, no. 55 (Boston), 1 July 1825; "Criticism on Female Beauty," *The Museum of Foreign Literature, Science, and Art* 7, no. 41 (Philadelphia), 1 November 1825; "Beauty," *Ladies' Magazine* (Boston), 1 January 1830; "A Chapter on Female Features," *The Albion, A Journal of News, Politics, and Literature* 3, no. 32 (New York), 8 August 1835; and "Chapter on Female Features," *Lady's Book* (New York), April 1836. These critiques increased in frequency during the 1840s, and especially during the 1850s (as the women's rights movement became increasingly prominent). For just some examples, see "An Intellectual Fashion," *The Universalist Palladium and Ladies' Amulet* 2, no. 9 (Portland, ME), 15 August 1840; J. C., "The Blue-Stocking," *The New World* 2, no. 11 (New York), 13 March 1841; "High Foreheads," *Boston Evening Transcript* (Boston), 2 December 1846; "Are We a Good-Looking People?" *Putnam's Monthly Magazine of American Literature, Science, and Art* 1, no. 3 (New York), March 1853; "Human Hair," *Littell's Living Age* 37, no. 469 (Boston), 14 May 1853; L. N. Morton, "A Chapter on Human Hair," *Peterson's Magazine* 24, no. 1 (Philadelphia), July 1853; "Beauty of Small Foreheads in Women," *Louisville Daily Courier*, 5 November 1853; "High Foreheads," *Evening Post* (New York), 26 June 1855; "High Foreheads," *American Union* 14, no. 11 (Boston), 14 July 1855; "The Ladies' Column. The Wardrobe, Toilette, Boudoir, etc.," *Porter's Spirit: A Chronicle of the Turf, Field Sports, Literature, and the Stage* 33, no. 7 (New York), 18 April 1857; "High and Low Foreheads," *Saturday Evening Post* 4 (Philadelphia), 27 June 1857; and "High and Low Foreheads," *The National Magazine, Devoted to Literature, Art, and Religion* 11 (New York), August 1857. For examples of authors and editors who defended women's high brows, see "Female Attraction," *Morning Star* (Limerick, ME), 11 May 1853; "Female Attraction," *Christian Era* (Boston), 7 July 1853; "Sketchings," *The Crayon* 2, no. 2 (New York), 11 July 1855; "Low Foreheads

and Beauty," *New York People's Organ* 15, no. 34 (New York), 23 February 1856; "High Foreheads, Beauty, and Intellect," *APJ* 23, no. 3 (New York), March 1856.

2. For a critique of "strong-minded bloomers," see Genio C. Scott, "For the Home Journal: Interesting to Ladies," *Home Journal* 26, no. 542 (New York), 28 June 1856. For "female brawlers," see "American Physician," Appendix to Alexander Walker, *Woman: Physiologically Considered as to Mind, Morals, Marriage, Matrimonial Slavery, Infidelity, and Divorce* (New York: J. & H. G. Langley, 1840), 374.

3. On women's public activism in the antebellum era, see Nancy A. Hewitt, *Women's Activism and Social Change: Rochester, New York, 1822–1872* (Ithaca, NY: Cornell University Press, 1984); Ellen Carol DuBois, "Outgrowing the Compact of the Fathers: Equal Rights, Woman Suffrage, and the United States Constitution, 1820–1878," *Journal of American History* 74, no. 3 (December 1987): 836–62; Judith Wellman, "Women's Rights, Republicanism, and Revolutionary Rhetoric in Antebellum New York State," *New York History* 69, no. 3 (July 1988): 352–84; Lori D. Ginzberg, *Women and the Work of Benevolence: Morality, Politics, and Class in the Nineteenth-Century United States* (New Haven, CT: Yale University Press, 1990); and Anne M. Boylan, *The Origins of Women's Activism: New York and Boston, 1797–1840* (Chapel Hill: University of North Carolina Press, 2002).

4. For two nineteenth-century critiques of coverture, see Sarah Josepha Hale, "Rights of Married Women," *Godey's Lady's Book* (New York), May 1837; and "The Legal Wrongs of Women," *United States Magazine and Democratic Review* 14, no. 71 (New York), May 1844. For secondary scholarship on the fight for married women's property legislation, see Suzanne D. Lebsock, "Radical Reconstruction and the Property Rights of Southern Women," *Journal of Southern History* 43 (May 1977), 197; Norma Basch, "Equity vs. Equality: Emerging Concepts of Women's Political Status in the Age of Jackson," *Journal of the Early Republic* 3, no. 3 (Autumn 1983): 297–318; Nancy Isenberg, *Sex and Citizenship in Antebellum America* (Chapel Hill: University of North Carolina Press, 1998); and Wellman, "Women's Rights, Republicanism, and Revolutionary Rhetoric in Antebellum New York State." For "new style of female politics," see Boylan, *The Origins of Women's Activism*, 137.

5. For foundational scholarship on how both Americans and Europeans have historically used science to enforce racial and gender hierarchies, see Anne Fausto-Sterling, *Myths of Gender: Biological Theories About Women and Men* (1985; New York: Basic Books, 1992); Thomas Laqueur, *Making Sex: Body and Gender from the Greeks to Freud* (Cambridge, MA: Harvard University Press, 1990); Londa L. Schiebinger, *Nature's Body: Gender in the Making of Modern Science* (Boston: Beacon Press, 1993); Cynthia Eagle Russett, *Sexual Science: The Victorian Construction of Womanhood* (Cambridge, MA: Harvard University Press, 1989); Stephen J. Gould, *The Mismeasure of Man* (1981; repr. New York: W. W. Norton & Company, 1996); Carroll Smith-Rosenberg and Charles Rosenberg, "The Female Animal: Medical and Biological Views of Woman and Her Role in Nineteenth-Century America," *Journal of American History* 60, no. 2 (September 1973): 332–56; George Fredrickson, *The Black Image in the White Mind: The Debate on Afro-American Character and Destiny, 1817–1914* (New York: Harper and Row, 1971); Winthrop Jordan, *White Over Black: American Attitudes toward the Negro, 1550–1812* (Chapel Hill: University of North Carolina Press, 1968); and Stuart Ewen and Elizabeth Ewen, *Typecasting: On the Arts & Sciences of Human Inequality* (New York: Seven Stories Press, 2006). Sharrona Pearl has also examined how people used physiognomy to make distinctions of race and ethnicity in the Victorian era. See Pearl, *About Faces: Physiognomy in Nineteenth-Century Britain* (Cambridge, MA: Harvard University Press, 2010). On the intersections between fashion and politics in early America, see Kate Haulman, *The Politics of Fashion in Eighteenth-Century America* (Chapel Hill: University of North Carolina Press, 2011); Linzy Brekke-Aloise, "'The Scourge of Fashion': Political Economy and the Politics of Con-

sumption in Post-Revolutionary America," *Early American Studies: An Interdisciplinary Journal* 33, no. 1 (Spring 2005): 111–39; Kathy Peiss, *Hope in a Jar: The Making of America's Beauty Culture* (New York: Metropolitan Books, 1998); and Catherine E. Kelly, *The Republic of Taste: Art, Politics, and Everyday Life in Early America* (Philadelphia: University of Pennsylvania Press, 2016). For scholarship on Britain in the same time period, see Erin Mackie, *Market á la Mode: Fashion, Commodity and Gender in* The Tatler *and* The Spectator (Baltimore, MD: Johns Hopkins University Press, 1997); and John Styles and Amanda Vickery, eds., *Gender, Taste, and Material Culture in Britain and North America, 1700–1830* (New Haven, CT: Yale University Press, 2006).

6. In recent years, scholars have begun to develop more nuanced understandings of the relationship between popular sciences and social inequality, pointing out that women's rights activists and abolitionists sometimes embraced phrenology. See Cynthia S. Hamilton, "'Am I Not a Man and a Brother?' Phrenology and Anti-Slavery," *Slavery and Abolition* 29, no. 2 (2008): 173–87; Carla Bittel, "Woman, Know Thyself: Producing and Using Phrenological Knowledge in 19th-Century America," *Centaurus* 55 (2013): 104–30; Britt Rusert, *Fugitive Science: Empiricism and Freedom in Early African American Culture* (New York: New York University Press, 2017); and James Poskett, *Materials of the Mind: Phrenology, Race, and the Global History of Science, 1815–1920* (Chicago: University of Chicago Press, 2019).

7. John B. M. D. Lafoy, *The Complete Coiffeur; or, An Essay on the Art of Adorning Natural, and Creating Artificial, Beauty* (New York, 1817), chapter 5 (for the "science of physiognomy," see pages 46–47). Richard Corson calls this work "one of the few books ever written (or at least published) by practicing hairdressers." See Corson, *Fashions in Hair: The First 5,000 Years* (London: Peter Owen, 1965), 464. Karen Halttunen suggests that physiognomy and phrenology might have influenced women's hairstyles, particularly between 1836 and 1856. During this period, middle-class women were expected to embody the "sentimental" ideal, telegraphing sincerity through open countenances and honest eyes. Halttunen suggests that physiognomy could have been responsible for the popularity of the cottage or "poke" bonnet, which carefully framed women's faces in a "U" shape without obscuring the countenance. With limited frippery and few distractions, observers could focus on the sentimental character traits being conveyed through a woman's visage. In her study of hairstyles in nineteenth-century France, Carol Rifelj similarly contends that physiognomy "informed contemporary hairdressing." See Karen Halttunen, *Confidence Men and Painted Women: A Study of Middle-Class Culture in America* (New Haven, CT: Yale University Press, 1982), 84–86; and Carol Rifelj, *Coiffures: Hair in Nineteenth-Century French Literature and Culture* (Newark: University of Delaware Press, 2010), 120.

8. "On Fashions," *The New-England Galaxy and United States Literary Advertiser* 8, no. 417 (Boston), 7 October 1825; and "On Fashions," *The Athenaeum; or, Spirit of the English Magazines* 4, no. 2 (Boston), 15 October 1825.

9. *First Annual Report of the Oneida Association: Exhibiting Its History, Principles, and Transactions to Jan. 1, 1849* (Oneida Reserve: Leonard & Company, 1849), 9. On the organ of "amativeness" and its connection to female sexuality, see Helen Lefkowitz Horowitz, *Rereading Sex: Battles over Sexual Knowledge and Suppression in Nineteenth-Century America* (New York: Vintage Books, 2002), 115–18, 328–31; and April R. Haynes, *Riotous Flesh: Women, Physiology, and the Solitary Vice in Nineteenth-Century America* (Chicago: University of Chicago Press, 2015), 145–46.

10. Hiram Powers to Mary Duncan, 20 August 1856, Hiram Powers Papers, Box 3, Folder 25, AAA. Charles Colbert has also argued that Powers deliberately emphasized desirable phrenological traits in his sitters. See Colbert, "'Each Little Hillock Hath a Tongue': Phrenology and the Art of Hiram Powers," *The Art Bulletin* 68, no. 2 (June 1986): 281–300.

11. This story first appeared in Margaret Howitt, *Twelve Months with Fredrika Bremer in Sweden* (London: Jackson, Walford, and Hodder, 1866), 2:242. Charlotte Bremer then told a similar version in her biography of her sister. See Charlotte Bremer Quiding, ed., *Life, Letters, and Posthumous Works of Fredrika Bremer* (Riverside, Cambridge: H. O. Houghton and Company, 1868), 17–18. True or not, other authors latched onto the anecdote and repeated it in their own descriptions of the feminist thinker. See Daniel G. Brinton, *Personal Beauty: How to Cultivate and Preserve It in Accordance with the Laws of Health* (Springfield, MA: W. J. Holland, 1870), 70; and Mrs. M. E. Sangster, "Little Fredrika Bremer," *Harpers Young People* 4, no. 180 (New York), 10 April 1883.

12. Catharine Maria Sedgwick to Katharine Maria Sedgwick Minot, 13 October 1849, in Mary E. Dewey, ed., *Life and Letters of Catherine M. Sedgwick* (New York: Harper & Brothers, 1872), 317; and Eliza Leslie, *Miss Leslie's Behaviour Book: A Guide and Manual for Ladies* (Philadelphia: T. B. Peterson and Brothers, 1839), 321.

13. *The Toilette of Health, Beauty, and Fashion* (1833; repr. Boston: Allen and Ticknor, 1834), 15–22; and John Bell, *Health and Beauty: An Explanation of the Laws of Growth and Exercise* (Philadelphia: E.L. Carey & A. Hart, 1838), 57–59.

14. Lynne Zacek Bassett, for instance, has argued that the "gothic" aesthetic of the mid-nineteenth century "looked mainly for its inspiration to the Renaissance, particularly the Elizabethan era, which was celebrated for its economic and cultural power." Bassett also contends that the Panic of 1837 led to "more restrained fashions," which might also explain the simplified hairstyles of the late 1830s and 1840s. See Lynne Zacek Bassett, *Gothic to Goth: Romantic Era Fashion and Its Legacy* (Hartford, CT: Wadsworth Atheneum of Art, 2016), 24, 30.

15. See "A Series of Papers on the Hair: Chapter IV," *Godey's Lady's Book* 50 (New York), April 1855; and "A Series of Papers on the Hair: Chapter V," *Godey's Lady's Book* 50 (New York), May 1855.

16. "An Intellectual Fashion," *Universalist Palladium & Ladies' Amulet* (Portland, ME), 15 August 1840.

17. The *Baltimore Sun* published advertisements for "Atkison's Depilatory" almost every day between late July and September of 1837. The ads counseled readers against using these products on their foreheads—warnings that would have only been necessary if people were, indeed, using them to attain high brows. For more on nineteenth-century beauty recipes, see Peiss, *Hope in a Jar*, esp. chapter 1.

18. In a divorce suit in 1860, Joseph W. Trust's wife, Mary Trust, declared that his cosmetics company earned him $6,000 per year (approximately $185,764 in today's currency). To avoid paying alimony, he denied these claims, insisting that "business was bad." On Joseph W. Trust, his alter ego, "Dr. Felix Gouraud," his family, and his business, see Joyce W. Warren, *Women, Money, and the Law: Nineteenth-Century Fiction, Gender, and the Courts* (Iowa City: University of Iowa Press, 2005), 19–43. For sample advertisements, see "To the Ladies," *New-York Tribune*, 19 April 1841; Advertisement, *New World* 4, no. 24 (New York), 11 June 1842; "Analysis of Beauty," *The Subterranean* 4, no. 39 (New York), 20 February 1847; "Intellectual Development of the Forehead," *The Subterranean* 4, no. 40 (New York), 27 February 1847; "Beauty and Fashion: The Forehead," *Spirit of the Times* 17, no. 37 (New York), 6 November 1847; "Gouraud's Poudres Subtiles," *Home Journal* 28, no. 126 (New York), 8 July 1848; "Advertisements," *Littell's Living Age* 37, no. 472 (Boston), 4 June 1853; and "Gouraud's Italian Medicated Soap," *New York Herald*, 3 December 1853. These ads were reprinted hundreds of times between the 1840s and the 1870s.

19. *The Toilette of Health, Beauty, and Fashion*, 51–54; and Daniel G. Brinton, *The Laws of Health in Relation to the Human Form* (Springfield, MA: W. J. Holland, 1870), 301–2.

20. Lydia Maria Child, *Letters from New York* (New York: C. S. Francis & Co., 1845), 248–49; and "To Correspondents," *Godey's Lady's Book* 47, no. 6 (Philadelphia), December 1853.

21. "Phrenology in a Bad Way," *Spirit of the Times: A Chronicle of the Turf, Agriculture, Field Sports, Literature and the Stage* 13, no. 51 (New York), 17 February 1844; "A High Forehead," *Phrenological Journal & Life Illustrated: A Repository of Science, Literature & General Intelligence* 40, no. 6 (New York), December 1864; and "To Eradicate the Hair," *APJ* 41, no. 5 (New York), May 1865.

22. For the critique of "strong-minded" women, see "Human Hair," *Littell's Living Age* 37, no. 469 (Boston), 14 May 1853; "Human Hair," *The Eclectic Magazine of Foreign Literature, Science, and Art* 29, no. 2 (New York), June 1853; "The Human Hair," *Pen and Pencil* 1, no. 25 (Cincinnati), 18 June 1853; and "A Series of Papers on the Hair: Chapter IV," *Godey's Lady's Book and Magazine* 50 (New York), April 1855. It seems that this quotation came from a review essay on "Human Hair" in the *Quarterly Review* 92, no. 184 (London), 1853. *Peterson's Magazine* published a similar critique in 1853, arguing that "Strong-minded women, who sweep the hair off the brow, so as to increase the apparent heighth [sic] of the forehead, only render themselves masculine-looking, and spoil whatever beauty of the face they may happen to have." See L. N. Morton, "A Chapter on Human Hair," *Peterson's Magazine* 24, no. 1 (Philadelphia), July 1853.

23. J. C., "The Blue-Stocking," *The New World; a Weekly Family Journal of Popular Literature, Science, Art and News* 2, no. 11 (New York), 13 March 1841.

24. "My Cousin Nell," *Journal of Agriculture* 2, no. 11 (Boston), 2 June 1852. For a similar example, see John Lofland's discussion of female beauty and intellect, which described female authors as ugly creatures with large foreheads and masculine features. This piece was reprinted numerous times between the 1840s and the 1870s: "Miscellaneous," *The Southern Recorder* 23, no. 17 (Milledgeville, GA), 10 May 1842; "The Gatherer," *American Masonic Register* 3, no. 41 (Albany, NY), 11 June 1842; John Lofland, *The Poetical and Prose Writings of John Lofland, MD, the Milford Bard* (Baltimore, MD: John Murphy, 1846); "Beauty and Intellect," *Family Journal: Literature, News, Romance, Science* 1, no. 25 (Baltimore, MD), 18 June 1859; "Men and Women," *Cape Girardeau Weekly Argus* 1, no. 24 (Cape Girardeau, MO), 26 November 1863; Untitled, *Tri-Weekly Kentucky Yeoman* 19, no. 170 (Frankfort, KY), 26 October 1871. For secondary scholarship on bluestockings and female pedants, see Mary Kelley, *Learning to Stand and Speak: Women, Education, and Public Life in America's Republic* (Chapel Hill: University of North Carolina Press, 2006), 99–102; and Lucia McMahon, *Mere Equals: The Paradox of Educated Women in the Early American Republic* (Ithaca, NY: Cornell University Press, 2012), 1–17.

25. George Cruikshank, *The Comic Almanack for 1847: An Ephemeris in Jest and Earnest* (London: David Bogue, Fleet Street, 1846), 22–25. For American reprintings, see Henry Mayhew, "My Wife Is a Woman of Mind," *Spirit of the Times* 19, no. 5 (New York), 24 March 1849; and "My Wife Is a Woman of Mind," *The Literary Union* 1, no. 17 (Syracuse), 28 July 1849.

26. Supporters of phrenology lamented how their discipline was "sneeringly called" bumpology. The artist Hiram Powers, for instance, complained that "almost every body is ridiculing it when perhaps there is not one out of a thousand—take the community at large—that knows the location of a single bump." See Hiram Powers, "Writings about Prejudice, circa mid-1800s," Box 10, Folder 31, Hiram Powers papers, AAA. For the sheet music to Blewitt's comic ballad, see J. L. Hatton, ed., *A Collection of New, Standard, and Popular Humorous Songs* (London: Boosey & Co., 1875), 124–25. For "Eccentric Song" and "quite a sensation," see "My Wife Is a Woman of Mind," *Illustrated London News*, 17 March 1849. For "racy style," see "Mr. Blewitt's Concert," *Belfast News-Letter*, 12 October 1849. For some advertisements and reviews of Blewitt's performances, see "Concerts," *The Era* (London), 28 January 1849 and 8 April 1849; "Mr. Blewitt's Benefit Night," *Freeman's Journal* (Dublin), 19 April 1849; "Reviews," *Liverpool Mercury* (Lancashire, England), 15 May 1849; "New Music," *Illustrated London News*, 14 December 1850; "Comic Songs," *Illustrated Lon-*

don News, 28 December 1850; "Mr. J. L. Hatton, Pianist," *Leeds Intelligencer* (Yorkshire, England), 16 April 1853.

27. For scholars who have analyzed the supposed conflict between women's maternal responsibilities and their intellectual enrichment in the nineteenth century, see Russett, *Sexual Science*; Nancy M. Theriot, *Mothers and Daughters in Nineteenth-Century America: The Biosocial Construction of Femininity* (Lexington: University Press of Kentucky, 1996), 86–88; and Kimberly A. Hamlin, *From Eve to Evolution: Darwin, Science, and Women's Rights in Gilded Age America* (Chicago: University of Chicago Press, 2017), esp. chapter 2. On the conflicts between women's intellectual cultivation and their duties as republican mothers in early America, see McMahon, *Mere Equals*, 139–63. In 1971, Madeleine B. Stern recognized connections between phrenologists and female activists. See Stern, *Heads and Headlines: The Phrenological Fowlers* (Norman: University of Oklahoma Press, 1971), esp. chapter 10. For a more recent examination of the intersections between phrenology and feminism, see Bittel, "Woman, Know Thyself." Besides Stern and Bittel, few scholars have explicitly analyzed the connections between female activism and the rise of cranial analysis in the United States. Historians of sexuality, however, have noted phrenology's influence on nineteenth-century attitudes toward sex. See Horowitz, *Rereading Sex*, 89–91, 115–18, 328–31; and Haynes, *Riotous Flesh*, 145–46.

28. In 1844, the Fowlers started advertising a series of lectures on "Woman—Her Character, Sphere, and Influence." In 1845, they began publishing regular columns on the "Nature" of woman.

29. On the cause of "noble-hearted women," see "The Worcester Female Convention," *APJ* 12, no. 9 (New York), 1 September 1850. For other positive phrenological reviews of women's rights conventions, see Mary, "Woman and Reform," *APJ* 12, no. 4, 1 April 1850; "Reform in the Condition of Woman: The Spirit and Means By Which It Is to Be Effected," *APJ* 12, no. 10, 1 October 1850; Peggoty, "Woman's Rights," *APJ* 14, no. 4, October 1851; "Woman's Rights Convention," *APJ* 14, no. 4, October 1851; "Mrs. E. Oakes Smith's Lecture," *APJ* 14, no. 5, November 1851; Anna, "Woman! Her Rights and Duties," *APJ* 14, no. 6, December 1851; "An Acceptable Present," *APJ* 14, no. 6, December 1851; "Woman's Rights Convention," *APJ* 15, no. 4, October 1852; "Woman's Rights Convention," *APJ* 17, no. 2, February 1853; "Miss Lucy Stone's Lectures on Woman's Rights," *APJ* 17, no. 6, June 1853; "Woman's Rights Convention," *APJ* 18, no. 3, September 1853; "Woman's Rights Convention," *APJ* 19, no. 1, January 1854; "Works on Woman's Rights," *APJ* 19, no. 3, March 1854. For examples of positive phrenological profiles of female activists, see "Character and Biography of Amelia Bloomer: Biographical Sketch," *APJ* 17, no. 3, March 1853; "Biography: Paulina Wright Davis," *APJ* 17, no. 6, June 1853; "Biography: Elizabeth Oakes Smith," *APJ* 18, no. 5, November 1853; "Grace Greenwood: A Portrait, Biography, and Phrenological Character," *APJ* 19, no. 1, January 1854; "The Champions of Social Reform: Elizabeth Cady Stanton and Susan B. Anthony," *APJ* 49, no. 3, March 1869.

30. For secondary scholarship on these issues, see Bittel, "Woman, Know Thyself."

31. "Phrenological Character of Mrs. E. Rider," Draper-Rice Family Papers, Box 3, Folder 6, AAS.

32. Madeleine Stern was one of the first scholars to highlight the connections between the Fowlers and early feminist activists. See Stern, *Heads and Headlines*, 166–79. Yet most scholarship on gender does not discuss phrenology, and most scholarship on phrenology does not treat gender as a central category of historical analysis. For an exception, see Bittel, "Woman, Know Thyself."

33. On Mott's juxtaposition of the "truths of phrenology" and "the dogmas & hidden mysteries" of religious enthusiasts, see Lucretia Coffin Mott to George Combe, 13 June 1839, in Beverly Wilson Palmer, ed., *Selected Letters of Lucretia Coffin Mott* (Urbana and Chicago: University of Illinois Press, 2002), 53; and Bittel, "Woman, Know Thyself," 120.

34. Orson Fowler, *Creative and Sexual Science; or, Manhood, Womanhood, and Their Mutual Interrelations* (New York: Fowler and Wells, 1870), 160.

35. Abby Kelley to Stephen S. Foster, 30 January 1843; Anna Breed to Abby Kelley, November 1838; Abby Kelley to Mr. and Mrs. Hudson, 12 April 1841; and W. Bassett to Abby Kelley, 12 November 1838; all in Abigail Kelley Foster Papers, AAS.

36. "The Phrenological and Water Cure Journals," *The Lily* 5, no. 9 (Seneca Falls, NY), 1 May 1853; Senex, "Harper's Editor and the Women. No. VI," *The Lily* 7, no. 15 (Mount Vernon, OH), 15 August 1854; and "What Woman Needs," *The Lily* 6, no. 12 (Mount Vernon, OH), July 1854.

37. Elizabeth Cady Stanton, *Eighty Years and More: 1815-1897* (New York: European Publishing Company, 1898), 43–44, 138–39; and Harriet Stanton Blatch and Theodore Stanton, eds. *Elizabeth Cady Stanton as Revealed in Her Letters, Diary and Reminiscences* (New York: Harper and Brothers, 1922), 2:46–47.

38. "Phrenological Reports, 1853," in Ellen Dubois, ed., *The Elizabeth Cady Stanton and Susan B. Anthony Reader: Correspondence, Writings, Speeches* (1981; repr. Boston: Northeastern University Press, 1992), 269–76. In a letter to her father, Stanton lightheartedly proclaimed that Fowler's analysis "often hits the nail on the head." See Blatch and Stanton, *Elizabeth Cady Stanton as Revealed in Her Letters, Diary and Reminiscences*, 2:47.

39. See, for example, "Sarah Kinson, or Margru," *APJ* (New York), 1 July 1850; and "Frederick Douglass: Portrait, Character, and Biography," *APJ* 43, no. 5 (New York), May 1866.

40. "Frederick Douglass: Portrait, Character, and Biography," *APJ* 43, no. 5 (New York), May 1866. Courtney Thompson points out that the Fowlers tried to foster a "good negro" stereotype, describing certain Black individuals as exceptions who nonetheless proved a racist rule. See Thompson, *An Organ of Murder: Crime, Violence, and Phrenology in Nineteenth-Century America* (New Brunswick, NJ: Rutgers University Press, 2021), 24.

41. Darcy Grimaldo Grigsby, *Enduring Truths: Sojourner's Shadows and Substance* (Chicago: University of Chicago Press, 2015), 13, 193. For excellent biographical analyses of Truth, see Nell Irvin Painter, *Sojourner Truth: A Life, a Symbol* (New York: W. W. Norton, 1997); and Margaret Washington, *Sojourner Truth's America* (Urbana: University of Illinois Press, 2009). Washington argued that Truth "embraced phrenology," but little evidence survives about Truth's relationship to cranial analysis, besides the phrenological reading of her head (179).

42. "Are We a Good-Looking People?" *Putnam's Monthly Magazine of American Literature, Science, and Art* 1, no. 3 (New York), March 1853.

43. "Female Attraction," *Morning Star* (Limerick, ME), 11 May 1853.

44. This quotation originated in a *Boston Post* article. I have found at least twenty-one reprintings of this exact quotation (and at least seven different phrenological rebuttals) in US newspapers and magazines between 1855 and 1877. This number does not include all the similar denunciations of women's high brows that simply used different phrasing. For just a few examples, see "High Foreheads," *The Daily Ohio Statesman* (Columbus, OH), 30 June 1855; "High Foreheads," *American Union* 14, no. 11 (Boston), 14 July 1855; and "High Foreheads Not Essential to Female Beauty," *Brother Jonathan* (New York), 4 August 1855. For some phrenological rebuttals, see: "Sketchings," *The Crayon* 2, no. 2 (New York), 11 July 1855; "Low Foreheads and Beauty," *New York People's Organ* 15, no. 34 (New York), 23 February 1856; and "High Foreheads, Beauty, and Intellect," *APJ* 23, no. 3 (New York), March 1856. Between 1855 and 1901, there were at least thirty-five reprintings of this article in English, Scottish, and Irish newspapers, and at least ten more articles that addressed the same theme in slightly different words, including this vigorous, pro-phrenology response: "Are High Foreheads a Mark of Beauty?" *Burnley Express* (Lancashire, England), 14 July 1883.

45. Genio C. Scott, "For the Home Journal: Interesting to Ladies," *Home Journal* 26, no. 542 (New York), 28 June 1856.
46. "High Foreheads, Beauty, and Intellect," *APJ* 23, no. 3 (New York), March 1856.
47. "Sketchings," *The Crayon* 2, no. 2 (New York), 11 July 1855.
48. "The Manly Brow Movement," *Harper's Bazaar* 2, no. 52 (New York), 25 December 1869.

CHAPTER FIVE

1. James Gotendorf, ed., *Love-Letters of Margaret Fuller, 1845–1846* (New York: D. Appleton and Co., 1903), 113.

2. Diary of Sara Jane Lippincott, July 1852, in *Haps and Mishaps of a Tour of Europe* (Boston: Ticknor & Co., 1854), 45.

3. For scholarship on shifts in criminality, poverty, and personal responsibility in early national America, see John Alexander, *Render Them Submissive: Responses to Poverty in Philadelphia, 1760–1800* (Amherst: University of Massachusetts Press, 1980); Michael Meranze, *Laboratories of Virtue: Punishment, Revolution, and Authority in Philadelphia, 1760–1835* (Chapel Hill: University of North Carolina Press, 1996); Peter Okun, *Crime and the Nation: Prison Reform and Popular Fiction in Philadelphia, 1786–1800* (New York: Routledge, 2002); Jonathan Simon, "Rise of the Carceral State," *Social Research* 74, no. 2 (Summer 2007): 476–81; and Jen Manion, *Liberty's Prisoners: Carceral Culture in Early America* (Philadelphia: University of Pennsylvania Press, 2015).

4. Several scholars have highlighted the contradictions between environmentalism and biological determinism that were inherent in phrenology, but few have pointed out that these inconsistencies first emerged in physiognomic discourses in the late eighteenth century. See Peter McCandless, "Mesmerism and Phrenology in Antebellum Charleston: 'Enough of the Marvellous,'" *Journal of Southern History* 58, no. 2 (May 1992): 204; Roger Smith, *The Norton History of the Human Sciences* (New York: W. W. Norton & Company, 1997), 395–96; Stuart Ewen and Elizabeth Ewen, *Typecasting: On the Arts and Sciences of Human Inequality* (New York: Seven Stories Press, 2006), 163–90; Cynthia S. Hamilton, "'Am I Not a Man and a Brother?' Phrenology and Anti-Slavery," *Slavery and Abolition* 29, no. 2 (2008): 173–87; Christopher J. Beshara, "Moral Hospitals, Addled Brains, and Cranial Conundrums: Phrenological Rationalisations of the Criminal Mind in Antebellum America," *Australasian Journal of American Studies* 29, no. 1 (July 2010): 36–60; and Carla Bittel, "Woman, Know Thyself: Producing and Using Phrenological Knowledge in 19th-Century America," *Centaurus: An International Journal of the History of Science and its Cultural Aspects* 55, no. 2 (May 2013): 104–30.

5. Johann Caspar Lavater, *Essays on Physiognomy; for the Promotion of the Knowledge and the Love of Mankind*, trans. Thomas Holcroft (London: G. G. J. and J. Robinson, 1789), 1:205.

6. "Physiognomy," *Weekly Museum* 9, no. 14 (New York), 24 September 1796.

7. R. W. James Mann, *An Oration: Addressed to the Fraternity of Free Masons* (Wrentham, MA, 1798), 9–10.

8. "(Inclosed)," *The Literary Mirror* 1, no. 10 (Portsmouth), 23 April 1808.

9. J. F. Daniel Lobstein, MD, *A Treatise upon the Semeiology of the Eye, for the Use of Physicians; and of the Countenance, for Criminal Jurisprudence* (New York: C. S. Francis, 1830), 35–36. See also pages 130–34.

10. For other discussions of the criminal physiognomy in the early republic, see Henry Bunbury, "The Wheelbarrow," *Daily National Intelligencer* (Washington, DC), 24 September 1816;

"Horrible Crime," *Ladies Port Folio* 1, no. 22 (Boston), 27 May 1820; "Physiognomy of Murderers," *Republican Star* 25, no. 19 (Easton, MD), 23 December 1823; Geryn, "Adventures of a Rambler, No. 1," *The American Monthly Magazine* 1, no. 4 (Philadelphia), May 1824; "Execution," *The National Advocate* (New York), 27 September 1824; Washington Irving, *Tales of a Traveller* (London: John Murray, 1824), 166; B. B. T., "The Stealing Propensity," *The Knickerbocker; or New York Monthly Magazine* 4, no. 3 (New York), September 1834; Emma Embury, "Newton Ainslie: A Sketch," *Lady's Book* (New York), August 1839; "Phrenological Examination of Prisoners," *American Phrenological Journal* 3, no. 2 (Philadelphia), 1 November 1840; and Lorenzo Niles Fowler, *The Illustrated Phrenological Almanac for 1852* (New York: Fowler and Wells, 1851), 17. In 1835, *Godey's Lady's Book* published a critique of those who too quickly drew physiognomic judgments when evaluating criminals. The author did not challenge the legitimacy of physiognomy, writing, "I quarrel not with the principles of this science, as they are laid down by learned professors." But the author did urge caution and asked people to interrogate their own assumptions before publishing them in the papers. See "Moral and Personal Deformity; A Hint to Those Who Frame Advertisements for Apprehending Offenders," *Godey's Lady's Book* (New York), September 1835.

11. Courtney E. Thompson, *An Organ of Murder: Crime, Violence, and Phrenology in Nineteenth-Century America* (New Brunswick, NJ: Rutgers University Press, 2021), 29. Thompson brilliantly shows how phrenology influenced elite understandings of crime and the people who committed it during the early nineteenth century, and also how phrenological ideas about crime continued to circulate in popular culture, even when physicians, lawyers, and scientists began to reject this science in the mid-nineteenth century.

12. In his notebook, Bartol compared physical beauty with "mental loveliness" or the "Beauty of Mind." He likewise debated the relative merits of physiognomy and phrenology, contemplating how they might be used together. See Cyrus A. Bartol Diary, 16 June 1834, Ms. N-1812, MHS.

13. "Mr. Barrett's Journal, containing Observations on Prisons and Asylums for the Insane, in the Western and Southern States," in *Annual Report of the Board of Managers of the Prison Discipline Society* (Boston: 63 Atkinson Street, 1844), 431. As Courtney Thompson has argued, "criminality . . . was not specifically racialized, at least in the first two thirds of the nineteenth century in the United States." See Thompson, *An Organ of Murder*, 23.

14. John Luckey, *Life in Sing Sing State Prison, as Seen in a Twelve Years' Chaplaincy* (New York: N. Tibbals & Co., 1860), 145. Luckey was no fan of phrenology, but even he used physiognomic language to describe incarcerated people.

15. Ansel D. Eddy, *Black Jacob, A Monument of Grace: The Life of Jacob Hodges, An African Negro, Who Died in Canandaiuga, NY* (Philadelphia: American Sunday-School Union, 1842), 37.

16. Embury, "Newton Ainslie: A Sketch." This story also had a religious message. Part of the reason that Ainslie allowed himself to be consumed by bitterness and swept up in a fit of passion was because he abandoned God to follow the blasphemous teachings of Thomas Paine. While in prison, however, he rekindled his Christianity.

17. Z. P., "Retribution," *The Advocate of Moral Reform* 3, no. 4 (New York), 15 February 1837. For scholarship on the New York Female Moral Reform Society and the *Advocate of Moral Reform*, see Therese L. Lueck, "Women's Moral Reform Periodicals of the 19th Century: A Cultural Feminist Analysis of *The Advocate*," *American Journalism* 16, no. 3 (Summer 1999): 37–52; Nicolette Severson, "'Devils Would Blush to Look': Brothel Visits of the New York Female Moral Reform Society, 1835 and 1836," *Journal of the History of Sexuality* 23, no. 2 (May 2014): 226–46; Lori D. Ginzberg, *Women and the Work of Benevolence: Morality, Politics, and Class in the 19th Century United States* (New Haven, CT: Yale University Press, 1990), 13–14, 19–21, 27, 56–59, 62, 113–14, and 128; and Christine Stansell, *City of Women: Sex and Class in New York, 1789–1860* (Chicago: University

of Illinois Press, 1987). For the role of women in prison reform, see Estelle Freedman, *Their Sisters' Keepers: Women's Prison Reform in America, 1830–1930* (Ann Arbor: University of Michigan Press, 1984). For moral reformers' interactions with sex workers in a different context, see Jeffery S. Adler, "Streetwalkers, Degraded Outcasts, and Good-for-Nothing Huzzies: Women and the Dangerous Class in Antebellum St. Louis," *Journal of Social History* 25, no. 4 (Summer 1992): 737–55.

18. C. L., "The Last Home of the Living Lost," *Advocate of Moral Reform and Family Guardian* 18, no. 16 (New York), 15 August 1852. Beginning in the mid-eighteenth century, new medical and social theories encouraged elite and middle-class individuals to cultivate a sense of "sympathy" for other people's suffering. But this often translated into voyeuristic and sensationalistic descriptions of individuals in anguish. See Karen Halttunen, "Humanitarianism and the Pornography of Pain in Anglo-American Culture," *American Historical Review* 100, no. 2 (April 1995): 303–34. See also William Hogarth, *The Harlot's Progress* (1732).

19. C. L., "The Last Home of the Living Lost."

20. Michel Foucault famously chronicled this transition in his pathbreaking work, *Discipline and Punish: The Birth of the Prison*, trans. Alan Sheridan (New York: Pantheon, 1977). For an examination of how this process played out in the American context, see Meranze, *Laboratories of Virtue*; Marvin E. Schultz, "'Running the Risks of Experiments': The Politics of Penal Reform in Tennessee, 1807–1829," *Tennessee Historical Quarterly* 52, no. 2 (Summer 1993): 86–97; Adam J. Hirsch, "From Pillory to Penitentiary: The Rise of Criminal Incarceration in Early Massachusetts," *Michigan Law Review* 80, no. 6 (May 1982): 1179–1269; and Jacqueline Thibaut, "'To Pave the Way to Penitence': Prisoners and Discipline at the Eastern State Penitentiary, 1829-1835," *The Pennsylvania Magazine of History and Biography* 106, no. 2 (April 1982): 187–222. For more recent work on prisons and prisoners in the early republic, see Manion, *Liberty's Prisoners*.

21. For the concept of "redemptive suffering," see Jennifer Graber, *The Furnace of Affliction: Prisons and Religion in Antebellum America* (Chapel Hill: University of North Carolina Press, 2011). For inmates' responses to this concept, see Graber, "Engaging the Trope of Redemptive Suffering: Inmate Voices in the Antebellum Prison Debates," *Pennsylvania History: A Journal of Mid-Atlantic Studies* 79, no. 2 (Spring 2012): 209–33.

22. See "Extract from Gov. Lincoln's Message, Jan. 1826," in *Reports of the Prison Discipline Society of Boston*, First Annual Report (1826; repr. Montclair, NJ, 1972), 4, quoted in Larry Goldsmith, "History from the Inside Out: Prison Life in Nineteenth-Century Massachusetts," *Journal of Social History* 31, no. 1 (Autumn 1997): 109. For Jared Curtis's commentary on prisoners at the Massachusetts State Prison in Charlestown, see "Jared Curtis Notebooks, 1829–1831," Ms. SBd-63, MHS. For a published version, see Philip F. Gura, ed., *Buried from the World: Inside the Massachusetts State Prison, 1829–1831* (Charlottesville: University of Virginia Press, 2005). I will cite the manuscript versions unless otherwise noted. Philadelphia's prison administrators were equally concerned about the failures of early-nineteenth-century prisons. After taking charge of the Eastern State Penitentiary, Samuel Wood criticized the Walnut Street Prison, which allowed prisoners to interact: "I assert, without fear of contradiction, that it is not possible for the Legislature to devise a system where men will be more completely contaminated, hardened, and depraved, than in that college of vice, the Walnut Street Prison." See Samuel R. Wood, "Sale of Walnut Street Prison," *Register of Pennsylvania* 7, no. 7 (Philadelphia), 12 February 1831, quoted in Manion, *Liberty's Prisoners*, 190.

23. "Jared Curtis Notebooks," vol. 1, MHS.

24. "Jared Curtis Notebooks," vol. 2, MHS.

25. "Jared Curtis Notebooks," vol. 1, MHS.

26. On Farnham, phrenology, and prison reforms at Sing Sing, see Janet Floyd, "Dislocations

of the Self: Eliza Farnham at Sing Sing Prison," *Journal of American Studies* 40, no. 2 (August 2006): 311–25; John Lardas Modern, "Ghosts of Sing Sing, or the Metaphysics of Secularism," *Journal of the American Academy of Religion* 75, no. 3 (September 2007): 615–50; and Jodi Schorb, *Reading Prisoners: Literature, Literacy, and the Transformation of American Punishment, 1700–1845* (New Brunswick, NJ: Rutgers University Press, 2014), 169–84.

27. See Marmaduke Blake Sampson, *Rationale of Crime, and Its Appropriate Treatment: Being a Treatise on Criminal Jurisprudence Considered in Relation to Cerebral Organization*, ed. Eliza Farnham (New York: D. Appleton & Company, 1846). For a discussion of the daguerreotyping process, see Madeleine B. Stern, "Mathew B. Brady and the *Rationale of Crime*: A Discovery in Daguerreotypes," *The Quarterly Journal of the Library of Congress* 31, no. 3 (July 1974): 126–35; Ewen and Ewen, *Typecasting*, 215; Modern, "Ghosts of Sing Sing," 631–33; and Floyd, "Dislocations of the Self," 314.

28. For descriptions of Farnham's radicalism, see Modern, "Ghosts of Sing Sing"; and Floyd, "Dislocations of the Self."

29. "State Penitentiary for the Eastern District of Pennsylvania Records, 1819–1955," Series 1, Admission Ledgers and Bound Volumes, Mss.365.P381p, APS. I am relying on volumes A, B, and D of Larcombe's notes. Volume C has been lost.

30. "State Penitentiary for the Eastern District of Pennsylvania Records, 1819–1955," Series 1, vol. A, APS.

31. "State Penitentiary for the Eastern District of Pennsylvania Records, 1819–1955," Series 1, vols. A and B, APS. As Larry Goldsmith has argued, "Prisoners may have been captives, but they were hardly passive" (see Goldsmith, "History from the Inside Out," 110).

32. "State Penitentiary for the Eastern District of Pennsylvania Records, 1819–1955," Series 1, vol. B, APS.

33. Monahan apparently insisted that he was a Presbyterian, but Larcombe was convinced by his appearance that he was Catholic. When Monahan left the prison, Larcombe wrote, "T. O. no hope." T. O. stood for "Time Out." "State Penitentiary for the Eastern District of Pennsylvania Records, 1819–1955," Series 1, vols. B and D, APS.

34. *Annual Report of the Board of Managers of the Prison Discipline Society* (Boston: Perkins and Marvin, 1836), 80.

35. "State Penitentiary for the Eastern District of Pennsylvania Records, 1819–1955," Series 1, vols. A, B, and D, APS. For an examination of how inmates navigated and responded to the "moral instruction" of prison reformers, see Graber, "Engaging the Trope of Redemptive Suffering," 209–33. Michel Foucault famously decried the "disciplinary gaze" of the panopticon: an architectural style that would have subjected prisons to reformers' constant surveillance. See Foucault, *Discipline and Punish*, 168.

36. "Notebooks Concerning Prisons and Prisoners," in William Parker Foulke Papers, 1840–1865, Box 7, Mss.B.F826, APS. The Pennsylvania System vied with the Auburn System for influence. At the Eastern State Penitentiary, prisoners were isolated in private cells and blocked from all interaction with fellow inmates. In Auburn, prison administrators forced inmates to work together, but in silence. The prison at Sing Sing followed the Auburn model. The goal in both systems was for prisoners to internalize punishment and engage in silent self-reflection. On the differences between these two systems, see Simon, "Rise of the Carceral State," 476–81; Modern, "Ghosts of Sing Sing," 625–28; and Meranze, *Laboratories of Virtue*, 254–59.

37. Foulke, "Notebooks Concerning Prisons and Prisoners," APS.

38. See Foulke, "Notebooks Concerning Prisons and Prisoners," APS. Jennifer Graber and Jen Manion have also shown how prisoners manipulated authority figures to attain their own goals, sometimes reinforcing the narratives of prison administrators and reformers. See Graber,

"Engaging the Trope of Redemptive Suffering"; and Manion, *Liberty's Prisoners*. Jodi Schorb and Larry Goldsmith have also demonstrated how inmates used reading and writing as strategies of resistance within antebellum penitentiaries. See Goldsmith, "History from the Inside Out"; and Schorb, *Reading Prisoners*, chapters 2 and 3.

39. Charles Dickens, *American Notes for General Circulation* (New York: Harper and Brothers, 1842), 42. Strangely, Dickens thought solitary confinement had a positive effect on women's appearances: "The faces of the women, as I have said, it humanises and refines. Whether this be because of their better nature, which is elicited in solitude, or because of their being gentler creatures, of greater patience and longer suffering, I do not know; but so it is. That the punishment is nevertheless, to my thinking, fully as cruel and as wrong in their case, as in that of the men, I need scarcely add" (42–43).

40. Dickens, *American Notes for General Circulation*, 40.

41. See *First Annual Report of the Board of Managers of the Prison Discipline Society, Boston, June 2, 1826*, 6th edition (Boston: Perkins and Marvin, 1830), 300.

42. Foulke, "Notebooks Concerning Prisons and Prisoners," APS.

43. Foulke, "Notebooks Concerning Prisons and Prisoners," APS. The problem of penitentiary-induced mental illness plagued prison administrators throughout the nation. In 1848, the Prison Discipline Society also wrote about the "Cases of Insanity, 'supposed to have originated' in the New Penitentiary in Philadelphia, and in the State Prison at Charlestown, Mass., in 1846." See the *Twenty-First Annual Report of the Board of Managers of the Prison Discipline Society, Boston, May 1846* (Boston: Damrell and Moore, 1848), 132–33.

44. See "British Penal Discipline," *Pennsylvania Journal of Prison Discipline and Philanthropy* 3, no. 3 (Philadelphia), July 1848. For Foulke's theory on masturbation and insanity, see Foulke, "Notebooks Concerning Prisons and Prisoners," APS (especially the entry on 14 November 1849). He presented this theory to the prison physician, Dr. Evans: "I then asked Dr. Evans whether he had formed any opinion of the extent to which the dementia was attributable to the practise [sic] of masturbation—He answered that he did not doubt that a portion of it was so—but not as much as was supposed by some persons." Dr. Parrish, by contrast, argued against Foulke's theory, citing medical testimony during a recent legal case. He argued that "the medical opinion then given was against the idea of the vice disqualifying the party" from serving as a testator on a will.

45. G. C., "Organized Charities," *Oneida Circular* 10, no. 48 (Brooklyn), 24 November 1873.

CHAPTER SIX

1. Frederick Douglass's review of Wilson Armistead's *A Tribute for the Negro*, published in the *North Star* (Rochester, NY), 7 April 1849. William Lloyd Garrison republished Douglass's criticism of white portrait painters, though not his review of Armistead's book. See Frederick Douglass, "Negro Portraits," *The Liberator* (Boston), 20 April 1849.

2. John Stauffer, *The Black Hearts of Men: Radical Abolitionists and the Transformation of Race* (Cambridge, MA: Harvard University Press, 2001), 51. For the most recent articulation of this argument, see John Stauffer, Zoe Trodd, and Celeste-Marie Bernier, eds., *Picturing Frederick Douglass: An Illustrated Biography of the Nineteenth Century's Most Photographed American* (New York: W. W. Norton & Company, 2015). Stauffer has argued that daguerreotypes in the nineteenth century were "thought to penetrate the sitter's soul as well as his mind." See Stauffer, "Frederick Douglass and the Aesthetics of Freedom," *Raritan: A Quarterly Review* 25, no. 1 (Summer 2005):

120; and Stauffer, "Creating an Image in Black: The Power of Abolition Pictures," in *Prophets of Protest: Reconsidering the History of American Abolitionism*, ed. Timothy Patrick McCarthy and John Stauffer (New York: The New Press, 2006), 256–67. Stauffer also published this essay in *Beyond Blackface: African Americans and the Creation of American Popular Culture*, ed. W. Fitzhugh Brundage (Chapel Hill: The University of North Carolina Press, 2011), 66–94. Other scholars have also written about Frederick Douglass's belief in the power of visual representation. See Ginger Hill, "'Rightly Viewed': Theorizations of Self in Frederick Douglass's Lectures on Pictures," in *Pictures and Progress: Early Photography and the Making of African American Identity*, ed. Maurice O. Wallace and Shawn Michelle Smith (Durham, NC and London: Duke University Press, 2012), 41; Laura Wexler, "'A More Perfect Likeness': Frederick Douglass and the Image of the Nation," in *Pictures and Progress*, 20–21; Eric Foner, "True Likenesses," in *Forever Free: The Story of Emancipation and Reconstruction* (New York: Vintage Books, 2013), 34–37; and Celeste-Marie Bernier, "A Visual Call to Arms against the 'Caracature [sic] of My Own Face': From Fugitive Slave to Fugitive Image in Frederick Douglass's Theory of Portraiture," *Journal of American Studies* 49, no. 2 (May 2015): 323–57. When Douglass thought, spoke, and wrote about portraits, he did so within an intellectual milieu that had been shaped by physiognomic and phrenological assumptions. In fact, Douglass himself once dined with the famous phrenologist George Combe, and also read his phrenological best-seller, *The Constitution of Man*. According to James Poskett, "Douglass singled out this book as one of the few phrenological works containing a fair portrayal of African character." See Poskett, *Materials of the Mind: Phrenology, Race, and the Global History of Science, 1815–1920* (Chicago: University of Chicago Press, 2019), 130.

3. I am building on recent works by scholars who have explored how African Americans responded to and shaped nineteenth-century racial science. See Bruce Dain, *A Hideous Monster of the Mind: American Race Theory in the Early Republic* (Cambridge, MA: Harvard University Press, 2002); Mia Bay, *The White Image in the Black Mind: African American Ideas about White People, 1830–1925* (New York: Oxford University Press, 2000); Patrick Rael, *Black Identity and Black Protest in the Antebellum North* (Chapel Hill: University of North Carolina Press, 2002); Rael, "A Common Nature, A United Destiny: African American Responses to Racial Science from the Revolution to the Civil War," in McCarthy and Stauffer, eds., *Prophets of Protest*; and Britt Rusert, "Delany's Comet: Fugitive Science and the Speculative Imaginary of Emancipation," *American Quarterly* 65, no. 4 (December 2013): 801. I am also building on a wealth of interdisciplinary scholarship that examines how African Americans used visual and literary culture to represent the meaning of freedom, claim citizenship, and challenge racism. In showing the impact of phrenology and physiognomy on Black intellectual thought, though, I am recovering the scientific logic that undergirded the artistic practices that are so well documented in the existing literature. See Nell Irvin Painter, *Sojourner Truth: A Life, A Symbol* (New York: W. W. Norton & Company, 1996), esp. chapters 20 and 26; Wallace and Smith, eds., *Pictures and Progress*, 4–5; Shawn Michelle Smith, *American Archives: Gender, Race, and Class in Visual Culture* (Princeton, NJ: Princeton University Press, 1999); Rusert, "Delany's Comet"; Jasmine Nichole Cobb, *Picture Freedom: Remaking Black Visuality in the Early Nineteenth Century* (New York: New York University Press, 2015); Sarah Blackwood, "'Making Good Use of Our Eyes': Nineteenth-Century African Americans Write Visual Culture," *MELUS: Multi-Ethnic Literature of the US* 39, no. 2 (Summer 2014): 42–65; and Sarah Blackwood, "Fugitive Obscura: Runaway Slave Portraiture and Early Photographic Technology," *American Literature* 81, no. 1 (2009): 93–125.

4. On the rise of scientific racism in the United States, see William R. Stanton, *The Leopard's Spots: Scientific Attitudes toward Race in America, 1815–1859* (Chicago: University of Chicago Press, 1960); George M. Fredrickson, *The Black Image in the White Mind: The Debate on Afro-*

American Character and Destiny, 1817–1914 (New York: Harper and Row, 1971); Stephen J. Gould, *The Mismeasure of Man* (1981; repr. New York: W. W. Norton, 1996); and Ann Fabian, *The Skull Collectors: Race, Science, and America's Unburied Dead* (Chicago: University of Chicago Press, 2010).

5. White phrenologists made deliberate attempts to spread phrenological knowledge to Black Americans, but their entreaties dripped with racist condescension. Despite his anti-slavery sympathies, Orson Fowler once used a phrenological lecture as an opportunity to critique the abolitionists for being "so denunciatory in their language," and for moving too fast. Then, he gave a separate lecture to a segregated Black audience. When *The Liberator* reviewed the event, the newspaper declared that because it was "almost impossible" to get Black audiences "to come out, and join in a class to hear lectures on any science, (owing, I suppose, to the degraded state in which they have been held,) it was thought best by friend Fowler to give them a lecture by themselves; which he did last evening." Essentially, the newspaper claimed that Black audiences were not as scientifically literate as white audiences, and thus in need of a segregated lecture. The newspaper insisted that the Black audience enjoyed the lecture, but then wrote, "I long to see the time when they [African Americans] will feel the same interest for the elevation of themselves, that their friends feel for them." Such statements reveal how white phrenologists practically (and arrogantly) reinforced racial inequality, even as they technically opposed slavery and encouraged Black people to be scientific practitioners. See "Reformatory: Lectures on Phrenology and Physiology," *The Liberator*, 4 October 1844.

6. Manisha Sinha writes that African Americans in the North were able to "develop a more vocal tradition of protest" than their enslaved counterparts, despite sharing ideological goals. See Sinha, "Coming of Age: The Historiography of Black Abolitionism," in McCarthy and Stauffer, eds., *Prophets of Protest*, 34.

7. I rely heavily on Patrick Rael's concept of "cofabrication." See Rael, *Black Identity and Black Protest*, 5, 10, 124, 174, and 283. For another examination of class divisions among African Americans in New York, see Leslie M. Harris, *In the Shadow of Slavery: African Americans in New York City, 1626–1863* (Chicago: University of Chicago Press, 2003), esp. chapter 6. On ideological hegemony: Audre Lorde, "The Master's Tools Will Never Dismantle the Master's House," in *Sister Outsider: Essays and Speeches* (Berkeley, CA: Crossing Press, 1984), 110–14. For a discussion of the historiographical debate surrounding Lorde's analysis within African American history, see Rael, "A Common Nature, A United Destiny," 185–86, and 195–99.

8. Cobb, *Picture Freedom*, 145–46.

9. Frederick Douglass, *Claims of the Negro, Ethnologically Considered* (Rochester, NY: Daily American Office, 1854), 20–21. William Lloyd Garrison republished part of Douglass's lecture: "The Negro Is a Man," *The Liberator* (Boston), 28 July 1854. William J. Wilson made similar arguments about white pictures of Black people, dismissing them as images that were "gotten up for the *American prejudice Market.*" See Wilson, "Afric-American Picture Gallery—Second Paper. By Ethiop," *Anglo-African Magazine* 1, no. 3 (New York), March 1859, 88.

10. Ginger Hill argues that Douglass used photographs to "claim the status of rights-bearing autonomy" and to "proclaim his complex interiority." Sarah Blackwood similarly argues that Americans thought daguerreotypes were "associated with the revelation of certain truths about their sitters." Neither scholar examines the more fundamental question of *why* Americans believed pictures reflected "truths" about people's character. See Hill, "'Rightly Viewed': Theorizations of Self in Frederick Douglass's Lectures on Pictures," 46; and Blackwood, "Fugitive Obscura," 97.

11. As George Price and James Brewer Stewart have argued, "Never before had white violence flared so intensely in northern cities," and never before "had disagreements in the North over

the meaning of 'race' carried such explosive potential." For a broad adumbration of these transformations, see James Brewer Stewart, "The Emergence of Racial Modernity and the Rise of the White North, 1790–1840," *Journal of the Early Republic* 18, no. 2 (Summer 1998): 181–217. See also Dain, *A Hideous Monster of the Mind*, 119–21; and Harris, *In the Shadow of Slavery*, esp. chapter 4.

12. Cynthia S. Hamilton, "'Am I Not a Man and a Brother?' Phrenology and Anti-Slavery," *Slavery & Abolition* 29, no. 2 (June 2008): 176.

13. Rael, "A Common Nature, A United Destiny," 190–91.

14. Britt Rusert, *Fugitive Science: Empiricism and Freedom in Early African American Culture* (New York: New York University Press, 2017), 4.

15. See, for example, "Anti-Phrenology," *The Colored American* (New York), 16 September 1837; "Phrenology Exploded," *The Colored American* (New York), 25 November 1837; "Phrenology," *The Liberator* (Boston), 29 November 1839; "Lectures on Phrenology," *The Liberator* (Boston), 13 November 1840; and "Practical Phrenology," *The Liberator* (Boston), 11 November 1853.

16. Harriet Beecher Stowe, *Uncle Tom's Cabin; or, Life among the Lowly* (Boston: John P. Jewett & Company; and Cleveland: Jewett, Proctor, and Worthington, 1852), 1:40–41 and 2:189.

17. For the description of Mr. Walters, see Frank J. Webb, *The Garies and Their Friends* (London: G. Routledge & Co., 1857), 121. For an articulation of the novel's political statement, see the Preface. For other anti-slavery works that employ physiognomy as a technology of character detection, see Sarah Josepha Hale, *Northwood; or, Life North and South* (Boston: Bowles and Dearborn, 1827); William Wells Brown, *Clotel; or, The President's Daughter* (London: Partridge & Oakey, 1853); Frederick Douglass, *The Heroic Slave*, in *Autographs for Freedom* (Boston: John P. Jewett, & Co., 1853); William Wells Brown, *The American Fugitive in Europe: Sketches of Places and People Abroad* (Boston: John P. Jewett and Company, 1855); William C. Nell, *The Colored Patriots of the American Revolution* (Boston: Robert F. Wallcut, 1855); Josephine Brown, *Biography of an American Bondman, by his Daughter* (Boston: R. F. Wallcut, 1856); and William Wells Brown, "Chapter IV: Slave Revolt at Sea," in *The Negro in the American Rebellion: His Heroism and His Fidelity* (Boston: Lee & Shepard, 1867).

18. For William Still's physiognomic descriptions, see "Journal C of the Underground Railroad in Philadelphia kept by William Still: containing notices of arrivals of fugitive slaves in Philadelphia with descriptions of their flight, 1852–1857," AmS.232, HSP. Lydia Maria Child also used physiognomy to suggest that certain individuals were particularly unsuited for slavery. When telling the story of a fictional slave who advocated for rebellion, she wrote: "His high, bold forehead and flashing eye indicated an intellect too active, and a spirit too fiery, for Slavery." See Child, "The Meeting in the Swamp," in *The Freedmen's Book* (Boston: Ticknor and Fields, 1866), 107.

19. William Wells Brown, *The Black Man, His Antecedents, His Genius, and His Achievements* (New York: Thomas Hamilton, 1863), 237. For the description of Forten, see pages 192–93.

20. Martin Robison Delany, *The Condition, Elevation, Emigration, and Destiny of the Colored People of the United States* (1852; repr. Frankfurt am Main: Outlook Verlag, 2020), 74, 84.

21. "Milton, O[hio]. June 18, 1848," *The North Star* (Rochester, NY), 7 July 1848; "Henry E. Lewis," *The North Star* (Rochester, NY), 22 December 1848; and "New York and Brooklyn News," *Frederick Douglass' Paper* (Rochester, NY), 2 February 1855. Britt Rusert mentions Lewis in *Fugitive Science*, 123–24. As early as 1831, *The Liberator* advertised that "A black clergyman, from the South" was giving "very successful" public lectures on phrenology. See Untitled, *The Liberator* (Boston), 30 July 1831.

22. Rusert, *Fugitive Science*, 123. For the images satirizing Black practical scientists, see "Free-knowledgy, or Black Bumpology," in *Crocketts Comic Almanac* (New York: Elton, 1839), AAS; and

"Professor Pompey Magnetizing an Abolition Lady," Digital Collections, AAS. James Poskett has pointed out that phrenologists of color traveled around the globe, facing mockery in the London press, just as they did in the United States. See Poskett, *Materials of the Mind*, 16.

23. Bruce Dain has called this work "the first major African American writing on record to address in an original and systematic manner racial differences and racial history." James Brewer Stewart and George R. Price have similarly argued that before Easton published his *Treatise*, "no American writer had ever attempted so comprehensive an analysis of 'black and white' in all of its ramifications." See Dain, *A Hideous Monster of the Mind*, 173; and Price and Stewart, ed., *To Heal the Scourge of Prejudice: The Life and Writings of Hosea Easton* (Amherst: University of Massachusetts Press, 1999), 26.

24. Dain, *A Hideous Monster of the Mind*, 173.

25. Hosea Easton, *A Treatise on the Intellectual Character, and Civil and Political Condition of the Colored People of the U. States* (Boston: Printed and Published by Isaac Knapp, 1837), 23. It was common in the nineteenth century for abolitionists to claim that slavery damaged both the minds and bodies of enslaved people. See Dea H. Boster, "An 'Epeleptick' Bondswoman: Fits, Slavery, and Power in the Antebellum South," *Bulletin of the History of Medicine* 83, no. 2 (Summer 2009): 285–87.

26. Harris, *In the Shadow of Slavery*, 200.

27. Frederick Douglass, *Claims of the Negro, Ethnologically Considered* (Rochester, NY: Daily American Office, 1854), 21.

28. Brown, *The Black Man, His Antecedents, His Genius, and His Achievements*.

29. "Afric-American Picture Gallery.—Fifth Paper," *Anglo-African Magazine* 1, no. 7 (New York), July 1859, 217.

30. Easton, *A Treatise on the Intellectual Character, and Civil and Political Condition of the Colored People*, 52–53. For "Mind acts on matter," see pages 6, 24, 44. For more information on Hosea Easton as an intellectual and activist, see Dain, *A Hideous Monster of the Mind*, 170–96; and Price and Stewart, *To Heal the Scourge of Prejudice*. William J. Wilson suggested a similar argument in a fictional series. See Wilson's description of the underground railroad and slavery's effect on Black bodies: "Afric-American Picture Gallery," *Anglo-African Magazine* 1, no. 2 (New York), February 1859.

31. Samuel Forry, "On the Relative Proportion of Centenarians, of Deaf and Dumb, of Blind, and of Insane in the Races of European and African Origin," *New York Journal of Medicine and the Collateral Sciences* 2, May 1844. For more on the rhetoric of disability within pro- and anti-slavery activism, see Baynton, "Disability and the Justification of Inequality in American History," in *The New Disability History: American Perspectives*, ed. Paul K. Longmore and Lauri Umansky (New York: New York University Press, 2001), 37–39; and Baynton, "Slaves, Immigrants, and Suffragists: The Uses of Disability in Citizenship Debates," *PMLA* 120, no. 2 (March 2005): 562. Easton was hopeful that freedom would invigorate the bodies and minds of formerly enslaved people. But Jim Downs has shown that Emancipation often left enslaved people with significant health problems and a lack of medical resources. See Downs, *Sick from Freedom: African-American Illness and Suffering during the Civil War and Reconstruction* (Oxford: Oxford University Press, 2012).

32. James McCune Smith, "Civilization: Its Dependence on Physical Circumstances," *Anglo-African Magazine* 1, no. 1 (New York), January 1859: 5–16. Nicholas P. Wood also describes Smith's opinion on the changing human form in "Jefferson's Legacy, Race Science, and Righteous Violence in Jabez Hammond's Abolitionist Fiction," *Early American Studies* 14, no. 3 (Summer 2016): 590. For the essay in which Smith makes these claims, see James McCune Smith, "On the Fourteenth Query of Thomas Jefferson's Notes on Virginia," *Anglo-African Magazine* 1, no. 8 (New York), Au-

gust 1859. For other instances in which Smith used physiognomy to argue for racial equality, see "For Frederick Douglass' Paper," *Frederick Douglass' Paper* (Rochester, NY), 18 December 1851; and Communipaw [James McCune Smith], "Nicaragua," *Frederick Douglass' Paper* (Rochester, NY), 8 January 1852.

33. For reviews of James McCune Smith's lecture on the "fallacy of phrenology," see "Phrenology," *The Colored American* (New York), 23 September 1837; and "Dr. Smith," *The Colored American* (New York), 30 September 1837.

34. Smith, "On the Fourteenth Query of Thomas Jefferson's Notes on Virginia," 228. In general, James McCune Smith was more dismissive of phrenology than he was of physiognomy. He titled his iconic series "Heads of the Colored People" as a critique to phrenological discourses. Even in these articles, though, he employs physiognomic descriptions of his subjects. See John Stauffer, *The Works of James McCune Smith: Black Intellectual and Abolitionist* (New York: Oxford University Press, 2006), 185–242.

35. Isaiah Weare kept a manuscript version of this document in his collection of "Notebooks, Letters, Financial Papers, and Newspaper Clippings (1855–1900)." See "Leon Gardiner Collection of American Negro Historical Society Records, 1790–1905," Collection 8, Box 9G, Folder 4, Document 2, HSP. For the published version, see William J. Wilson, "For Frederick Douglass' Paper. From Our Brooklyn Correspondent," *Frederick Douglass' Paper* (Rochester, NY), 9 November 1855.

36. After Wilson's physiognomic profile, Isaiah Weare angrily wrote to *Frederick Douglass' Paper*, claiming that he never spoke to the anonymous "Brooklyn Correspondent" who had nevertheless "volunteered to be my daguerreotypist, and to give to the public, pictures *gratis*." It appears that Weare was far less radical than both Douglass and Wilson, for he made sure to insist that he *was not* the leader of the Philadelphia delegation, and that "no word or sentence, indicating a sectional idea or preference, escaped [his] lips." See Isaiah C. Weare, "Letter from Isaiah C. Weare," *Frederick Douglass' Paper* (Rochester, NY), 9 November 1855. Weare maintained a low public profile in the 1850s, but he did serve on the Executive Committee of the Mother Bethel AME Church and became an advocate for Black voting rights. He forged relationships with both William Still and James Forten, two prominent members of the Philadelphia Black community. The historical record is ambiguous on the proper spelling of his last name; it is listed sometimes as "Weare" and other times as "Wears." I used "Weare" because this is how both he and Wilson spelled it in the 1850s (though it appears he later used "Wears"). For a closer examination of Weare's political activism in the decades following the Civil War, see Harry C. Silcox, "The Black 'Better Class' Political Dilemma: Philadelphia Prototype Isaiah C. Wears," *Pennsylvania Magazine of History and Biography* 113, no. 1 (January 1989): 45–66.

37. Wilson, "For Frederick Douglass' Paper."

38. Wilson, "For Frederick Douglass' Paper."

39. "Afric-American Picture Gallery—Second Paper. By Ethiop," *Anglo-African Magazine* 1, no. 3, March 1859. Though writing in the twentieth century, bell hooks echoed Wilson's concerns. During the era of segregation, she argues, African Americans needed a space to portray images of their race: "Since no 'white' galleries displayed images of black people created by black folks, spaces had to be made within diverse black communities." See bell hooks, "In Our Glory: Photography and Black Life," in *Art on my Mind: Visual Politics* (New York: The New Press, 1995), 59. On the Afric-American Picture Gallery, see John Ernest, *Liberation Historiography: African American Writers and the Challenge of History, 1794–1861* (Chapel Hill: University of North Carolina Press, 2004), 321–29; and Ivy Wilson, *Specters of Democracy: Blackness and the Aesthetics of Politics in the Antebellum US* (New York: Oxford University Press, 2011), 145–68.

40. Wilson, *Specters of Democracy*, 148.

41. "Afric-American Picture Gallery.—Fifth Paper. By Ethiop," *Anglo-African Magazine* 1, no. 7 (New York), July 1859. For the history of Phillis Wheatley's portrait, see Gwendolyn Dubois Shaw, "'On Deathless Glories Fix Thine Ardent View': Scipio Moorhead, Phillis Wheatley, and the Mythic Origins of Anglo-African Portraiture in New England," in Gwendolyn DuBois Shaw and Emily K. Shubert, eds., *Portraits of a People: Picturing African Americans in the Nineteenth Century* (Seattle and London: University of Washington Press, and Andover, MA: Addison Gallery of American Art, 2006), 26–40; and Eric Slauter, "Looking for Scipio Moorhead: An 'African Painter' in Revolutionary North America," in *Slave Portraiture in the Atlantic World*, ed. Agnes Lugo-Ortiz and Angela Rosenthal (Cambridge: Cambridge University Press, 2013), 89–111. For the theory of the facial angle, see Petrus Camper, *The Works on the Connexion between the Science of Anatomy and the Arts of Drawing, Painting, Statuary, &c. &c.*, translated from the Dutch by T. Gogan, MD (London, 1794), 42.

42. Nell Irvin Painter, *History of White People* (New York: W. W. Norton & Company, 2010), 66–67. Miriam Meijer argues that nineteenth-century racial theorists purposefully distorted Petrus Camper's work. She contends that Camper was arguing for the universal humanity of mankind. See Meijer, *Race and Aesthetics in the Anthropology of Petrus Camper* (Amsterdam: Rodopi, 1999).

43. Robert Chambers, *Chamber's Information for the People: A Popular Encyclopedia*, first American edition (Philadelphia: G. B. Bieber & Co., 1848), 1:68.

44. For descriptions of Wheatley's "intellectual" countenance, see Abigail Mott, *Narratives of Colored Americans* (1826; repr. New York, 1875), 8; Wilson Armistead, *A Tribute for the Negro* (New York: William Harned, 1848), 346; and Brown, *The Black Man, His Antecedents, His Genius, and His Achievements*, 231.

45. James McCune Smith, "Nicaragua," *Frederick Douglass' Paper* (Rochester, NY), 8 January 1852; Frederick Douglass, "A Tribute for the Negro," *The North Star* (Rochester, NY), 7 April 1849.

46. M. H. Freeman, "The Educational Wants of the Free Colored People," *Anglo-African Magazine* 1, no. 4 (New York), April 1859, 115–19. Freeman raised issues that Black Americans continue to face. As scholars like Susannah Walker have argued, "forty years after Stokely Carmichael declared that 'black is beautiful' the phrases 'good hair' and 'bad hair' still have meaning for African Americans." See Walker, *Style and Status: Selling Beauty to African American Women, 1920–1975* (Lexington: University of Kentucky Press, 2007), 2.

47. Freeman, "The Educational Wants of the Free Colored People." See also bell hooks, *Black Looks: Race and Representation* (Boston: South End Press, 1992), 1.

48. Mia Bay, "The Battle for Womanhood Is the Battle for Race: Black Women and Nineteenth-Century Racial Thought," in Mia E. Bay, Farah J. Griffin, and Martha S. Jones, eds., *Toward an Intellectual History of Black Women* (Chapel Hill: University of North Carolina Press, 2015), 76–77. Scholars have pointed out that the Black female educator, Sarah Mapps Douglass, taught her students an anti-racist form of anatomy. April Haynes and Britt Rusert suggest that Douglass's lessons provided alternatives to the claims of phrenologists and craniologists like Samuel Morton (though Haynes hints that phrenologists and craniologists came to the same derogatory conclusions about Black minds, which was not always true). See Haynes, *Riotous Flesh*, 160–62; and Rusert, *Fugitive Science*, 181–218.

49. Thomas Jefferson, *Notes on the State of Virginia* (London: John Stockdale, 1787), 230; and Charles White, *An Account of the Regular Gradation in Man, and in Different Animals and Vegetables; and from the Former to the Latter* (London: C. Dilly, 1799), 134–35, quoted in Painter, *History of White People*, 70–71. Londa Schiebinger has argued that white male scientific thinkers spent most of their time talking about Black men and white women—rather than Black women—because they saw the first two groups as "contenders for power." Black women, she posits, seemed

less threatening. See Schiebinger, "The Anatomy of Difference: Race and Sex in Eighteenth-Century Science," *Eighteenth-Century Studies* 23, no. 4 (Summer 1990): 389. On the history of Black female beauty, see Janell Hobson, *Venus in the Dark: Blackness and Beauty in Popular Culture* (New York: Routledge, 2005); and Stephanie M. H. Camp, "Black Is Beautiful: An American History," *The Journal of Southern History* 81, no. 3 (August 2015): 675–90.

50. Journal of Ida B. Wells, 21–28 January 1886, in Dorothy Sterling, ed., *We Are Your Sisters: Black Women in the Nineteenth Century* (New York: W. W. Norton, 1997), 481–82. On Wells's Cabinet portraits and her exchanges with Charles Morris, see James West Davidson, *"They Say": Ida B. Wells and the Reconstruction of Race* (New York: Oxford University Press, 2007), 87–89.

51. On the slow decline of phrenology in the 1860s and 1870s, see Ronald G. Walters, *American Reformers, 1815–1860* (1978; repr. New York: Hill and Wang, 1997), 163–65; and Daniel Patrick Thurs, *Science Talk: Changing Notions of Science in American Culture* (New Brunswick, NJ: Rutgers University Press, 2007), 52.

52. For Douglass's lectures from the 1860s, see Stauffer, Trodd, and Bernier, eds., *Picturing Frederick Douglass*, 123–73. Direct quotations are from pages 130–34.

53. Celeste-Marie Bernier, John Stauffer, and Zoe Trodd suggest that Douglass valued photographs for their "truth value, or objectivity." See *Picturing Frederick Douglass*, xi. For Douglass's complex meditations on the politics of vision, see his lectures on photography in *Picturing Frederick Douglass*, 130, 134, 143, 148, 165. See also Hill, "'Rightly Viewed': Theorizations of Self in Frederick Douglass's Lectures on Pictures," 54–58.

54. For almost physiognomic efforts to read Douglass's character in his likeness, see Stauffer, *The Black Hearts of Men*, 46; and Bernier, Stauffer, and Trodd, eds., *Picturing Frederick Douglass*, xxiv–xxvi. Nell Irvin Painter also reads traits like strength, intelligence, and maturity in images of Sojourner Truth. See Painter, "Representing Truth: Sojourner Truth's Knowing and Becoming Known," *Journal of American History* 81, no. 2 (September 1994): 485. Celeste Bernier has argued that when interpreting portraits of Black Americans, scholars should "return not only to earlier black authors' and artists' strategies of self-imaging in general but to Douglass's endorsement of an alternative theoretical language in particular." See Bernier, "A Visual Call to Arms against the 'Caracature [sic] of My Own Face,'" 356.

55. bell hooks, *Black Looks*, 115. Both Walter Johnson and Jasmine Nichole Cobb have argued that white men positioned themselves as viewers in the nineteenth century, priding themselves on their ability to "see" and scrutinize Black bodies. See Walter Johnson, *Soul by Soul: Life Inside the Antebellum Slave Market* (Cambridge, MA: Harvard University Press, 1999), 137; and Cobb, *Picture Freedom*, 38–43. Christopher Lukasik has similarly pointed out that Lavater's physiognomic system rested on the premise that "the physiognomist needed to possess a number of traits—beauty, education, leisure, and, by extension, capital among them—in order to read faces accurately." See Lukasik, *Discerning Characters: The Culture of Appearance in Early America* (Philadelphia: University of Pennsylvania Press, 2011), 35. See also Christopher Rivers, *Face Value: Physiognomical Thought and the Legible Body in Marivaux, Lavater, Balzac, Gautier, and Zola* (Madison: University of Wisconsin Press, 1994), 94.

CONCLUSION

1. Gilbert Haven, "Cleopatra and Sybilla," *National Anti-Slavery Standard* (New York), 4 October 1862.

2. Haven, "Cleopatra and Sybilla."

3. As Sharrona Pearl has contended, physiognomy constituted a "slippery and flexible concept that changed with every interrogation, every elucidation, every user, and every use." See Pearl, *About Faces: Physiognomy in Nineteenth-Century Britain* (Cambridge, MA: Harvard University Press, 2010), 2.

4. Audre Lorde, "The Master's Tools Will Never Dismantle the Master's House," in *Sister Outsider: Essays and Speeches* (Berkeley, CA: Crossing Press, 1984), 110–14.

5. See Francis Galton, *Hereditary Genius: An Inquiry into its Laws and Consequences* (London: Macmillan and Co., 1869); Cesare Lombroso, *Criminal Man*, trans. Mary Gibson and Nicole Hahn Rafter (1876; Durham, NC: Duke University Press, 2006); Lombroso, *The Man of Genius* (London: W. Scott, 1891); and Lombroso, *Criminal Woman, the Prostitute, and the Normal Woman*, trans. Mary Gibson and Nicole Hahn Rafter (1893; Durham, NC: Duke University Press, 2004).

6. See Herbert Spencer, "Personal Beauty," *Leader* 5, nos. 212 and 216 (15 April and 13 May 1854). Both essays discuss the intersections of race, gender, and physiognomy.

7. On the rise of anthropology, evolutionary biology, and "sexual science" in the late nineteenth century, see Cynthia Eagle Russett, *Sexual Science: The Victorian Construction of Womanhood* (Cambridge, MA: Harvard University Press, 1989); Rachel Malane, *Sex in Mind: The Gendered Brain in Nineteenth-Century Literature and Mental Sciences* (New York: Peter Lang Publishing, Inc., 2005); Carla Bittel, *Mary Putnam Jacobi and the Politics of Medicine in Nineteenth-Century America* (Chapel Hill: University of North Carolina Press, 2009); and Kimberly A. Hamlin, *From Eve to Evolution: Darwin, Science, and Women's Rights in Gilded Age America* (Chicago: University of Chicago Press, 2014).

8. Carl Senior, at the National Institute of Mental Health, has argued that facial beauty conveys one's reproductive fitness, writing that "facial beauty is an honest signal of the genotypic and phenotypic quality of the bearer." See Senior, "Beauty Is in the Brain of the Beholder," *Neuron* 38 (22 May 2003): 525–28. For other studies that attempt to discern internal traits from facial features, see Michael P. Haselhuhn and Elaine M. Wong, "Bad to the Bone: Facial Structure Predicts Unethical Behavior," *Proceedings of the Royal Society* 279, no. 1728 (7 July 2012): 571–76; Justin M. Carré, Cheryl M. McCormick and Catherine J. Mondloch, "Facial Structure Is a Reliable Cue of Aggressive Behavior," *Psychological Science* 20, no. 10 (October 2009): 1194–98; Clare A. M. Sutherland, Lauren E. Rowley, Unity T. Amoaku, Ella Daguzan, Kate A. Kidd-Rossiter, Ugne Maceviciute, and Andrew W. Young, "Personality Judgments from Everyday Images of Faces," *Frontiers in Psychology* 6 (October 2015): 1–11; Leslie A. Zebrowitz and Joann M. Montepare, "Social Psychological Face Perception: Why Appearance Matters," *Social and Personality Psychology Compass* 2/3 (2008): 1497–1517; R. Thora Bjornsdottir and Nicholas O. Rule, "The Visibility of Social Class from Facial Cues," *Journal of Personality and Social Psychology* 113, no. 4 (2017): 530–46; and Yilun Wang and Michal Kosinski, "Deep Neural Networks Are More Accurate than Humans at Detecting Sexual Orientation from Facial Images," *Journal of Personality and Social Psychology* 114, no. 2 (2018): 246–57. Some scholars have pointed out that people *perceive* traits in faces, even if they do not explicitly argue that faces indeed convey character. See Mirella Walker and Thomas Vetter, "Changing the Personality of a Face: Perceived Big Two and Big Five Personality Factors Modeled in Real Photographs," *Journal of Personality and Social Psychology* 110, no. 4 (2016): 609–24; and Raluca Petrican, Alexander Todorov, and Cheryl Grady, "Personality at Face Value: Facial Appearance Predicts Self and Other Personality Judgments among Strangers and Spouses," *Journal of Nonverbal Behavior* 38 (2014): 259–77.

9. Todorov, *Face Value*, 39, 64–69.

10. Alexander Todorov, *Face Value: The Irresistible Influence of First Impressions* (Princeton, NJ: Princeton University Press, 2017), 5, 30, 42; and Courtney Thompson, "Rediscovering 'Good' and 'Bad' Heads in the Phrenological Present," *Nursing Clio*, 8 December 2020, https://nursingclio.org/2020/12/08/rediscovering-good-and-bad-heads-in-the-phrenological-present/.

11. Deanna Pai, "From Beyoncé to Taylor Swift, Decoding Celebrity Faces," *Allure*, 11 July 2016.

12. Todorov, *Face Value*, 154. For "gaining credibility," see David Robson, "How Your Face Betrays Your Personality and Health," *BBC Future*, 12 March 2015.

SELECTED BIBLIOGRAPHY

MANUSCRIPT COLLECTIONS

AMERICAN ANTIQUARIAN SOCIETY (AAS)

Abigail Kelley Foster Papers
Allen-Johnson Family Papers
Chase Family Papers
Draper-Rice Family Papers
Gale Family Papers
Georgiana Souther Barrows Diary
Ruth Henshaw Bascom Diaries

AMERICAN PHILOSOPHICAL SOCIETY (APS)

Eastwick Collection
Gratz Family Papers
Peale-Sellers Family Collection
State Penitentiary for the Eastern District of Pennsylvania Records
William Parker Foulke Papers

ARCHIVES OF AMERICAN ART (AAA)

Hiram Powers Papers

BOSTON ATHENAEUM (BA)

Ann Greene Chapman, Notebooks and Diary
Clara P. Balch, Apollonian wreath [commonplace book]
Cornelia Wells Walter, Extract Book
Lucretia Fiske Farrington, "Notebook for 1834"
Papers of Cephas Thompson and Family

COLLEGE OF PHYSICIANS OF PHILADELPHIA (CPP)

Elm Hill Private School and Home for the Education of Feeble-Minded Youth Records
Nathaniel Chapman Papers
"Likenesses of the Insane, taken for Dr. Jn. K. Mitchell during his Service at the Pennsylvania Hospital."
Nathaniel Chapman Papers

CORNELL UNIVERSITY LIBRARY, DIVISION OF RARE AND MANUSCRIPT COLLECTIONS

Fowler and Wells Families Papers
Mary Ferguson Diaries

HISTORICAL SOCIETY OF PENNSYLVANIA (HSP)

Caroline H. Hance Long Diary
Elizabeth C. Clemson, Autograph Album and Commonplace Book
Leon Gardiner Collection of American Negro Historical Society Records
William Still, "Journal C of Station No. 2 of the Underground Railroad"
Journal of Hannah Margaret Wharton
Lydia and Mary Thomas, Commonplace Book
Mary C. Smith, Autograph Album and Poem Book
Mary G. Billmeyer, Collection of Poetry and Clippings
Mary L. Baldwin, Album
Mary McShane, Autograph Album
Peirce Family Papers
Powel Family Papers
Rebecca F. Taylor, Autograph Album
Rhoda Ann Hampton, Autograph Book
Sally Bridges, Autograph Book
Sarah Logan Fisher Wister, Diary
Susan L. Wattson, Autograph Book
Susanna Longstreth, Autograph Book

HUNTINGTON LIBRARY

Georgiana Bruce Kirby Journal
James M. Monroe Diary

MASSACHUSETTS HISTORICAL SOCIETY (MHS)

Adams Family Papers
Adams Family Papers: An Electronic Archive, http://www.masshist.org/digitaladams/
Anna Cabot Lowell Papers
Anna Cabot Lowell II Diaries
Cary Family Diaries and Commonplace Books

Charles Sedgwick Papers
Cyrus A. Bartol Diary
Dana Family Papers
Eliza M. Spencer Diary
Elizabeth Coombs Adams, Commonplace Books
Jared Curtis Notebooks
Lowell Family Papers
Lucinda Read, Journal and Commonplace Books
Richardson Family Papers
Samuel Gray Ward, Account of a Meeting with Jones Very
Wigglesworth Family Papers II

RUTHERFORD B. HAYES PRESIDENTIAL LIBRARY

Charles E. Frohman Collections

VIRGINIA HISTORICAL SOCIETY (VHS)

Byrd Family Papers
Early Family Papers
Frances Cornelia Barbour Collins, Scrapbook
Harriet Cary Christian, Album
Mary Ann Caruthers, Album
Mary Anna McGuire Claiborne
Mary Virginia Early Brown, Autograph Album
Rosina Ursula Young Mordecai, Commonplace Book
Sophia Coutts, Album

WINTERTHUR LIBRARY

Emma Buckley Howard Edwards, Scrapbooks
James Terry Papers
Moore-Wells Family Papers

PUBLISHED WOMEN'S LETTERS AND DIARIES

Addison, Daniel Dulany, ed. *Lucy Larcom: Life, Letters, and Diary.* Boston: Houghton, Mifflin, and Co., 1894.
Anderson, James House, ed. *Life and Letters of Judge Thomas J. Anderson and Wife.* Columbus, OH: F. J. Heer, 1904.
Ballstadt, Carl, Elizabeth Hopkins, and Michael Peterman, eds. *Letters of a Lifetime.* Toronto: University of Toronto Press, 1985.
Birney, Catherine H., ed. *The Grimké Sisters: Sarah and Angelina Grimké, the First American Women Advocates of Abolition and Woman's Rights.* Boston: Lee and Shepard, 1885.

Blackwell, Elizabeth. *Pioneer Work in Opening the Medical Profession to Women: Autobiographical Sketches.* London: Longmans, Green, and Co., 1895.

Blatch, Harriot Stanton, and Theodore Stanton, eds. *Elizabeth Cady Stanton as Revealed in Her Letters, Diary and Reminiscences.* New York: Harper and Brothers, 1922.

Brady, Patricia, ed. *George Washington's Beautiful Nelly: The Letters of Eleanor Parke Custis Lewis to Elizabeth Bordley Gibson, 1794–1851.* Columbia: University of South Carolina Press, 1991.

Bunkers, Suzanne L., ed. *All Will Yet Be Well: The Diary of Sarah Gillespie Huftalen, 1873–1952.* Iowa City: University of Iowa Press, 1989.

Burr, Virginia Ingraham, ed. *Secret Eye: The Journal of Ella Gertrude Clanton Thomas, 1848–1889.* Chapel Hill: University of North Carolina Press, 1990.

Butler, Frances Anne. *Journal of a Residence in America*, vol. 1. Philadelphia: Carey, Lea, and Blanchard, 1835.

Cary, Virginia. *Letters on Female Character, Addressed to a Young Lady on the Death of Her Mother.* Richmond, VA: Ariel Works, 1830.

Chadwick, John White, ed. *A Life for Liberty: Anti-Slavery and Other Letters of Sallie Holley.* New York: G. P. Putnam's Sons, 1899.

Chronicles of a Pioneer School from 1792 to 1833, Being the History of Miss Sarah Pierce and Her Litchfield School. Compiled by Emily Noyes Vanderpoel. Edited by Elizabeth C. Barney Buel. Cambridge, MA: Harvard University Press, 1903.

Clark, Allen C., ed. *Life and Letters of Dolly [sic] Madison.* Washington, DC: Press of W. R. Roberts Co., 1914.

Cook, Clarence, ed. *A Girl's Life Eighty Years Ago: Selections from the Letters of Eliza Southgate Bowne.* New York: Charles Scribner's Sons, 1887.

Cooper, Susan Fenimore. *Journal of a Naturalist in the United States.* London: Richard Bentley & Son, 1855.

Crafts, Hannah. *The Bondwoman's Narrative.* Edited by Henry Louis Gates. New York: Warner Books, 2002.

Crane, Bathsheba H. Morse. *Life, Letters, and Wayside Gleanings, for the Folks at Home.* Boston: J. H. Earle, 1880.

Crane, Elaine Forman, ed. *The Diary of Elizabeth Drinker.* Boston: Northeastern University Press, 1991.

Curtis, Caroline G., ed. *The Cary Letters.* Cambridge: Riverside Press, 1891.

Cushing, Caroline Elizabeth Wilde. *Letters, Descriptive of Public Monuments, Scenery, and Manners in France and Spain.* Newburyport, MA: E. W. Allen & Co., 1832.

Daniel Webster in England: Journal of Harriette Story Paige, 1839. Boston: Houghton, Mifflin, & Co., 1917.

Dawson, Francis Warrington, ed. *A Confederate Girl's Diary: Sarah Morgan Dawson.* Boston and New York: Houghton Mifflin Company, 1913.

Degen, Maria Kittredge Whitney. *Diary of a Grand Tour of Europe and the Middle East, 1850–1852.* 3 vols. Alexandria, VA: Alexander Street, 2002.

Dewey, Mary E., ed. *Life and Letters of Catharine M. Sedgwick.* New York: Harper & Brothers, 1871.

Diary of Annie L. Van Ness, 1864–1881. Alexandria, VA: Alexander Street, 2004.

Diary of Sarah Connell Ayer. Portland, ME: Lefavor-Tower Co., 1910.

"Diary of Sarah Pugh." In *Memorial of Sarah Pugh: A Tribute of Respect from Her Cousins.* Philadelphia: J. B. Lippincott & Co., 1888.

Drinkwater, Anne T., ed. *Memoir of Mrs. Deborah H. Porter.* Portland, ME: Sanborn & Carter, 1848.

Edmonds, Sarah Emma. *Nurse and Spy in the Union Army.* Hartford, CT: W. S. Williams & Co., 1865.

Emerson, Sarah Hopper Gibbons, ed. *Life of Abby Hopper Gibbons: Told Chiefly through Her Correspondence.* New York: G. P. Putnam's Sons, 1897.

Farrand, Max, ed. *A Journey to Ohio in 1810: As Recorded in the Journal of Margaret Van Horn Dwight.* New Haven: Yale University Press, 1912.

Fergenson, Laraine R. "Margaret Fuller as a Teacher in Providence: The School Journal of Ann Brown." *Studies in the American Renaissance* (1991): 59–118.

Freiberg, Malcolm, ed. *Journal of Madam Knight.* New York: Wilder and Campbell, 1825.

Fuller, Margaret. *Summer on the Lakes.* Boston: Charles C. Little and James Brown, 1844.

Gotendorf, James, ed. *Love-letters of Margaret Fuller: 1845–1846.* New York: Appleton and Company, 1903.

Hallowell, Anna. *James and Lucretia Mott: Life and Letters.* Boston: Houghton, Mifflin and Company, 1884.

Hare, Catherine, comp. *Life and Letters of Elizabeth L. Comstock.* Philadelphia: John C. Winston & Co., 1895.

Harris, Robin, and Terry Harris, eds. *The Eldon House Diaries: Five Women's Views of the 19th Century.* Toronto: Champlain Society, 1994.

Harris, Sharon M., ed. *Selected Writings of Judith Sargent Murray.* New York: Oxford University Press, 1995.

Harwell, Richard Barksdale, ed. *Kate: The Journal of a Confederate Nurse.* Baton Rouge: Louisiana State University Press, 1998.

Hassal, Mary. *Secret History; or, The Horrors of St. Domingo, in a Series of Letters, Written by a Lady at Cape Francois, to Colonel Burr, Late Vice-President of the United States, Principally during the Command of General Rochambeau.* Philadelphia: Bradford & Inskeep, 1808.

Hawthorne, Sophia Peabody. *Notes in England and Italy.* New York: G. P. Putnam & Son, 1869.

Hilen, Andrew, ed. *Diary: A European Tour with Longfellow, 1835–1836.* Seattle: University of Washington Press, 1956.

Hillard, Katherine, ed. *My Mother's Journal: A Young Lady's Diary of Five Years Spent in Manila, Macao and the Cape of Good Hope from 1829–1834.* Boston: G. H. Ellis, 1900.

Holland, Rupert Sargent, ed. *Letters and Diary of Laura M. Towne Written from the Sea Islands of South Carolina, 1862–1884.* Cambridge, MA: Riverside Press, 1912.

Hopkinson, Christina, ed. *Diary and Letters of Josephine Preston Peabody.* Boston: Houghton, Mifflin, & Co., 1925.

Hudspeth, Robert N., ed. *The Letters of Margaret Fuller.* Ithaca, NY: Cornell University Press, 1983.

Hunt, Gaillard, ed. *The First Forty Years of Washington Society in the Family Letters of Margaret Bayard Smith.* New York: Frederick Ungar Publishing, 1906.

Kendall, Phebe Mitchell, comp. *Maria Mitchell: Life, Letters, and Journals.* Boston: Lee & Shepard, 1896.

Kopacz, Paula. "The School Journal of Hannah (Anna) Gale." *Studies in the American Renaissance* (1996): 67–113.

La Corbinière, Clémentine de, ed. *The Life and Letters of Sister St. Francis Xavier.* St. Louis: B. Herder Books Co., 1917.

Larcom, Lucy. *A New England Girlhood, Outlined from Memory.* Boston and New York: Houghton Mifflin Company, 1889.

Lee, R. H., ed. *Memoir of the Life of Harriet Preble: Containing Portions of Her Correspondence, Journal and Other Writings, Literary and Religious.* New York: G. P. Putnam's Sons, 1856.

Lensink, Judy Nolte, ed. *"A Secret to Be Buried": The Diary and Life of Emily Hawley Gillespie, 1858–1888*. Iowa City: University of Iowa Press, 1989.

Letters of Anna Seward: Written Between the Years 1784 and 1807. 6 vols. Edinburgh: George Ramsay & Company, 1811.

The Life and Letters of Elizabeth Payson Prentiss. New York: A. D. F. Randolph, 1882.

Lippincott, Sara. *Haps and Mishaps of a Tour of Europe*. Boston: Ticknor & Co., 1854.

Loughborough, Mary Ann. *My Cave Life in Vicksburg: With Letters of Trial and Travel*. New York: D. Appleton & Co., 1864.

Lyman, Arthur T., ed. *Arthur Theodore Lyman and Ella Lyman: Letters and Journals*. Menasha, WI: George Banta Publishing Co., 1932.

Macleod, Margaret, ed. *The Letters of Letitia Hargrave*. Toronto: Champlain Society, 1947.

MacMullen, Ramsay, ed. *Sisters of the Brush: Their Family, Art, Lives & Letters*. New York: Past Times, 1997.

Marcus, Jacob R., comp. *The American Jewish Woman: A Documentary History*. New York: Ktav Publishing House, 1981.

McDowell, Amanda, and Lela McDowell Blankenship, eds. *Fiddles in the Cumberland*. New York: Richard R. Smith, 1943.

McGuire, Judith Brockenbrough. *Diary of a Southern Refugee During the War*. 3rd ed. Richmond, VA: J. W. Randolph & English, 1889.

McKee, Ruth K., ed. *Mary Richardson Walker: Her Book*. Caldwell, ID: Caxton Printers, 1945.

McMahon, Lucia, and Deborah Schriver, eds. *To Read My Heart: The Journal of Rachel Van Dyke, 1810–1811*. Philadelphia: University of Pennsylvania Press, 2015.

Memoir of Mrs. Harriet L. Winslow, Thirteen Years a Member of the American Mission in Ceylon. New York: American Tract Society, 1840.

Mighels, Ella Sterling. *Life and Letters of a Forty-Niner's Daughter*. San Francisco: Harr Wagner Publishing Company, 1929.

Mohr, James C., and Richard E. Winslow, eds. *Cormany Diaries: A Northern Family in the Civil War*. Pittsburgh: University of Pittsburgh Press, 1982.

Morton, Sarah Wentworth. *My Mind and Its Thoughts, in Sketches, Fragments, and Essays*. Boston: Wells and Lilly, 1823.

Mott, Richard F., ed. *Memoir and Correspondence of Eliza P. Gurney*. Philadelphia: J. B. Lippincott & Co., 1884.

Mug, Mary Theodosia, ed. *Journals and Letter of Mother Theodore Guerin*. Saint-Mary-of-the-Woods, IN: Providence Press, 1937.

Myers, Albert Cook, ed. *Sally Wister's Journal: A True Narrative Being a Quaker Maiden's Account of Her Experiences with Officers of the Continental Army, 1777–1778*. Philadelphia: Ferris & Leach Publishers, 1902.

Myers, Robert M., ed. *The Children of Pride: A True Story of Georgia and the Civil War*. New Haven: Yale University Press, 1972.

Nelson, David T., ed. *The Diary of Elisabeth Koren, 1853–1855*. Northfield, MN: Norwegian-American Historical Association, 1955.

Noyes, Emily Hoffman Gilman. *A Family History in Letters and Documents*. St. Paul: Self-published, 1919.

Palmer, Beverly Wilson, ed. *Selected Letters of Lucretia Coffin Mott*. Urbana and Chicago: University of Illinois Press, 2002.

Peake, Elizabeth. *Pen Pictures of Europe*. Philadelphia: J. B. Lippincott & Co., 1874.

Philipson, David, ed. *Letters of Rebecca Gratz*. Philadelphia: Jewish Publication Society, 1929.

Pierce, Fay, ed. *Music Study in Germany: From the Home Correspondence of Amy Fay.* New York: Macmillan & Co., 1913.

A Place in Thy Memory. New York: J. F. Trow, 1850.

Powers, Elvira J. *Hospital Pencillings: Being a Diary While in Jefferson General Hospital, Jeffersonville, Ind., and Others at Nashville, Tennessee, as Matron and Visitor.* Boston: E. L. Mitchell, 1866.

Putnam, Ruth, ed., *Life and Letters of Mary Putnam Jacobi.* New York: G. P. Putnam's Sons, 1925.

The Remembrancer: Diary of S. A. Morton, 1800–1809. Alexandria, VA: Alexander Street, 2004.

Richards, Laura E., Maud Howe Elliot, and Florence Howe Hall. *Julia Ward Howe, 1819–1910.* 2 vols. Boston: Houghton, Mifflin, & Co., 1915.

Robinson, Harriet Jane Hanson. *Loom and Spindle; or, Life among the Early Mill Girls.* New York: Thomas Y. Crowell & Company, 1898.

Robinson, Sara Tappan Doolittle Lawrence. *Kansas: Its Interior and Exterior Life.* 4th ed. Boston: Crosby, Nichols, & Co., 1856.

Salls, Helen Harriet, ed. *Pamela Savage of Champlain, Health Seeker in Oxford.* Raleigh: North Carolina Division of Archives and History, 1952.

Shartle, E. H., ed. *Emma Cullum Cortazzo, 1842–1918.* Meadville, PA: Self-published, 1919.

Shuffelton, Frank. "Margaret Fuller at the Greene Street School: The Journal of Evelina Metcalf." *Studies in the American Renaissance* (1985): 29–46.

Sigourney, Lydia Huntley. *Letters to My Pupils, with Narrative and Biographical Sketches.* 2nd ed. New York: Robert Carter & Brothers, 1853.

Stanton, Elizabeth Cady. *Eighty Years and More: 1815–1897.* New York: European Publishing Company, 1898.

Steele, Eliza R. *Summer Journey in the West.* New York: John S. Taylor, 1841.

Sterling, Dorothy, ed. *We Are Your Sisters: Black Women in the Nineteenth Century.* New York: W. W. Norton & Company, 1984.

Stevenson, Brenda, ed. *The Journals of Charlotte Forten Grimké.* New York: Oxford University Press, 1988.

Strout, Richard Lee, ed. *Maud.* New York: Macmillan & Co., 1939.

"'Treasure in My Own Mind': The Diary of Martha Lawrence Prescott, 1834–1836." *The Concord Saunterer* 11 (2003): 92–152.

Van Doren, Mark, ed. *Correspondence of Aaron Burr and His Daughter Theodosia.* New York: Covici-Friede, 1929.

Wallis, Mary David. *Life in Feejee; or, Five Years among the Cannibals.* Boston: W. Heath, 1851.

Willard, Emma. *Journal and Letters from France and Great-Britain.* Troy, NY: N. Tuttle, 1833.

Willard, Frances E. *Glimpses of Fifty Years: The Autobiography of an American Woman.* Chicago: Woman's Temperance Publication Association, 1889.

Willard, Frances E. *Nineteen Beautiful Years; or, Sketches of a Girl's Life.* New York: Harper and Brothers, 1864.

Willson, Ann. *Familiar Letters of Ann Willson.* Philadelphia: Wm. D. Parrish & Co., 1850.

Wilson, W. Emerson, ed. *Phoebe George Bradford Diaries.* Wilmington: Historical Society of Delaware, 1975.

Wisner, Benjamin B., ed. *Memoirs of the Late Mrs. Susan Huntington, of Boston.* 2nd ed. Boston: Cocker & Brewster, 1826.

Worthington, Miriam Morrison, ed. *Diary of Anna R. Morrison, Wife of Isaac L. Morrison.* Springfield: Illinois State Historical Society, 1914.

INDEX

Page numbers in italics refer to figures.

abolitionism: belief in human perfectibility and, 137–38, 164; Black female beauty and, 186, 193–94; facial analysis and, 145, 173–74, 177–79, 250n25; support for popular sciences and, 6, 9–10, 15, 45, 132–33, 141–42, 172, 196–97, 211n18, 237n6; white supremacy and, 75–76. *See also* Brown, John, Jr.; Brown, John, Sr.; Brown, William Wells; Chapman, Maria; Douglass, Frederick; Easton, Hosea; Foster, Abby Kelley; Foulke, William Parker; Garrison, William Lloyd; Goodell, William; Grimké, Charlotte Forten; Haven, Gilbert; Holley, Sallie; Kirby, Georgiana Bruce; Lippincott, Sara Jane; Mott, Abigail; Mott, Lucretia; Sizer, Nelson; Smith, James McCune; Still, William; Truth, Sojourner; Weld, Angelina Grimké; Wilson, William Joseph

Adams, Abigail, 17, 20–21, 33
Adams, John, 4, 20–21
Adams, John Quincy, 22, 33–34, 61, 221n68
Adams, Louisa, 22–23
African Americans: appeal of popular sciences and, 6, 9–10, 87–89, 136, 168, 174–79, 181, 188, 249n21; beauty and, 56–57, 70–71, 82–83, 111–15, 169–70, 173–74, 179, 186, 188–89, 193–94, *195*; brows and, 136–37, 173, 178–79, 182–86, *184*, 193–94, *195*; exclusion from republican citizenship, 9, 12–14, 24, 66, 167; intelligence and, 66–67, 70–71, 83, *90*, 90–91, 136, 174; racist depictions of, 166, 170, 176; use of popular sciences to justify exploitation of, 9, 65–66. *See also* Bias, J. J. Gould; Bond, Hannah; Brown, William Wells; Delany, Martin; Douglass, Frederick; enslaved people; Freeman, Martin H.; gender; Green, Sarah Margru Kinson; Grimké, Charlotte Forten; Montgomery, Mary Virginia; Paul, Sarah; race; Shadd, Mary Ann; Smith, James McCune; Still, William; Turner, Henry McNeal; Weare, Isaiah; Wells, Ida B.; Wells, Lewis G.; Wheatley, Phillis; Wilson, William Joseph; women, Black

Allen, Mary Ware, 110
American Female Moral Reform Society, 151
American Medical Association, 7
American Phrenological Journal, 48, 55–56, 70, 89–90, 110, 136–37, 222nn15–16
Anthony, Susan B., 9–10, 45, 131, 135, 137, 138, 140

Baron de Beaujour, Louis-Auguste Félix, 25–26
Barrows, Georgiana, 51
Bartlett, Joseph, 21
Bartol, Cyrus, 147–48
Bascom, Ruth Henshaw, 86

beauty: African Americans and, 56–57, 70–71, 82–83, 111–15, 169–70, 173–74, 179, 186, 188–89, 193–94, *195*; ancient Greeks and, 36, 56; brows and, 40–42, 55, 100, 115–27, *121*, *122*, *123*, *124*, *125*, 141; children and, 55, 187–88; craniums and, 81–82, 117–18, 120, 138, 196; "expressive beauty," 82, 93, 96–97, 101–6, 111, 229; perfect beauty, 71, 81–82, 96–97, 100; scientific beauty, 81–82, 97, *98*, 99–103, 107, 111–12, 126, 229; self-improvement and, 6, 46, 55, 74, 82, 107, 109, 163–64, 181, 196–97. *See also* fashion
Beecher, Henry Ward, 45, 50
Bell, John, 28–29; *Health and Beauty*, 120
Bell, Thomas, 71
Bias, J. J. Gould, 175
Blackwell, Elizabeth, 59
Blewitt, Jonathan, 130
Bloomer, Amelia, 131, 134, 140
Blumenbach, Johann Friedrich, 10
Bond, Hannah, 87–88
Bondwoman's Narrative, The, 87–88
Brackett, Anna, 96
Branagan, Thomas, 40–42
Breed, Anna, 133–34
Bremer, Fredrika, 118–19
Broadhurst, Thomas, 107
Brontë, Charlotte, 92
Brown, John, Jr., 44–45
Brown, John, Sr., 44–45, 194
Brown, William Wells, 174–75, 178, 186, 190
brows: beauty and, 40–42, 55, 100, 115–27, *121*, *122*, *123*, *124*, *125*, 138–41; Black men and, 136–37, 178–79; Black women and, 173, 182–86, *184*, 193–94, *195*; character and, 11, 46–47, *58*, 82, 102–6, 173, 187, 198; criminality and, 147–52; gender conservatives and, 141–43; "highbrow" vs. "lowbrow," 1, 2–3, 56–60; intelligence and, 4, 11, 27, 32, 55, 57–59, 69, 90, 115, 119, 173–75, 182; mutability of, 176–81; politicians and, *30*, *31*, 60–65, *63*; scientific racism and, 68, 166–67, 169–70, 175–76, 186; women intellectuals and, 128–30, 138–43, *142*; women's inferiority and, 69–70
Buffon, Comte de, 10, 25
Burns, Robert, 92

Caldwell, Charles, 61, 70
Calhoun, John C., 179
Camper, Petrus, 10, 36, *38*, *39*, 185–86, 219n60, 252n42. *See also* facial angles
Carey, Mathew, 35
Cary, Margaret, 86
Chambers, Robert, 101, 186
Chapman, Maria, 112
Chapman, Nathaniel, 22
character: class and, 35, 195–97; criminality and, 144–45, 147–52, 154–64; ethnicity and, 71–74, *72*, 154–55; facial expressions and, 82, 102–6; fictional characters and, 20, 35, 41, 45, 83, 173–74; gender and, 36–43, 68–71, *73*, 127–28, 131–38, 195–97; historical tradition of facial analysis and, 10, 19–20; politicians and, 60–65, *62*, *63*; race and, 35–36, 65–68, 166–67, 170, 191–92, 195–97; reading and, 91–95; republican virtue and, 18, 23–34; scientific study of, 11, 46–47, *58*, 173, 187, 198; self-improvement and, 52–53, 75–79, 108–9, 152, 168, 172; sketches and, 18, 20, 34, 91–93, *97*, 173
Chase, Lucy, 51, 80, 229n1
Child, Lydia Maria, 126, 249n18
children: beauty and, 55, 187–88; intelligence and, 51; Orson Fowler and, 85; rearing of, 11, 35, 54–55, 83–84, 88; scientific education and, 11, 22; Spurzheim and, 85
citizenship. *See* republican citizenship
class: character and, 35, 195–97; craniums and, 53; eyes and, 59; gender and, 51, 59–60, 85, 102–3, 109–11, 232n36; phrenology and, 51–53, 84–85; physiognomy and, 9, 12–13, 18, 35, 59–60, 75–79, 91, 140–41, 164, 177–78, 194–97; republican citizenship and, 13–14, 24, 32, 51
Clay, Edward Williams, 169–70
Clay, Henry, 61
Clemson, Elizabeth, 110
Colfax, Richard H., 66
Combe, George, 52, 60–61, 67, 86, 105, 133, 229n2; *The Constitution of Man*, 47, 84, 246n2
Crafts, Hannah. *See* Bond, Hannah
Crane, Bathsheba, 108–9

craniometry, 9, 67–68, 167, 173. *See also* Morton, Samuel
craniums: beauty and female fashion, 81–82, 117–18, 120, 138, 196; character and, 51, 54, 96, *98*, *99*, 145, 150–51, 158; childrearing and, 70, 87; class and, 53; famous men and, 60–64, 105–6; fictional characters and, 83; gender and, 13, 135; intelligence and, 67, 132; marriage and, 54; personality and, *62*; race and, 89, 136–37, 167, 170, 189; scientific study of, 3, 9, 11–12, 44, 47–49; self-improvement and, 53, 55, 75; tripartite division of, 57–59, *58*, 218n46
criminology, 9, 198. *See also* Galton, Francis; Lombroso, Cesare; moral reform; penitentiaries
Cruikshank, George, 128–30, *129*
Curtis, Harriot, 51
Curtis, Jared, 154–55, 163

Dall, Caroline Healey, 96, 232n27
Darwin, Charles, 89, 198, 220n64
Delany, Martin, 170, 175, 178
Delaplaine, Joseph, 26–27, 29
depilatories, 122–26, 138
Dickens, Charles, 60, 161, 225n50, 246n39
Douglass, Frederick: appeal of popular sciences and, 9–10, 178; *Frederick Douglass' Paper*, 181–82; *The North Star* and, 166, 175–76; photography and, 136–37, *171*; visual representations and race, 166, 170, 187, 190–91, 246n2
dress reform movement, 140. *See also* Bloomer, Amelia
Drinker, Elizabeth, 21–22

Easton, Hosea, 176–80
Eddy, Ansel D., 149
Edwards, Emma Howard, 33–34
Embury, Emma C., 97, 99–100, 149–50
Encyclopaedia Britannica, 37, 39
enslaved people: appeal of popular sciences and, 6, 9–10, 87–89, 174, 177–79; beauty and, 111–15, 179; use of popular sciences to justify exploitation of, 9, 65–66. *See also* Bond, Hannah; Montgomery, Mary Virginia

enslavers, 6, 14, 65–66, 87–88, 113
ethnicity: beauty and, 70; Indigenous Americans and, 66; phrenology and, 53, 59–60, 66, 71–79, 84–85, 91, 140–41, 154–55, 164, 177–78, 194–97; physiognomy and, 9, 12–13, 45, 71–74, *72*, *73*, 154–55, 236n5. *See also* immigrants; Irish Americans
ethnology: challenges to white ethnologists, 167–69, 170–73, *171*, 175–76, 180–81, 183, 186–88, 190–91; as scientific discipline, 9, 67; scientific racism and, *39*, 68, 175, 193. *See also* Combe, George; craniometry; Douglass, Frederick; Morton, Samuel; Prichard, James Cowles
eyes: beauty and, 41–42, 70, 82, 101, 112–13, 193; character and, 29–30, 32–33, 40, 92–93, 108, 134, 144–46, 158; class and, 59; Fuller and, 96; intelligence and, 27, 32, 47–48, 57, 81, 84, 107, 109–10, 149; race and, 60, 63, 70, 87–88, 112–13, 162, 174, 178–79, 182, 185; scientific study of, 3–4, 49, 111

facial angles: Camper's theory of, 36; Fuller and, 96, *99*; race and, 76, 180; scientific racism and, *38*, *39*, 185; Wheatley and, 184–85. *See also* Camper, Petrus; ethnology
factory workers, 6, 59, 64, 85. *See also* Curtis, Harriot; Larcom, Lucy; Robinson, Harriet
Farnham, Eliza, 84, 155–56, 163
Farrington, Lucretia Fiske, 91–92, 231n19
fashion: dress reform movement and, 140; hairstyles and, 115–22, *121*, *123*, *124*, 126–30, *129*, 138, *142*, 143
feminism. *See* women's rights
Ferguson, Mary, 51
fine arts: beauty and, 56–57, *121*, 122, *123*, *124*, *125*, 150, 166, 169–70, 187, 190–91; gender and, 57, 118, 193–94, *195*, 237n10; nationalism and, 24–25, 29; politicians and, 24–25, 27–30, *30*, *31*; race and, 166, 169–70, 187, 190–91. *See also* Clay, Edward Williams; Douglass, Frederick; Hogarth, William; Peale, Charles Willson; Peale, Rembrandt; Powers, Hiram; Story, William Wetmore; Stuart, Gilbert; Washington, George; Webster, Daniel

Foley, Margaret, 60
foreheads. *See* brows
Foster, Abby Kelley, 45, 54, *125*, 133–34
Foulke, William Parker, 159–63, 246n44
Fowler, Charlotte. *See* Wells, Charlotte Fowler
Fowler, Lorenzo: ambivalence toward female intelligence, 80, 131–33, 135, 240n28; collaboration with Farnham, 155; "Difference between the Sexes," 69; establishment of American phrenology and, 48–56, *49*; *Illustrated Phrenological Almanac for 1850*, 68, 151; Garrison and, 221n6
Fowler, Lydia Folger, 132, 135, 141
Fowler, Orson: critique of capitalism, 53; establishment of American phrenology and, 48–56, *49*; *Fowler's Practical Phrenology*, 53; scientific racism and, 68; women and, 84, 131, 133, 140, 240n28
Fowler and Wells, 44, 73, 222nn15–16, 228n77. *See also* Fowler, Lorenzo; Fowler, Orson; Sizer, Nelson; Wells, Charlotte Fowler; Wells, Samuel
Franklin, Benjamin, 17–18, 24, 29
Freeman, Martin H., 187–88, 190
Fuller, Margaret: beauty and, 95–96, *98*, *99*; Fowlers and, 48, 84; *The History of Women's Suffrage* and, 141; as popular scientist, 9–10, 84, 101, 133; women prisoners and, 144–45

Gale, Anna, 92–93, *94*, 110
Gall, Franz Joseph, 11, 46–47, 52, 61, 212n22, 224n27
Galloway, Samuel, 107–8
Galton, Francis, 198
Garrison, William Lloyd, 9–10, 45, 112, 221n6, 248n9
gender: anatomy and physiology, 9–10, 18, 70, 72, 147; character and, 36–43, 68–71, 73, 127–28, 131–38, 195–97; intelligence and, 5, 41–42, 59, 69–71, 81, 95–97, 100–101, 126–28, 130–33, 138–41, 174–75, 189; print culture and, 127–30, *129*, *142*, 142–43; republican motherhood, 12–13, 18, 55, 83–84, 127; "true woman," 64–65, 72, 121, 127–28, 131, 138–39, 198. *See also* women, Black; women, white

gender conservatives: appeal of popular sciences and, 9, 45–46, 56, 138; critiques of female beauty and fashion, 115–17, *124*, 130, 139, 141–42, *142*; as opponents of popular sciences, 114, 143. *See also* Blewitt, Jonathan; Cruikshank, George; Mayhew, Henry; Scott, Genio C.
gender equality. *See* women's rights
Gliddon, George, 67–68, 227n65
Godey's Lady's Book, 96, 120, 126, 149
Goodell, William, 76
Graham, Sylvester, 55–56
Gratz, Rebecca, 83–84, 108
Green, Sarah Margru Kinson, 89–91, *90*, 136
Grimké, Charlotte Forten, 59, 92, 112–14, 174–75

Hale, Sarah Josepha, 107
Hance, Caroline, 86–87
Haven, Gilbert, 193–94, *195*
health reform, 52, 55–56, 140
Hillard, Harriet Low, *123*
Hodges, Jacob, 149
Hogarth, William, 150
Holley, Sallie, 96, 99
hooks, bell, 188, 192, 251n39

immigrants, 45, 64, 70–74, *72*, *73*, 79, 154–55, 172, 198
imperialism, 10, 66. *See also* scientific racism
Indigenous Americans, 26–27, 61, 65–66, 68, 75–76. *See also* scientific racism
intellectuals: challenges to white ethnologists and, 167–69, 170–73, *171*, 175–76, 180–81, 183, 186–88, 190–91; depictions of female intellectuals, 95–96, *98*, *99*, 126–30, *129*; "moral uplift" and, 177–78. *See also* Bias, J. J. Gould; Brown, William Wells; Delany, Martin; Douglass, Frederick; Easton, Hosea; Foulke, William Parker; Freeman, Martin H.; Fuller, Margaret; Grimké, Charlotte Forten; Laundrey, Simon Foreman; Lewis, Henry; Shadd, Mary Ann; Smith, James McCune; Still, William; Weare, Isaiah; Wells, Lewis G.; Wilson, William Joseph
intelligence: African Americans and, 66–67, 70–71, 83, *90*, 90–91, 136, 174; brows and,

INDEX

4, 11, 27, 32, 55, 57–59, 69, 90, 115, 119, 173–75, 182; children and, 51; craniums and, 67, 132; ethnicity and, 51, 71–73, 72; eyes and, 27, 32, 47–48, 57, 81, 84, 107, 109–10, 149; fine arts and, 24–25; gender and, 5, 41–42, 59, 69–71, 81, 95–97, 100–101, 126–28, 130–33, 138–41, 174–75, 189; jaws and, 182; race and, 90, 136, 167, 174–75, 178, 186, 189, 191; republican citizenship and, 18–19, 40, 65; scientific study of, 3–4, 47, 50, 65, 177; self-improvement and, 55, 82, 103, 107, 109, 149–50; white women and, 69–71, 138–43, *142*

Irish Americans, 70–74, 72, 73, 154–55. *See also* immigrants

Jackson, Andrew, 61
Jacobi, Mary Putnam, 108
Jacques, Daniel, 78
jaws: animal propensities and, 32, 58, 70, 182; intelligence and, 182; race and, 66, 70, 177, 181–82; self-improvement and, 79–80
Jefferson, Thomas, 4–5, 24, 27, 32, 111, 180; daughters of, 100, 110

Kelley, Abby. *See* Foster, Abby Kelley
Kirby, Georgiana Bruce, 84

Larcom, Lucy, 85
Larcombe, Thomas, 156–59, 163
Laundrey, Simon Foreman, 175
Lavater, Johann Caspar: brows and, 58; criminality and, 146; *Essays on Physiognomy*, 20, 22, 28–29, 42, 106; facial features vs. expressions, 103–5; Gall's critiques of, 11, 47; gender and, 37, 39–40, 42, 120; *Physiognomiche Fragmente*, 19; physiognomic principles and, 10–11, 19–21, 32, 54, 182; racism and, 33, 35–36; transatlantic and lasting popularity of, 19, 22, 86–87, 110–11; Washington and, 29–30, *30*, *31*
Lawrence, William, 186
Leslie, Eliza, 119
Lewis, Henry, 175–76
Lippincott, Sara Jane, 144–45
literature and poetry: novels, 7, 11, 20, 45, 83, 87–88, 92–93, 107, 113–14, 155, 173–74; poems, 83, 92–93, 95, 106, 110–11, 133–34. *See also* Bond, Hannah; Brontë, Charlotte; More, Hannah; Morton, Sarah Wentworth; print culture; Scott, Walter; Sedgwick, Catharine Maria; Stowe, Harriet Beecher; Turnbull, Robert; Webb, Frank; Weems, Mason Locke; Whitman, Sarah Helen; Whitman, Walt

Lobstein, J. F. Daniel, 146
Lombroso, Cesare, 198
Low, Harriet. *See* Hillard, Harriet Low
Lowell, Anna Cabot, 102–3, 109–11, 232n36
Luckey, John, 148–49, 156

Madison, James, 61
Mann, Horace, 45
Mann, R. W. James, 146
Mansfield, Mary Ann, 86
market economy, 8, 12, 34, 64–65, 75, 77, 79
Mayhew, Henry, 128–30
meritocracy: beauty and, 99–100, 106; Blacks and, 186; cultural influence of, 4–5, 52–53; Daniel Webster and, 64; myth of, 23–24, 46, 78–79
Mighels, Ella Sterling, 87
Mitchell, John Kearsley, 22
Montgomery, Mary Virginia, 88–91, 114, 231n18
moral reform: connection to popular sciences, 7, 9–10, 196; nineteenth century as "age of reform," 8, 75; rise of penitentiaries and, 152–53; self-improvement and, 145, 148–49, 164, 197; sex workers and, 149–52; women and, 116. *See also* Bartol, Cyrus; Curtis, Jared; Farnham, Eliza; Foulke, William Parker; Larcombe, Thomas; Nightingale, Florence
More, Hannah, 92
Morton, Samuel, 67, 167, 173
Morton, Sarah Wentworth, 106–7
Mott, Abigail, 186
Mott, Lucretia, 9–10, 45, 112–13, 116, 131, 133

naturalists, 6, 7, 10, 68, 166–67, 170, 192. *See also* Blumenbach, Johann Friedrich; Buffon, Comte de; Camper, Petrus; White, Charles

Neal, John, 95
New York Female Moral Reform Society, 149–52
Nightingale, Florence, 73, 73–74
Nott, Josiah, 67–68, 227n65

Oberlin College, 89, 114, 175
Oneida utopian community, 118

Paige, Harriet Story White, 100
Paul, Sarah, 85
Peale, Charles Willson, 24–25
Peale, Rembrandt, 29
penitentiaries: Eastern State Penitentiary, 156, 159–62; Massachusetts State Prison, 154–55; rise of, 152–53; Sing Sing State Prison, 148–49, 155–56; social reform and, 53, 75; solitary confinement and, 158–62. *See also* Curtis, Jared; Farnham, Eliza; Foulke, William Parker; Larcombe, Thomas
Phelps, Almira, 107
phrenology: class and, 51–53, 84–85; craniometry and, 9, 67–68, 167, 173; ethnicity and, 9, 12–13, 45, 53, 59–60, 66, 71–79, 84–85, 91, 140–41, 154–55, 164, 177–78, 194–97; European origins, 46–47; general rules of, 11, 47, 52; popularization in the US, 47–51, 49; relationship to physiognomy, 10–11, 44, 46–50, 53–54, 56–57; scientific racism and, 68, *138*, *139*; tripartite division of cranium and, 57–59, *58*. See also *American Phrenological Journal*; Combe, George; Fowler, Lorenzo; Fowler, Orson; Gall, Franz Joseph; popular sciences; Sizer, Nelson; Spurzheim, Johann Gaspar; Wells, Charlotte Fowler; Wells, Samuel
physicians, 6–8, 14, 43, 47–48, 131, 162, 179. *See also* Bell, John; Bias, J. J. Gould; Blackwell, Elizabeth; Chapman, Nathaniel; Fowler, Lydia Folger; Jacobi, Mary Putnam; Lewis, Henry; Lobstein, J. F. Daniel; Mitchell, John Kearsley; Ray, Isaac; Rush, Benjamin; Smith, James McCune
physiognomy: beauty and, 40–42, 55, 100, 115–27, *121*, *122*, *123*, *124*, *125*, 141; class and, 9, 12–13, 18, 35, 59–60, 75–79, 91, 140–41, 164, 177–78, 194–97; ethnicity and, 9, 13, 45, 71–74, 154–55, 236n5; European origins of, 10–11, 19–21; facial angles and, 36, *38*, *39*, 76, 96, *99*, 180, 184–85; general rules of, 20–21, 30, 32, 43; medical physiognomy, 146–47, 161–62; in the modern world, 6–7, 199–201; popularization in the US, 11, 19–24, 51, 111; relationship to phrenology, 10–11, 44, 46–50, 53–54, 56–57; tripartite division of countenance and, 32, 57. *See also* Camper, Petrus; Lavater, Johann Caspar; popular sciences
Pike, Mary Hayden, 11
Poe, Edgar Allan, 11, 48, 96
politics: phrenology and, 60–65; physiognomy and, 18–19, 23–34; slavery and, 14, 23, 45, 66–68; suffrage and, 9, 12–14, 66, 167. *See also* abolitionism; Clay, Henry; Jackson, Andrew; Madison, James; meritocracy; republican citizenship; scientific racism; Van Buren, Martin; Washington, George; Webster, Daniel; white supremacy; women's rights
popular sciences: accessibility of, 5, 13, 21, 43, 48, 50, 78, 91, 167, 172–73, 196; appeal to nineteenth-century Americans, 9–10, 23–24, 34–35, 46, 49–50, 64–65, 77–79, 172, 178; Black practitioners, 167, 172–76, 249n21; as challenge to oppression, 135, 167–69, 170–73, *171*, 175–76, 181–92; critiques of, 87, 140–41, 176, 178, 180–81, 187, 251n34; ethnicity and, 9, 13, 45, 71–74, *72*, *73*; European origins of, 10–11, 19–21, 46–47; human perfectibility and, 6, 15, 46, 52–53, 55, 74–80, 82, 103, 107–9, 145, 148–50, 152, 163–64, 168, 172, 176–81, 196–97; ideological flexibility of, 46, 50, 54, 56, 78–79, 82, 137, 145, 164, 167, 172, 196; lawyers and, 6–7, 11, 159–60, 174, 216n15; long-term significance of, 195–201; as tools of oppression, 13–14, 24, 35, 43, 64, 74, 76, 115–17, *124*, 130, 139, 141–42, *142*, 167–68, 190, 198, 201; widespread popularity of, 11, 19–20, 22, 47, 86–87, 110–11; women practitioners, 51, 80, 83–84, 87–92, 104, 108, 114, 132, 135, 141, 155–56, 163, 231n19. *See also* phrenology; physiognomy; science

Powel, Samuel, 28
Powers, Hiram, 57, 118, 237n10
Prichard, James Cowles, 186
print culture: gender and, 127–30, *129*, *142*, 142–43; promotion of popular sciences and, 11, 22, 26, 28–29, 61, 83, 91, 117, 178; race and, 181–85, *184*; rise of, 48; scientific texts and, 19–20, 37, 50; women's fashion and, 126–27. See also *American Phrenological Journal*; *Godey's Lady's Book*; literature and poetry; *Putnam's Monthly Magazine*
Prior, Margaret, 151
prisons. See penitentiaries
pseudoscience, as problematic term, 6–7, 133, 141
Putnam, Mary. See Jacobi, Mary Putnam
Putnam's Monthly Magazine, 138–39

Quakers, 6, 21, 33, 152, 186

race: character and, 35–36, 65–68, 166–67, 170, 191–92, 195–97; craniums and, 89, 136–37, 167, 170, 189; eyes and, 60, 63, 70, 87–88, 112–13, 162, 174, 178–79, 182, 185; facial angles and, 76, 180; fine arts and, 166, 169–70, 187, 190–91; intelligence and, 90, 136, 167, 174–75, 178, 186, 189, 191; print culture and, 181–85, *184*; republican citizenship and, 9, 12–14, 24, 66, 167; as social construct, 187–88; visual representations of, 136–37, 166, 170, *171*, 182–83, 187, 190–91, 246n2; whiteness, 32, 60, 71–72, 114. See also African Americans; Indigenous Americans; scientific racism; white supremacy; women, Black; women, white
Ray, Isaac, 22
Redfield, James W., 22, 84, 216n20
republican citizenship: Black intellectuals and, 177–79; class and, 13–14, 24, 32, 51; facial analysis and, 17–19, 24–25, 32, *171*; gender and, 13, 18, 32, 65, 79, 198; intelligence and, 66–67, 70–71, 83, 90, 90–91, 136, 174; motherhood and, 12–13, 18, 55, 83–84, 127; race and, 9, 12–14, 24, 32, 66–67, 76, 167, 171, 178; virtue and, 12–13, 65, 152
Rider, Lucy, 132

Robinson, Harriet, 51, 60, 85
Runkle, Mary, 151
Rush, Benjamin, 21, 24, 27

Sanderson, John, 27
science: anatomy, 4, 10–11, 13–15, 18, 24, 28, 36, 41, 46–47, 50, 56, 67, 69, 71, 78, 96, 108, 115, 133, 175, 180, 195, 198, 201, 207n2; evolution, 89, 198, 220n64; exclusiveness of, 5; historiography of, 8–9, 194, 199; Linnaean classificatory system, 26; magnetism, 50, 88; mesmerism, 50, 88, 176; modern, 6–7, 9, 167, 199–201; natural history, 6, 7, 10, 68, 166–67, 170, 192; physiology, 49, 71, 133–34; relationship to pseudoscience, 6–7, 50, 87, 133, 140–41, 200; as tool of oppression, 4, 13–14, 24, 35, 43, 64, 74, 76, 115–17, *124*, 130, 139, 141–42, *142*, 167–68, 190, 198, 201. See also craniometry; ethnology; phrenology; physiognomy; popular sciences; scientific racism
scientific racism: brows and, 68, 166–67, 169–70, 175–76, 186; Camper and, 36, *38*, *39*, 185–86, 219n60, 252n42; craniometry and, 167, 173; Fowlers and, 68; Lavater and, 36; Morton and, 67; polygenesis and, 167; use of popular sciences to challenge or support, 9, 13–14, 168–69, 175–76, 181, 190–91; rise of, 35–36, 65–68, 172; Spurzheim and, 224n27. See also Clay, Edward Williams; Colfax, Richard H.; Morton, Samuel; white supremacy
Scott, Genio C., 140
Scott, Walter, 92, 111, 231n20
Sedgwick, Catharine Maria, 22, 110, 119, 233n37
Shadd, Mary Ann, 182–83
Sizer, Nelson, 44–45, 135, 137
Smith, James McCune, 178, 180–81, 187, 251n34
Smith, Margaret Bayard, 100, 110
Smith, Samuel Stanhope, 177
Spencer, Herbert, 198
Spurzheim, Johann Gaspar: Bartol and, 147–48; collaboration with Gall, 47; death of, 61; popularity of, 47, 111; publications, 212n22; scientific racism and, 85, 224n27; self-improvement and, 52; visits to schools, 85

Stanton, Elizabeth Cady: cranial examination of, 135, 138; dress reform and, 140; interest in phrenology, 9–10, 45, 131, 134–35; Seneca Falls Convention and, 116
Still, William, 174, 251n36
Story, William Wetmore, 193–94; *Cleopatra*, 195
Stowe, Harriet Beecher, 11; *Uncle Tom's Cabin*, 45, 113, 173
Stuart, Gilbert, 24

temperance movement: Black women and, 136; Fowlers and, 48, 53, 132; as moral reform, 75, 116, 164
Thompson, William, 158–59
Trust, Joseph W. (pseud. Dr. Felix Gouraud), 123–26
Truth, Sojourner, 45, 137, 241n41
Turnbull, Robert, 92
Turner, Henry McNeal, 13–14

Van Buren, Martin, 60–61
Van Dyke, Rachel, 104

Walker, Alexander, 71
Washington, George: facial analysis of, 17–18, 24, 28–33, *30*, *31*, 61, 213n1; familiarity with physiognomy, 217n39
Weare, Isaiah, 182–83, 251n36
Webb, Frank, 173–74
Webster, Daniel, 61, *63*, 63–64
Webster, Noah, 34–35
Weems, Mason Locke, 11, 41–42
Weld, Angelina Grimké, 76
Weld, Theodore, 76
Wells, Charlotte Fowler, 73, 240n28; establishment of American phrenology and, 48–49, *49*, 52–54, 56; women's rights and, 135, 141
Wells, Ida B., 189–90
Wells, Lewis G., 175
Wells, Samuel, 44, 49, 73–74, 93; *How to Read Character*, *58*, 62; *New Physiognomy*, 39, 73, 109
Wharton, Hannah Margaret, 91–92, 111
Wheatley, Phillis, 183–86, *184*
White, Charles, 111

white supremacy: anti-slavery authors and, 114; popular sciences as challenge to, 56, 90–91, 167, 186–87, 190–92, 194–96; popular sciences as evidence for, 35–36, *38*, 56, 67–68, 70–71, 85, 89–90, *90*, 137, 172, 185–87, 190, 194–96. See also Caldwell, Charles; Calhoun, John C.; Camper, Petrus; Clay, Edward Williams; Colfax, Richard H.; Gliddon, George; Lavater, Johann Caspar; Morton, Samuel; Nott, Josiah; Spurzheim, Johann Gaspar
Whitman, Sarah Helen, 95
Whitman, Walt, 11, 48
Whittier, John Greenleaf, 112–13, 133
Willard, Emma, 104–6
Wilson, William Joseph, 170, 178, 181–86, 190
Winslow, Harriet Wadsworth, 77
Wirt, William, 92
women, Black: beauty and, 56–57, 70–71, 82–83, 111–15, 169–70, 173–74, 179, 186, 188–89, 193–94, *195*; character and, 70, 136, 174–75; "double jeopardy" and, 189; education and, 81, 85; enslaved, 87–91, *90*, 114, 136, 231n18; middle class, 89–92, 112–14; as intellectuals, 59, 92, 112–14, 174–75, 182–83; intelligence and, 70–71, 83, *90*, 90–91, 136, 174; as mothers and wives, 112. See also Bond, Hannah; gender; Green, Sarah Margru Kinson; Grimké, Charlotte Forten; Montgomery, Mary Virginia; Paul, Sarah; race; Shadd, Mary Ann; Truth, Sojourner; Wheatley, Phillis
women, white: beauty and, 70–71, 81–82, 93, 96–107, *98*, 111–14, 126, 229; education and, 40, 81, 84–85, 93, 95, 103, 127, 131; elite, 102–3, 109–11, 232n36; exclusion from republican citizenship, 18, 65, 79, 198; as facial and cranial analyzers, 45, 80–95, 114, 137–38, 197–98; intelligence and, 69–71, 138–43, *142*; Jewish, 83–84; as mothers and wives, 9, 12–13, 18, 37, 40–41, 55, 70, 72, 83–84, 108, 118, 127–28, *129*, 130–31, 139, 147–48, 198; "true womanhood" and, 64–65, 72, 121, 127–28, 131, 138–39, 198; working class, 6, 51, 59–60, 85. See also Anthony, Susan B.; Barrows, Georgiana; Bascom, Ruth Henshaw;

beauty; Blackwell, Elizabeth; Bloomer, Amelia; Brackett, Anna; Breed, Anna; Bremer, Fredrika; Brontë, Charlotte; Cary, Margaret; Chapman, Maria; Chase, Lucy; Child, Lydia Maria; Clemson, Elizabeth; Crane, Bathsheba; Curtis, Harriot; Dall, Caroline Healey; Drinker, Elizabeth; Edwards, Emma Howard; Embury, Emma C.; Farnham, Eliza; Farrington, Lucretia Fiske; Ferguson, Mary; Foley, Margaret; Foster, Abby Kelley; Fowler, Lydia Folger; Fuller, Margaret; Gale, Anna; gender; Hillard, Harriet Low; Holley, Sallie; Jacobi, Mary Putnam; Kirby, Georgiana Bruce; Larcom, Lucy; Lippincott, Sara Jane; Lowell, Anna Cabot; Mott, Abigail; Mott, Lucretia; Robinson, Harriet; Stanton, Elizabeth Cady; Stowe, Harriet Beecher; Weld, Angelina Grimké

women's rights: backlash to, 115–17, *124*, 127–30, *129*, 139, 141–42, *142*; suffrage and, 12–13, 75, 116; support for popular sciences and, 9–10, 14–15, 45, 53, 114, 117, 131–32, 134–36, 138–39, 141, 143, 196–97. *See also* Anthony, Susan B.; Blackwell, Elizabeth; Bloomer, Amelia; Brackett, Anna; Farnham, Eliza; Foster, Abby Kelley; Fowler, Lydia Folger; Jacobi, Mary Putnam; Kirby, Georgiana Bruce; Lippincott, Sara Jane; Mott, Lucretia; Stanton, Elizabeth Cady; Truth, Sojourner; Weld, Angelina Grimké; Wells, Charlotte Fowler